Springer Series in
MATERIALS SCIENCE 24

Springer
Berlin
Heidelberg
New York
Barcelona
Hong Kong
London
Milan
Paris
Singapore
Tokyo

Physics and Astronomy ONLINE LIBRARY

http://www.springer.de/phys/

Springer Series in
MATERIALS SCIENCE

Series Editors: R. Hull · R. M. Osgood, Jr. · H. Sakaki · A. Zunger

The Springer Series in Materials Science covers the complete spectrum of materials physics, including fundamental principles, physical properties, materials theory and design. Recognizing the increasing importance of materials science in future device technologies, the book titles in this series reflect the state-of-the-art in understanding and controlling the structure and properties of all important classes of materials.

Series homepage – http://www.springer.de/phys/books/ssms

Volumes 1–30 are listed at the end of the book.

Klaus Graff

Metal Impurities in Silicon-Device Fabrication

Second, Revised Edition

With 47 Figures

 Springer

Dr. Klaus Graff
Telefunken Electronic
Theresienstrasse 2
74072 Heilbronn, Germany

Present Address:
Hegelmaierstrasse 46
74076 Heilbronn, Germany

Series Editors:

Prof. Robert Hull
University of Virginia
Dept. of Materials Science and Engineering
Thornton Hall
Charlottesville, VA 22903-2442, USA

Prof. H. Sakaki
Institute of Industrial Science
University of Tokyo
7-22-1 Roppongi, Minato-ku
Tokyo 106, Japan

Prof. R. M. Osgood, Jr.
Microelectronics Science Laboratory
Department of Electrical Engineering
Columbia University
Seeley W. Mudd Building
New York, NY 10027, USA

Prof. Alex Zunger
NREL
National Renewable Energy Laboratory
1617 Cole Boulevard
Golden Colorado 80401-3393, USA

Library of Congress Cataloging-in-Publication Data.
Graff, Klaus, 1931-
Metal impurities in silicon-device fabrication / Klaus Graff.- - 2nd, rev. ed.
p. cm. - - (Springer series in materials science; 24)
Includes bibliographical references and index.
ISBN 3540642137 (hardcover : alk. paper)
1. Semiconductors- -Defects. 2. Silicon- -Inclusions. 3. Silicon- -Defects. 4.
Contamination (Technology) I. Title. II. Springer series in materials science ; v. 24.
TK7871.852.G73 1999
621.3815'2- -dc21   99-058100

ISSN 0933-033X

ISBN 3-540-64213-7 2nd Edition Springer-Verlag Berlin Heidelberg New York

ISBN 3-540-58317-3 1st Edition Springer-Verlag Berlin Heidelberg New York

Springer-Verlag is a company in the specialist publishing group BertelsmannSpringer
© Springer-Verlag Berlin Heidelberg 1995, 2000
Printed in Germany

Typesetting: PS™ Technical Word Processor
Cover concept: eStudio Calamar Steinen
Cover production: *design & production* GmbH, Heidelberg

Printed on acid-free paper   SPIN: 10673156   57/3144/Xo   5 4 3 2 1 0

Preface

At the end of the first edition of this monograph it was mentioned that the current data should be corrected and completed within reasonable periods of time. Although this period was rather short, the time has come to prepare a second edition with additional and more recent results taken from almost 100 selected papers that appeared in the meantime. In addition to a large number of minor corrections, further details and comments have been added.

Two main advances in knowledge and understanding have enabled us to solve problems, on the one hand, and fill blanks, on the other hand. They are

(i) the detection and investigation of several hydrogen complexes that form during the etching of silicon samples and can considerably deteriorate the detection of the characteristic deep energy levels of the impurity to be investigated by means of DLTS. Only avoiding the formation of hydrogen complexes by avoiding etching enables one to record the true DLT spectra, for example, of palladium and platinum. Both exhibit three deep energy levels which formerly could not be clearly detected or identified because of the reduced concentrations.

(ii) The investigation of many rare 4d and 5d transition metals by DLTS measurements performed after the growth of contaminated crystals. The new results fill almost all blanks and hence enabled a systematic overview of the behaviour of transition metals in silicon. A large number of these transition metals can hardly be diffused into silicon because of their high reactivity with the ambient atmosphere or their extremely low diffusivity and solubility. As a consequence of the filled blanks a conspicuous similarity between the electrical properties of the metals within one group of equal number N of s plus d electrons could be observed. This led to the tentative conclusion that the boundary between stable interstitial and substitutional defects might be located between the groups of $N = 8$ and $N = 9$. However, this assumption has been proven only for the 3d metals.

New data are presented in Chap. 5 in the sequence of the 3d, 4d, and 5d transition metals. In addition, all tables have been revised with respect to recently published data. Identified complexes of the corresponding metal are also included in the tables. The electrical properties of all metals and complexes presented in the text are summarized in two appendices at the end of the text, one in alphabetic order of the chemical symbols and the other in numeric order of their actication energies.

In conclusion, we have tried to make this monograph even more suitable as a reference for scientists and engineers working in laboratories or at their desks. For convenience, we have summarized in the listing to follow the symbols used throught the text and, in particular, in the tables.

I am deeply grateful to Prof. J. Weber, Max-Planck Institute in Stuttgart (FRG) for providing me with several preprints and selected suitable recent literature and to Prof. W. Schröter and Dr. M. Seibt, University of Göttingen, who spent extended time discussing various problems and reviewing collected data. I am also grateful to Dr. H. Lemke, Technical University of Berlin, for private telephone communications and for preprints. I am obliged to many colleagues, meeting regularly in workshops on point defects in silicon at the Max-Planck Institute in Stuttgart, for presenting their recent results and sending me preprints. In addition I want to thank my editor Dr. H.K.V. Lotsch, Springer-Verlag, Heidelberg, for giving me the opportunity to revise this monograph. Finally, I thank my wife Ursula for her patience during the frequent absence of her husband.

Heilbronn
August 1999   *Klaus Graff*

Preface to the First Edition

This monograph deals with the physical principles and properties of transition-metal impurities dissolved or precipitated in monocrystalline silicon. It is intended as an introduction to the important field of defect engineering, which is of growing interest for the fabrication of modern electronic devices. On the other hand, it is intended as a reference for scientists and engineers actively involved in silicon device research and development, in research institutes and industrial laboratories. It is assumed that the reader already has a basic understanding of the physics of silicon and of the technological processing of silicon wafers in present device fabrication.

Since the well-known investigations performed by G.W. Ludwig and H.H. Woodbury in the early 60's, our knowledge of the behaviour of metal impurities in silicon, on contamination sources, their impact on device performance, and on detection facilities has increased tremendously. Although this development has not yet ceased, the multitude of highly specialized publications which appeared in the last decades calls for a critical review of the manifold details. The present monograph provides a survey of the basic physical processes which lead to contamination, dissolution, diffusion and precipitation of metal impurities in silicon. It also gives an overview of modern analytical facilities and specific data on the properties of transition metals in silicon as far as they are known to date. Finally, gettering phenomena are discussed since they are directly coupled to the presence of unwanted metal impurities. So far, gettering remains an indispensible technological process in modern fabrication techniques. It is hoped that this monograph can serve as a basis for a better understanding of the sophisticated reactions observed in silicon samples contaminated with metal impurities, and taking place during the multitude of technological processes which are required for manufacturing electronic devices.

I am deeply grateful to many of my former colleagues at TEMIC and other scientific and industrial laboratories in central Europe. First of all I am obliged to my former supervisors Drs. G. Goldbach and D. Wolff for giving me the opportunity to write this monograph alongside my engagement in defect engineering, and ingot and process control at TEMIC Heilbronn, FRG. In addition, I want to thank H. Pieper for numerous discussions and ideas for solving problems in measurement and preparation techniques and for his technical assistance over many years. I am grateful to Mrs. C. Huger and Miss P. Heim for preparing numerous samples and making measurements which enabled most of our own results to be obtained. I want to thank D. Wallis for drawing all the figures and Mrs. A. Müller for performing the photographic work. In addition, thanks are expressed to all colleagues met during many workshops on "Point Defects in Silicon" initiated by Profs. E. Sirtl and J. Schneider in 1979. Since then these workshops have taken place twice a year at the Max-Planck Institute in Stuttgart/FRG under the guidance of Prof. H.-J. Queisser. These meetings were always a source of useful information and a possibility to become acquainted with experts from the industrial and university laboratories of central Europe, their methods and their most recent results. Finally, I am obliged to Dr. M. Seibt, University of Göttingen for critically reading this manuscript and for providing me with the TEM micrographs from precipitates of the main impurities in silicon. In addition, he suggested many corrections and improvements, as well as additional references. Last but not least, I especially thank my wife Ursula for her understanding that writing and its subsequent corrections require much time and patience.

Heilbronn
June 1994   *Klaus Graff*

Contents

Abbreviations and Notation

A	Area of a divice [cm^2], see (2.5)
a/aa	Acceptor / double acceptor
AAS	Atomic absorption spectrometer (Sect.6.1.3)
a.u.	Arbitrary units (Sect.6.3.1)
BSD	Backside damage (Sect.8.2.2)
C^{eq}	Equilibrium concentration [cm^{-3}], see (4.31)
CZ	Czochralski-grown silicon crystal
c_{pk}	Process capability (sect.7.1.1)
$c_{\mathrm{p/n}}$	Capture rate for holes / electrons [cm^3/s], see (6.1)
CV	Capacitance-voltage (measurement) (Sect.8.2.3)
D/D^{eff}	Diffusivity / effective diffusivity [cm^2/s], see (2.3) / (4.2)
D_0	Temperature-independent preexponential factor of D, see (2.3)
d/dd/ddd	Donor / double donor / triple donor
DLTS	Deep level transient spectroscopy (Sect.6.2.1)
e	(Capture cross-section) for electrons (Table 3.4)
E_{b}	Binding energy of pairs [eV]
$e_{\mathrm{n}}/e_{\mathrm{p}}$	Emission rates for electrons and holes [1/s], see (2.5)
E_τ	Energy of the trap level [eV], see (3.2)
E_∞	Correction term to change entropy values of activation energies in enthalpy values equal activation energy of temperature-dependent capture cross-sections [eV], see (3.1)
el. inact.	Electrically inactive
FZ	Float-zone-grown silicon crystal
h	(Capture cross-section) for holes (Table 3.4)
H_{M}	Migration enthalpy [eV], see (2.3)
H_{S}	Solution enthalpy [eV], see (2.2)
HT	High temperature
H_{x}	Unknown number of hydrogen atoms
I	Silicon selfinterstitial, see (5.27)
IG	Internal gettering (Sect.8.1)
i/s	Interstitial / substitutional

i_s	Generation current [A], see (2.5)
$k = 8.617 \cdot 10^{-5}$	Boltzmann constant [eV/K]
L	Diffusion length [cm], see (2.1 and 6)
Me/Me$_i$/Me$_s$	Metal / Metal on interstitial site / metal on substitutional site, see (4.32)
Me$^+$/Me^0/Me$^-$	Positively charged / neutral / negatively charged metal atom, see (3.5)
metast.	Metastable
N	Majority-carrier concentration in n-type Si [cm^{-3}], see (2.4)
n	Sum of s and d electrons
NAA	Neutron activation analysis (Sect.6.1.1)
Orient.	Orientation (Table 4.1)
P	Majority-carrier concentration in p-Si [cm^{-3}], see (6.4)
$q = 1.60 \cdot 10^{-19}$	Electronic charge [C], see (2.5)
Ref.	Paper listed under "References"
rel	(Metal)-related (defect) (Table 4.10)
r.u.	Relative units (Table 2.1)
Rem.	Remarks
RT	Room temperature
S	Solubility [cm^{-3}], see (2.2)
S_0	Temperature-independent preexponential factor of S [cm^{-3}], see (2.2)
S_s	Solution entropy [eV], see (2.2)
SIMS	Secondary ion mass spectrometer (Sect.6.1.2)
SPC	Statistical process control (Sect.7.11)
SRV	Surface recombination velocity [cm/s] (Sect.6.2.2)
t	Time, or diffusion duration [s], see (2.1)
T	Temperature, measured in [K] or [°C], as indicated
T_{DIS}	Dissociation temperature of pairs in [K] or [°C]
T_{DLTS}	Temperature of a DLTS peak [K] (Sect.6.2.1)
TEM	Transmission electron microscopy
TXRF	Total reflection X-ray fluorescence analysis (Sect.6.1.6)
UCL	Upper control limit (Sect.7.1.1)
USL	Upper specification limit (Sect.7.1.1)
V	Vacancy, see (5.25)
VPD	Vapor-phase decomposition (Sect.6.1.4)
$v/v_e/v_h$	Thermal velocity of charge carriers / of electroncs / of holes [cm/s], see (2.4) and (3.2)
$x_s - \lambda$	Energy difference between trap level and Fermi level [eV], see (2.5)
α	Capture coefficient [cm$^{-1} \cdot$ s^{-1}], see (4.12)

ΔE_A	Actication energy of impurity [eV] measured from the nearest band edge
ΔE_A !	Enthalpy value of the activation energy [eV], see (3.3)
$\sigma / \sigma_e / \sigma_h$	Charge-carrier capture cross-section / for electrons / for holes [cm^2], see (2.4), (3.4) and (6.2)
σ^*	Temperature-dependent capture cross-section [cm^2], see (3.1)
σ_M	Majority-charge-carrier capture cross-section [cm^2], see (2.4)
τ	Minority-charge-carrier recombination lifetime [s], see (2.4)
τ_0	Low injection-level minority-charge-carrier lifetime [s], see (2.4)
τ_{cap}	Capture time constant [1/s], see (6.1)
$+0.2/-0.2$	Value of ΔE_A [eV], the sign denotes the distance to valence-band edge (+) / conduction-band edge (−)

1. Introduction

The increasing complexity and miniaturization of modern integrated circuits requires higher yields and hence a decreasing density of defects in the electrically active zone of a silicon device. In a submicrometer structure a single metal precipitate may cause a distortion of electrical properties and consequently results in a faulty integrated circuit. Therefore, economic production of devices requires materials and processes of utmost purity. Better knowledge of the behavior of the main impurities which are usually introduced into the silicon material during device production helps to reduce additional contamination. As a consequence the subsequent purification of the silicon material gained through gettering processes requires less effort. Since modern dislocation- and swirl-free silicon crystals exhibit less grown-in gettering centers, the requirements for cleanness or for high gettering effectiveness are enhanced. The impurity problem is rendered even more serious by the tendency in modern technology to lower the temperature of the diffusion processes or to replace phosphorus diffusion, which is a well-known gettering process by phosphorus implantation and rapid thermal processing.

On the other hand, the technological processes for device manufacturing become more involved due to an increasing complexity of the tools and the application of modern techniques such as ion implantation, rapid thermal processing, plasma etching, etc. As a consequence, the device engineer has become highly specialized and is not able to follow the progress in special branches of semiconductor physics and technology even though he should apply this knowledge to increase device performance and enhance yields.

In the past decade materials science and the knowledge about the behavior of metal impurities in silicon during the sequence of technological processes have increased considerably. This transpired mainly due to the enhanced technological requirements as well as being a consequence of the extremely rapid development of modern detection methods starting with the application of Deep Level Transient Spectroscopy (DLTS) [1.1] as a common and convenient tool to detect impurities even in low concentrations. Hence, a multitude of relevant papers were published in this period. Today the number of articles which report on the properties of transition metals in silicon by far exceeds 1000. Therefore, even experts on silicon defect eng-

1

ineering are often not familiar with the recent literature and use, for example, data for the properties of deep energy levels, which have been meanwhile found faulty and were replaced by improved values in recent years.

Many investigations concerning the properties of deep energy levels in silicon have been performed in Europe in the last decade and the results are not as wide-spread as they should be. Therefore, a review on the behavior of deep energy level impurities in silicon may be helpful for defect and device engineers and even for materials scientists, if it provides an overview on the actual knowledge on this special branch of technology. It should help to close the information gap and may, in addition, serve, as a useful reference on the properties, behavior, and reliable data concerning defined metal impurities in silicon.

The present monograph reviews the knowledge on the characteristics of the transition-metal impurities in silicon, giving special attention to the 3d transition metals since the main impurities found in silicon during device fabrication belong to this group. With the exception of the alkaline and earthalkaline metals, most of the metals forming deep energy levels in silicon belong to the transition metals (88 metals including the rare-earth metals and actinides). The symbol 3d (4d, 5d, respectively) specifies the outer electron configuration of the neutral atom. The transition metals are characterized by an increasing number of electrons in the orbit with increasing atomic number ($3d^1$ = Sc, $3d^2$ = Ti...) exhibiting the same electron configuration. Figure 1.1 lists the transition metals as a section of the periodic table of elements including the atomic numbers, the abbreviation for the respective element, and the symbol for the outer-electron configuration.

21	22	23	24	25	26	27	28	29	30
Sc	Ti	V	Cr	Mn	Fe	Co	Ni	Cu	Zn
$3d^1$ $4s^2$	$3d^2$ $4s^2$	$3d^3$ $4s^2$	$3d^5$ $4s^1$	$3d^5$ $4s^2$	$3d^6$ $4s^2$	$3d^7$ $4s^2$	$3d^8$ $4s^2$	$3d^{10}$ $4s^1$	$3d^{10}$ $4s^2$
39	40	41	42	43	44	45	46	47	48
Y	Zr	Nb	Mo	Tc	Ru	Rh	Pd	Ag	Cd
$4d^1$ $5s^2$	$4d^2$ $5s^2$	$4d^4$ $5s^1$	$4d^5$ $5s^1$	$4d^6$ $5s^1$	$4d^7$ $5s^1$	$4d^8$ $5s^1$	$4d^{10}$ –	$4d^{10}$ $5s^1$	$4d^{10}$ $5s^2$
57	72	73	74	75	76	77	78	79	80
La	Hf	Ta	W	Re	Os	Ir	Pt	Au	Hg
$5d^1$ $6s^2$	$4f^{14}$ $5d^2$ $6s^2$	$4f^{14}$ $5d^3$ $6s^2$	$4f^{14}$ $5d^4$ $6s^2$	$4f^{14}$ $5d^5$ $6s^2$	$4f^{14}$ $5d^6$ $6s^2$	$4f^{14}$ $5d^9$ –	$4f^{14}$ $5d^9$ $6s^1$	$4f^{14}$ $5d^{10}$ $6s^1$	$4f^{14}$ $5d^{10}$ $6s^2$

Fig. 1.1. Section of the Periodic Table showing the group of transition metals, their atomic numbers and the electron configurations

This study is confined to a presentation of experimental facts, useful knowledge on the behavior of the impurities during technological processes, and reliable data relevant for technological use in device manufacturing and for preparing samples for scientific investigations. In general, this study does not present the theoretical background of deep energy levels in semiconductors. For a detailed discussion of the associated theoretical problems the reader is refered to the literature [1.2-4].

After a brief summary of the general behavior of transition metals in silicon in Chap.2 the different methods of impurity contamination are outlined followed by a discussion of the impact of impurities on device performance. In Chap.3 the important properties of the transition metals in silicon are presented such as their solubilities and diffusivities as a function of the temperature, the activation energies of the deep energy levels of the metals and of various complexes, and finally the structure of their precipitates, as deduced from recent Transmission Electron Microscopy (TEM) investigations [1.5, 6]. The properties of the technologically important impurities in silicon-device fabrication are presented in Chap.4, including data on iron, nickel, copper and several other metals which can be repeatedly detected in processed silicon. In Chap.5 the peculiarities of most of the remaining rare impurities are reported for completeness. Detection methods for dissolved and precipitated impurities, as far as they can be applied to routine measurements, are presented in Chap.6. Several routine techniques to monitor the amount of impurity contamination in silicon wafers before and after the application of technological processes are discussed in Chap.7. In Chap.8 problems arising with impurity gettering are treated. A short overview on various gettering mechanisms is given and suitable recently developed techniques to monitor the gettering efficiency are presented. Finally, Chap.9 offers some general conclusions together with an outlook on expected future trends in the development of defect monitoring and engineering. This chapter also contains a list of missing or unreliable data, and the required investigations to improve our knowledge about the behavior of metal impurities in silicon.

The references cited in this study are restricted to the most important papers presenting new methods, new results and reliable data which were used to form the average values listed in the tables. Repeatedly references utilized for one table are collected under one reference number, and it may happen that publications are cited under different numbers if, for instance, the paper reports solubilities, diffusivities, and activation energies for the electrically active defects of a metal. The comprehension of all data in one reference number is profitable to include recent results by adding a new paper to the listing of earlier ones without changing all the following reference numbers.

The activation energies of the transition metals in silicon and their complexes, as far as they are known today, are compiled in two tables at the end of the book. Appendix A.1 presents the metals or complexes in alphabetical order of their chemical symbols to provide a fast overview of the known electrical data of these impurities. Appendix A.2 lists the activation energies in numerical order, starting from the highest negative values (difference to the conduction-band edge) and ending with the highest positive values (difference to the valence-band edge). This table may serve to quickly identify deep energy levels detected by DLTS and to correlate them to defects which have already been studied and characterized.

Only those deep energy levels are listed which appeared to be reliably attributed to well-defined transition-metal impurities or compounds. The multitude of other published results which were probably correlated to an impurity and contradict other more reliable results were not considered to avoid confusion as much as possible. Deep energy levels originating from other lattice distortions, other elements or radiation-induced defects are also not considered in order to keep the list easy to survey. Nevertheless an identification of an unknown defect by measuring only its activation energy is, in general, rather difficult regarding the actual measurement precision of DLTS measurements of about 0.1 eV and the number of possible defects listed in Appendix A.2 increasing up to 7.

Hence for a reliable indentification of an unknown defect further informations are required such as the majority-carrier capture cross-sections or the limitation of the possible contaminations. One reason for data erroneously published in the first edition of this monograph, for example, could be meanwhile enlightened. During sample preparation for the DLTS measurement the specimen were etched to improve the electrical performance of the evaporated Schottky contacts. However, etching of silicon generates hydrogen which can diffuse into the surface region of the sample even at room temperature. Hydrogen may then react not only with the dopant element but also with some transition-metal impurities by forming electrically active or inactive compounds. By this way deep energy levels can disappear and others may appear, which should be correlated not to the isolated metal but to a hydrogen complex. More details will be discussed in Sects.2.2.2, 3.3.2 and Chaps.4,5).

2. Common Properties of Transition Metals

Before treating the specific properties of transition metals in silicon, their common properties should briefly be reviewed in order to provide the background for the following two sections on impurity contamination and on their impact on the device performance.

2.1 General Behavior

As will be discussed in the next section, there are many ways to introduce impurities into silicon wafers or onto their surfaces by means of a solid-, liquid-, or vapor-phase contamination. Even today it is still a serious problem to remove especially noble-metal impurities from silicon surfaces if they have been contaminated once. So far, the most reliable technique consists of etching the wafer in order to remove about 1 μm of the wafer thickness at the surface. This, however, deteriorates the quality of the polished silicon-wafer front side by affecting mainly its flatness and by causing the formation of etch pits.

During the high-temperature process a metal-impurity contamination layer deposited on the surface of a wafer frequently changes to a bulk contamination since most of the impurity metals diffuse into the bulk of the wafer. In the case of contaminated surfaces there are only two possibilities to avoid impurity diffusion:

- By an almost quantitative evaporation of the impurity metal from the silicon surface, preferably before diffusion of the impurity into the bulk of the sample begins.
- By a chemical reaction of the impurity metal with the ambient atmosphere forming stable and immobile oxides, nitrides, or other compounds, which remain on the silicon surface or change to a vapor phase and are carried off by the ambient atmosphere.

After diffusion at high temperatures the concentrations of transition metals in silicon samples can be related to their respective solubilities S in silicon at the diffusion temperature T if the amount of the surface contamination is sufficiently high to form an inexhaustible diffusion source. In thermal equilibrium the solubility is coupled to the equilibrium boundary phase which is formed at the diffusion temperature between the metal and the silicon substrate. For the 3d transition metals (M) this phase is a silicide of the composition MSi_2 (M:Ti...Ni), with the exception of copper which forms the copper-rich silicide Cu_3Si. The extension of the contaminated volume within the silicon wafer depends upon the diffusion length of the respective impurity metal. The diffusion length L is quadratically related to the product of the respective diffusivity D of the impurity in silicon and the diffusion duration t

$$L^2 = Dt .$$ (2.1)

Both the diffusivity D of the transition metal and its solubility S depend exponentially on the diffusion temperature T according to the two following Arrhenius equations

$$S = S_0 \exp\left[S_S - \frac{H_S}{kT}\right] \quad \text{for} \quad T < T_{eut} ,$$ (2.2)

$$D = D_0 \exp\left[\frac{-H_M}{kT}\right] ,$$ (2.3)

where S_S and H_S symbolize the solution entropy and enthalpy, respectively, H_M the migration enthalpy, k the Boltzmann constant, and S_0 and D_0 are the temperature-independent preexponential factors. With respect to the solubility the Arrhenius equation only holds for temperatures below the eutectic temperature T_{eut} if this exists well below the melting point of silicon. The solubilities are retrograde with maximum values at temperatures above the eutectic temperature (Mn, Fe, Co, Ni, Cu) [2.1].

As will be shown in Sects. 3.1 and 2, the values of both the solubilities and the diffusivities differ by several orders of magnitude between room temperature and common diffusion temperatures, as well as between different transition metals at the same temperature. As a consequence the extension of the contaminated volume after a high-temperature treatment can differ considerably in size for different impurities and for different processes. Several examples will be presented in Chap. 3.

During cooling of the sample at the end of the heat treatment the solubility of the impurity drastically decreases according to (2.2) and the dissolved metal becomes increasingly supersaturated. The following reactions are possible to overcome this supersaturation. They depend on the supersaturation which is a function of the process conditions, such as the temperature and the cooling rate adopted. In addition, the superasturation is a function of the properties of the impurity such as its solubility and diffusivity at the process temperature and at lower temperatures.

(i) The impurity precipitates within the bulk of the silicon wafer during the cooling of the sample. The precipitation requires a sufficiently high impurity concentration and a sufficiently high diffusivity of the impurity atoms in order to form nuclei for homogeneous precipitation. In the presence of foreign (heterogeneous) nuclei formed by lattice defects or by other impurity precipitates a sufficiently high diffusivity of the impurity atom is likewise needed. The impurity atoms must diffuse to the nuclei within the cooling period, and nuclei may be present in higher or lower densities.

(ii) The impurity atoms diffuse to the surfaces of the silicon wafer during cooling of the sample and form precipitates at both wafer surfaces. Again, a high diffusivity of the impurity atoms in silicon is required. As will be discussed later in more detail these surface precipitates cause the so-called **haze** following preferential etching of the samples.

(iii) The impurity atoms remain dissolved within the volume of the wafer because of low diffusivities of the impurity atoms coupled with the high cooling rates applied to the sample. In this case they are electrically active and form deep energy levels within the band gap of silicon. Frequently, the energy levels exhibit different charge states such as double donors, donors, acceptors, and double acceptors. As a consequence of supersaturation, the quenched-in impurities are metastable and give rise to various structural changes at room temperature or slightly elevated temperatures. Thus, dissolved impurities tend to form complexes with other impurities, e.g., in the form of donor-acceptor pairs. Pairs can be formed with the shallow acceptors of the doping elements, for example, boron (MeB), with deep acceptors of other transition metals ($MeAu_s$), with hydrogen (MeH), or with isolated substitutional atoms of the same metal ($Me_i Me_s$). Sometimes the impurities form electrically active complexes with each other (e.g., manganese). During subsequent low-temperature annealing processes they may also form precipitates.

For high and moderately high cooling rates (in the range between 1 and 250 K/s) impurity precipitation in the volume (Item i above) and outdiffusion to the surfaces (Item ii above) are generally observed in the same sample if the impurity belongs to the haze-forming transition metals. The haze-forming metals are characterized by high solubilities coupled with high diffusivities. However, even all three phenomena can be observed in the same sample: precipitation, outdiffusion, and electrically active dissolved impurities, for example, in the case of a simultaneous contamination of the wafer with different impurities. If a sample is contaminated by iron and copper, and a rather high cooling rate was applied after the diffusion process, the haze-forming copper forms nucleation centers for iron. But if the cooling period is too short for a quantitative precipitation of all the iron atoms, then a fraction of the iron atoms remains electrically active and dissolved.

In general, precipitated impurities will redissolve during a subsequent heat treatment at high temperatures. The concentration of the redissolved impurities depends on the respective solubility which, in turn, is a function of the temperature of the second process. For a quantitative dissolution of precipitated impurities, for example, for gettering purposes, it is recommended to apply a slightly higher diffusion temperature compared to the former process. Only the cooling rate at the end of the last temperature process is crucial for the amount of precipitated, gettered, or the electrically active fraction of the impurity concentration. If the original impurity diffusion source has been removed from the surfaces before performing the second heat treatment, an outdiffusion of the impurities during the subsequent heat treatment may be observed. The outdiffusion continues until a steady state has been achieved between in- and outdiffusion. An equilibrium cannot be reached if the outdiffusing impurity atoms evaporate from the sample surfaces or react with the ambient atmosphere, forming immobile compounds such as oxides or nitrides. In this case, outdiffusion continues and the annealing process acts as a gettering process. Additional gettering effects are observed in the presence of highly doped (extrinsic) zones in silicon samples. They are discussed in more detail in Chap. 8.

In conclusion, the behavior of metal impurities in silicon samples during technological processes may be of a different kind. It depends on the properties of the respective impurity metal, on the one hand, and on the parameters of the technological process, on the other hand, including conditions at the sample surfaces. For a better understanding of the behavior of the impurity reactions of the transition metals in silicon at elevated temperatures, one has to consider the following parameters:

- The respective solubilities and diffusivities of the impurities as a function of the sample temperature to estimate the diffusion lengths of the impurities.
- The boundary conditions at the sample surfaces which determine the rate of in- and outdiffusion of the impurities into and out of the silicon sample.
- The cooling rate applied at the end of the thermal processing which is an important parameter for most of the impurity gettering mechanisms and for impurity precipitation.
- Finally, the doping concentrations in the silicon sample which may affect the solubilities and diffusivities of the impurities in extrinsic silicon. So high doping concentrations may yield gettering effects in the case of locally formed, highly doped zones in the silicon sample, for instance, after phosphorus diffusion (Sect. 8.1).

8

2.2 Contamination of Silicon Wafers

Unintentional impurity contamination of clean silicon wafers is still a serious problem. The freshly polished or epitaxially grown silicon-wafer surface attracts all kinds of metal and nonmetal impurities to saturate its free valences. The surface remains active even after forming a layer of "native oxide" of about 1 nm in thickness within the first hour after its exposure to the atmosphere. It would be helpful if a crystal producer could cover clean silicon wafers with a thermal oxide before packing them into boxes for shipment. Any particle or metal contamination would only affect the oxide layer and could easily be removed before starting the device production. However, this procedure, although proposed repeatedly throughout the years, is still not common practice.

There are many kinds of likely impurity contaminations and contamination sources during the manufacturing of silicon devices, starting with the polycrystalline ingot to grow the single crystals. For a more systematic presentation, the impurity contamination can be divided into three different mechanisms: the solid-phase, the liquid-phase, and the vapor-phase contaminations. Contamination is discussed in this chapter in general terms, whereas possible contamination sources for specific impurities will be treated in connection with the discussion of the main impurities in Chap. 4 and the other impurities in Chap. 5.

2.2.1 Solid Phase

Solid-phase metal contamination takes place by any mechanical contact of the silicon wafer with a metal such as a susceptor; by a handling facility, for instance, in the form of tweezers; or by a contaminated carrier. Even Teflon [PTFE, Poly(TetraFluorEthene)] tweezers after long-time use can be contaminated and then will transmit metal impurities to clean wafer surfaces (Sect. 7.1).

On the other hand, solid-phase contamination can preferentially be applied to an intentional "clean"-metal contamination of a wafer to study the specific properties or reactions of the respective metal [2.2]. The rougher back side of a wafer is scratched by a piece of an etched pure-metal wire after having etched the surface of the wafer to remove unwanted impurities and the native oxide. The resulting contamination is as pure as the wire, and cross contamination can be avoided. However, the evaporation of the same metal onto a silicon-sample surface can introduce various other interfering impurities, as known from the literature [2.3, 4]. Finally sputtering and ion implantation can introduce unwanted contaminations and cause, in addition, different kinds of intrinsic defects and defect clusters. These defects may affect the diffusivity of the metal to be investigated and may induce complexing which changes the behavior of the impurity metal in silicon to be studied [2.5, 6].

Due to the different degrees of hardness of the various transition metals, a mechanical contamination can give rise to lattice distortions in the surface region of the silicon sample if the metal is harder than silicon. The distortion of the silicon surface may, however, cause the formation of dislocations and stacking faults in the silicon sample during a subsequent heat treatment which can, in turn, cause unwanted gettering effects. Many metals which have been investigated were deposited on silicon wafers by mechanical contamination and could be detected in the silicon sample after applying a suitable diffusion process. An intentional mechanical contamination by means of palladium, for example, was used for the development of a new method to determine the efficiency of gettering processes, and will be reported in Sect. 8.2.2.

2.2.2 Liquid Phase

Liquid-phase contamination can occur during any contact of the silicon wafer with contaminated liquids, especially solutions for a wafer cleaning purpose. Problems may arise especially in cleaning procedures when wafers in various states of device production pass the same cleaning facility several times.

For the electrochemical segregation of transition metals from solutions onto silicon surfaces the respective electronegativities of the metals in relation to that of silicon are important. The data for the transition metals are listed in Table 2.1. Since the electronegativity of silicon amounts to 1.8, all transition metals with electronegativities exceeding this value segregate preferably on silicon surfaces.[1] About one half of all transition metals listed in Table 2.1 exhibit electronegativities of ≥ 1.8. The respective metals are printed in bold in the table. The segregation of impurity metals on silicon wafers can even be utilized for purification of the solution by inserting a large number of silicon wafers or silicon samples with large surfaces such as cracked wafers.

The replating of impurities can play an essential role in cleaning procedures. Especially after the cleaning of contaminated wafers in inorganic acids which dissolve, at least partly, the metals adsorbed on their surfaces,

---

[1] The terminology "electronegativity" is commonly used in anorganic chemistry, but its definition differs slightly for different schools. In the context of the present book it is convenient to state the electronegativity relative to the value of silicon.

Table 2.1. Electronegativities in relative units (r.u.) of the transition metals. Bold fonts denote metals with values ≥ 1.8 r.u., being the electronegativity of Si (for an explanation, see Sect. 2.2.2)

Sc	Ti	V	Cr	Mn	Fe	Co	Ni	Cu	Zn
1.3	1.5	1.6	1.6	1.5	**1.8**	**1.8**	**1.8**	**1.9**	1.6
Y	Zr	Nb	**Mo**	Tc	**Ru**	**Rh**	**Pd**	**Ag**	Cd
1.3	1.4	1.6	**1.8**		**2.2**	**2.2**	**2.2**	**1.9**	1.7
La	Hf	Ta	W	**Re**	**Os**	**Ir**	**Pt**	**Au**	**Hg**
1.1	1.3	1.5	1.7	**1.9**	**2.2**	**2.2**	**2.2**	**2.3**	**1.9**

replating effects can be detrimental for the subsequent wafers which are treated in the same solution. Even fractions of impurity monolayers deposited on silicon wafer surfaces are sufficient to cause rather high impurity concentrations in the volume of the wafer after their diffusion. As a consequence even residual metal contaminations adsorbed on the cleaning-vessel's side walls after renewing the solution and rinsing the vessel can still be detrimental for device performance. These phenomena are well-known for gold contamination which can only be avoided by exchanging contaminated vessels. Other impurity metals exhibiting electronegativities higher than 1.8 will affect similar replating.

An intentional liquid-phase contamination with transition metals is possible if the respective electronegativity exceeds 1.8. However, the resulting adsorption layers can be discontinuous and inhomogeneously distributed on the wafer surface forming a multiplicity of islands. This can affect the purpose of the intentional contamination. Nevertheless, this technique is sometimes applied even in production lines, for example, for the purpose of carrier lifetime doping.

In order to avoid liquid-phase contamination during technological processes such as surface cleaning or oxide etching it is proposed to use utmost clean chemicals suitably purified for semiconductor industrial application. This is particularly important for the selection of nitric acid, hydrochloric acid, ammonium chloride, sodium- or potassium hydroxide which all may contaminate silicon surfaces with unwanted impurities mainly iron, nickel, and copper.

A peculiar contamination of silicon takes place during the chemical etching of samples by means of acid or alkaline solutions which both remove silicon from the sample surfaces. During the dissolution of silicon the formation of hydrogen can be observed by the generation of many smaller or larger bubbles. Hydrogen, however, belongs to the very fast diffusing elements in silicon [2.7] especially in form of positively charged ions [2.8] exhibiting extremely small ion diameters. As a consequence, hydrogen diffuses into the surface region of the sample even at room temperature where it can react with doping elements [2.9, 10] and with metal impurities [2.8, 11-14] by forming various hydrogen complexes. These complexes can be electrically inactive but passivating the dopant and impurity atoms. Other complexes are electrically active and generate deep-energy levels within the band gap of silicon.

Although hydrogen is not a metal impurity it should be mentioned here in connection with metal impurity contamination because of its influence on the detection of metal impurities. The wide-spread detection method Deep Level Transient Spectroscopy (DLTS) discussed in Sect.6.2.1, mainly uses Schottky barriers which, in general, require an etch process before evaporating the rectifying metal contact onto the sample surface. Due to the diffusion or drift of the ionized hydrogen atoms into the space charge region at the sample surface and the reaction with the impurity metal the deep energy levels of the isolated impurity atom to be detected can vanish at least partly because of the passivation of its electrical activity by hydrogen. On the other hand, additional electrically active metal-hydrogen complexes can be created which change completely the spectrum. Although the contamination with hydrogen takes place in a liquid etch solution the proper reaction is a vapor-phase diffusion process or ion drift taking place at room temperature in the near-surface space-charge region where DLTS measurements take place.

2.2.3 Vapor Phase

Vapor-phase contamination of silicon wafers with transition-metal impurities can take place during various processes such as oxidation or diffusion in furnace tubes, dry plasma etching, ion implantation, chemical vapor deposition, as well as evaporation and sputtering processes for contact formation. The transition-metal impurity can be present in the vapor phase (e.g., in the atmosphere of the furnace tube) or it can be sputtered from any metal surrounding in the equipment during the process (e.g., in ion implantation- or sputtering equipment). In order to avoid sputtering of transition-metal impurities during processing, all transition-metal components in the respective equipment must be shielded by means of quartz or aluminum which exhibits low sputtering rates combined with low diffusivities in silicon.

It is still difficult to keep the furnace tubes free of impurity-metal contamination since resistor heaters are unlimited impurity sources. With increasing diffusion temperatures, the evaporation of metals from the resistor heater increases drastically. This causes an enhanced vapor pressure and an increasing impurity diffusion through the quartz wall into the interior of the furnace tube. Since the impurities also penetrate the silicon samples, high impurity contaminations in the bulk of the silicon wafers can be observed after performing high-temperature processes. After diffusion at 1170°C the following impurity concentrations have been reported in the literature [2.15]: $6.6 \cdot 10^{16}$ iron atoms/cm^3, $3.8 \cdot 10^{15}$ nickel atoms/cm^3, $1.5 \cdot 10^{15}$ copper atoms/cm^3 and other impurities (Cr, Co, Ta) of minor concentrations. The amount of impurity content in the furnace tube can be reduced by using double-wall quartz tubes floated by oxygen with the addition of small amounts of HCl (1%) in the annulus [2.15]. But also single-wall quartz tubes can be kept much cleaner if high-temperature oxidation processes are carried out in the idle time (without the prime wafers) using slightly higher temperatures than in the normal process. With the exception of gold the transition metals oxidize, and the oxides are immobile or will be floated with the atmosphere. Gold contamination in the furnace tube can be reduced only by evaporating the gold at high temperatures. However, this requires longer times on the order of days, depending on the amount of gold contamination and the height of the diffusion temperature which can be applied.

Another source for metal contaminations in furnace tubes is the impurity content which is introduced with contaminated silicon wafers. Again iron, nickel, and copper are the main impurities observed on silicon-wafer surfaces of as-received ingots. As illustrated in Fig. 2.1, iron evaporates from a sample in position 2 in the boat which was intentionally contaminated by a cross-shaped scratch of iron. The iron vapor mainly contaminates the neighboring sample in position 3 which is situated in the direction of the gas stream. The cross-shaped contamination can be found again in position 3 but the strength of the contamination drops toward the edge of the sample. In the edge region the sample in position 1 is contaminated, although it is situated opposite to the direction of the floating atmosphere. The temperature process is performed at 1150°C for 30 minutes in an inert-gas atmosphere. At the end of the heat treatment the samples are cooled to room temperature within 5 minutes and the iron haze is revealed by a subsequent preferential etching (Sect. 3.4.2).

Although the amount of impurities adsorbed on the surfaces of a single contaminated silicon wafer is low, the total amount of impurities introduced into the furnace tube can be considerably high, taking into account the large number of wafers which are processed simultaneously and in sequence. Therefore the contamination, even of incoming wafers, should be checked before the processing starts in order to avoid contamination of all of the peripheral facilities such as cleaning solutions, furnace tubes, and other equipment (Chap. 7).

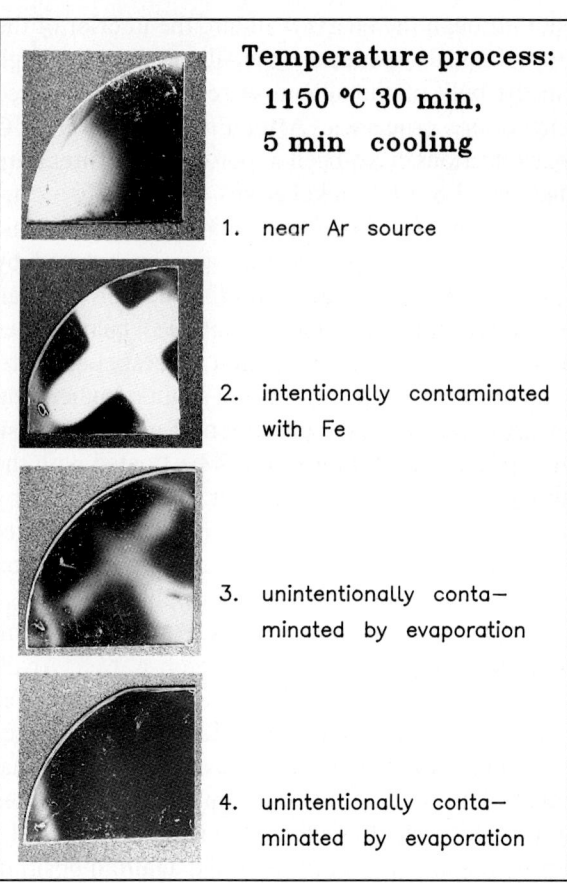

Temperature process:
1150 °C 30 min,
5 min cooling

1. near Ar source

2. intentionally contaminated
   with Fe

3. unintentionally conta—
   minated by evaporation

4. unintentionally conta—
   minated by evaporation

Fig. 2.1. Contamination of neighboring samples during heat treatment in a furnace tube. Evaporation of iron from an intentionally cross-shaped iron-contaminated sample in position 2. Detection of iron by the haze test

2.3 Impact on Device Performance

Little has been published in the literature about transition-metal contamination of silicon wafers during device manufacturing. In contrast, much information has been published on the deterioration of device performance due to metal and other impurities. A review has recently appeared by *Kolbesen* et al. [2.16].

The impact of various defects on device performance and on the yield depends strongly on the type of device and its specifications. The large number of different types of electronic devices with the multiplicity of various specifications cannot be discussed here since it is beyond the scope of this monograph. Therefore, we must restrict ourselves to report on the general influence of transition-metal induced defects on the electrical properties of silicon devices and to present some characteristic examples.

As discussed in Sect.2.1, transition-metal impurities can be dissolved or they can be precipitated after the last heat treatment performed on the silicon wafer during the manufacturing process. The state of a defect depends on the properties of the respective impurity, on the one hand, and the temperature and cooling rate applied during the last heat treatment, on the other hand. Although the influence of both types of defects, the dissolved and the precipitated transition metals, on the device performance is similar, the physical mechanisms of both are different. Therefore the impact of the two different types of defects upon the electrical device properties will be discussed seperately. Their influence on specific devices will be illustrated by a few examples without being required to give a complete listing of possible faults. For details the reader is referred to specialized publications [2.16] and the references therein.

2.3.1 Dissolved Transition Metals

The transition metals dissolved in silicon are electrically active and exhibit deep energy levels which act as donor or acceptor states. They may affect the doping concentration in the silicon sample if the impurity concentration is sufficiently high, for example, a few percent of the original charge-carrier concentration. In modern technology, impurity concentrations in processed silicon wafers rarely exceed concentrations of 10^{13} cm^{-3}. Therefore, a possible change in the doping concentration is limited to high-resistivity silicon wafers as used, for example, for the production of photodiodes, high-voltage transistors, or thyristors. Even in these devices, failures due to altered charge-carrier concentrations are rarely found since the main impurities in processed silicon wafers are iron, copper, and nickel. They usually exhibit the highest impurity concentrations and tend to precipitate quantitatively during the extended cooling periods after common high-temperature processing. Furthermore, diffusion processes are followed by low-temperature processes for contact annealing and welding purposes where even iron precipitates quantitatively.

A severe deterioration of the electrical properties is expected from dissolved impurities with high minority-carrier capture cross-sections. A high capture cross-section causes a drastic reduction of the minority-carrier lifetime in the silicon sample, which still depends on the respective impurity concentration. The low-injection-level minority-carrier recombination lifetime τ_0 is connected to the minority-carrier capture cross-section by the following equation which is valid for a single deep energy level in the band gap [2.17]

$$\tau_0 = \frac{1}{\sigma v N} .$$

(2.4)

Here the carrier capture cross-section for electrons, σ_e [cm^2], must be inserted for p-type silicon and the capture cross-section for holes, σ_p, for n-type silicon. The thermal diffusion velocity of electrons is denoted by v in p-type silicon ($2 \cdot 10^7$ cm/s at 300 K) and of holes in n-type silicon ($1.6 \cdot 10^7$ cm/s at 300 K). The impurity concentration per cm^3 is given by N.

The capture cross-sections for different transition metals can differ by several orders of magnitude. As a consequence the carrier lifetime of a silicon sample can even be determined by an impurity of minor concentration if this is a lifetime killer with a high minority-carrier capture cross-section. Therefore, the tolerable impurity concentration which achieves a carrier lifetime exceeding a defined minimum value depends upon the chemical nature of the respective impurity, its carrier capture cross-section for electrons in p-type silicon and for holes in n-type silicon. Both parameters can differ by orders of magnitude and consequently the tolerable concentration for a defined impurity can be quite different in p- and n-type silicon.

Another parameter which depends on the impurity concentration in the device and which is often specified, for example, for photodiodes, is the generation leakage current. The generation current i_S is calculated using the equation [2.18]

$$i_S = q(x_S - \lambda) A N \frac{e_n e_p}{e_n + e_p} ,$$

(2.5)

where q is the electronic charge, $(x_S - \lambda)$ is the energy difference between the trap level and the Fermi level, A is the area of the device, N is the impurity concentration, and $e_{n/p}$ denotes the emission rates for electrons and holes, respectively. The emission rates are the inverse carrier lifetimes, and therefore they can be calculated using (2.4) if the respective data for σ, v and N are known. So the leakage current is proportional to the impurity concentration N but it also depends upon the carrier capture cross-sections of the transition metal for electrons and for holes, σ_n and σ_p. Since the minority-carrier lifetime is a function of the minority carrier capture cross-section, the leakage current will be different for different transition metals even if the same lifetime is adjusted. This is important for lifetime doping where various impurities can be applied to adjust the required minority-carrier lifetime, but they will result in different leakage currents [2.18]. The relations become more complicated if an impurity exhibits more than a single deep energy level. This is the case for lifetime killers that are commonly applied for lifetime doping such as gold and platinum.

There is still a lack of reliable experimental data for the carrier capture cross-sections of the various transition metals. Most frequently only the capture cross-sections for the majority-charge carriers have been measured but even these results scatter by orders of magnitude (Sect. 3.3). Thus, a calculation of the generation leakage current is, in general, not possible or may yield unreliable results. Furthermore, there may be other reasons for an enhanced leakage current. A correlation of measured impurity concentrations to the specification limit of faulty devices due to enhanced leakage currents can, however, yield useful information. An appropriate example is shown in Fig. 2.2. The gold impurity contents in photodiodes on a silicon wafer were determined by means of DLTS (Sect. 6.2) and correlated to the area of faulty devices due to leakage current which exceeded the specified limits. The maximum gold concentration to achieve a dark current which does not yet exceed the specified limit for these photodiodes amounts to $(0.7 \div 1.0) \cdot 10^{12}$ cm^{-3}.

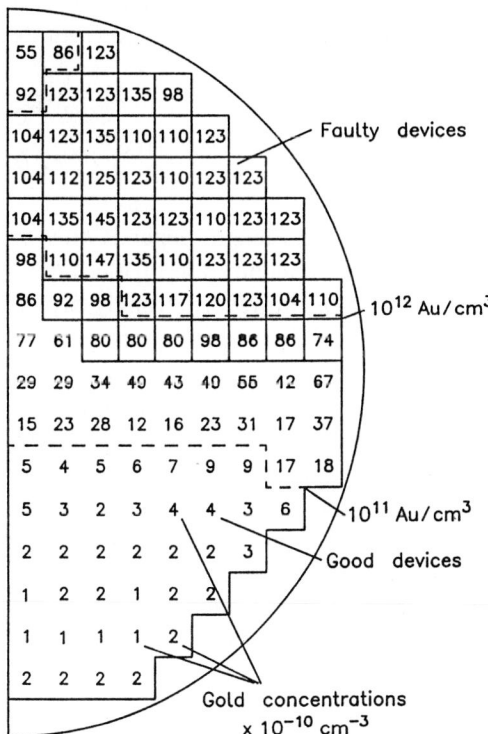

Fig. 2.2. Gold concentrations measured by DLTS in the bulk of photodiodes on a half wafer. The area of faulty devices due to enhanced dark currents is indicated by full lines bordering the gold concentrations. The faulty devices coincide with gold contents beyond about 10^{12} cm^{-3}

2.3.2 Precipitated Transition Metals

Although the physical mechanism is different the deterioration of device performance due to precipitates is quite similar to that for dissolved impurities: The diffusion length of excess minority-charge carriers will be reduced, which is equivalent to the reduction of the carrier lifetime (2.1), and the leakage current will be enhanced. Devices containing precipitates within the space-charge region of the rectifying contact exhibit typical soft reverse current-voltage characteristics. The reason is the shortening of the pn-junction by precipitates which exhibit dielectric constants different from those of silicon. Thus the breakdown voltage is locally reduced and the leakage current is locally enhanced, depending on the concentration, size, and chemical nature of the respective precipitates. The appearance of a soft reverse current-voltage characteristic is always connected to the presence of precipitates. Therefore, it can be employed to detect precipitates in faulty devices.

The reduction of the minority-carrier diffusion length is caused by the recombination of the excess carriers at the phase boundaries of the precipitates, which are present in the volume of the sample. A quantitative relationship between the minority carrier diffusion length L and the density of $NiSi_2$ precipitates, N, in a silicon sample was recently deduced from experiments [2.19]

$$L = \frac{0.7}{N^{1/3}} .$$
(2.6)

This relationship clearly reveals that the diffusion length depends only on the density of the precipitates and not on the concentration of the impurities. The density and the size of the precipitates, however, strongly depend on the cooling rate applied at the end of the last high-temperature process. The recombination at the precipitates affects the volume properties of the device and not only the properties of the space charge region as in the case of the soft reverse current-voltage characteristics. Therefore, precipitates must be avoided in opto-electronic devices because their electrical behavior is, in general, determined by the volume properties of the silicon sample. Since internal gettering is based on an oxygen precipitation mechanism (Sect. 8.1) it cannot be applied to devices where high carrier lifetimes are required in the volume, for instance, in the manufacturing of solar cells.

On the other hand, the limitation of the minority-carrier diffusion length by precipitates can be utilized to determine the depth of the oxygen-free zone in the electrically active surface region of wafers where internal gettering is applied [2.20]. The maximum diffusion length which can be measured is correlated to the thickness of the denuded zone. Beneath this zone the excess-charge carriers recombine at the oxygen precipitates which exhibit mostly high concentrations beneath the denuded zone. The minority-carrier diffusion length depends mainly upon the concentration of the precipitates and less on their sizes since the free-path length of the charge carriers is limited by the mean distance between the precipitates, as was shown by (2.6). Therefore, a suitable temperature process can increase the excess charge-carrier diffusion length if during this process large precipitates grow at the expense of the smaller ones in the neighboring regions increasing the free distance between the precipitates. This ripening process of precipitates is observed repeatedly during heat treatments performed subsequently at reduced temperatures.

However, the main reason for faulty devices due to precipitates is the appearance of soft reverse current-voltage characteristics, which also serves as a suitable detection method for precipitates in the space-charge region of the device. Usually the precipitates consist of iron, copper, or nickel silicides (Sect. 3.4) since these are the main impurities in silicon device production. Copper and nickel precipitate by a homogeneous nucleation mechanism, i.e., that they do not require foreign nuclei. However, iron is expected to precipitate mainly via a heterogeneous nucleation mechanism which requires lattice defects or other impurity precipitates for the generation of nuclei where iron can be segregated.

Many impurities precipitate preferably at the silicon-silicon dioxide interface and cause reduced gate-oxide breakdown voltages [2.16]. As could be shown by TEM, the transition-metal silicides intersect the original silicon-wafer surface. This is due to the enhanced lattice parameters of the silicides compared to those of the surrounding host lattice [2.21]. As a consequence, the thickness of the gate-oxide layer is reduced, resulting in a lowering of the local breakdown voltage [2.16]. The lowering of the breakdown voltage is a further detrimental effect of impurity precipitates situated near the silicon-silicon dioxide interface. Like other effects it can be reduced or omitted by reducing the impurity content in the wafer as a consequence of cleaner materials, processes or of effective gettering processes.

3. Properties of Transition Metals in Silicon

In spite of their regular, outer-electron configuration, the properties of the transition metals in silicon differ considerably even within the same sequence of 3d, 4d, or 5d transition metals, resulting in quite different features during and after heat treatments. The main properties which determine the behavior of transition metals in silicon and their impact on device performance are the solubilities and the diffusivities as a function of the sample temperature, and, as a consequence, their electrical activity and their precipitation behavior. Although the respective properties of the specific metals are different, there are general chemical trends which enable a simultaneous discussion of the properties. This holds at least for the 3d transition metals where most of the parameters of interest have been determined in the last decade or before. In contrast to the 3d transition metals the properties of the 4d and 5d transition metals are less known. Because of the lack of data on their solubilities and diffusivities, the chemical features for these two sequences cannot be demonstrated in the same way. Only the activation energies of most of the 4d and 5d transition metals were published recently. Therefore the solubilities, diffusivities, electrical activities and the precipitation behavior in the bulk of a silicon sample and at its surface will be discussed together in Sects.3.1-4, mainly for the sequence of the 3d transition metals. The properties of the 4d and 5d transition metals cannot be treated in similar details with the exception of several, technologically important metals of these groups. Chapter 4 lists specific peculiarities of the main impurities in silicon and of several other technological important impurities of the 4d and 5d transition metals such as molybdenum, palladium, platinum and gold. In Chap.5 most of the residual rare impurities in silicon are reviewed to the extent that their data have been determined, or special properties are known thus far.

The data presented in the tables of the subsequent chapters have been taken from the literature and have been critically selected with regard to their reliability. Doubtful data have mostly been omitted to avoid any confusion. Some corrections mainly concerning the accuracy of the listed values can again become desirable in a few years, but they should have a small effect on the conclusions drawn from the chemical trends or the physical mechanism of the processes. Whereas models for calculating the activation energies of the deep energy levels have been developed in the past

[3.1-3], models for the diffusivities appeared recently [3.4], and models to calculate the solubilities of the transition metals are still missing. This further development could generate new and more accurate measurements which may yield improved results and might help to fill in the blanks in the tables.

3.1 Solubilities

The solubility of an impurity is defined as the maximum impurity concentration which can be dissolved in thermal equilibrium in a sample at a given temperature. It depends on the surface conditions of the sample since the thermal equilibrium is adjusted by the equality of the in- and outdiffusion of the impurity atoms at the sample surfaces. The source of impurity at the surface should be inexhaustible within the diffusion period applied, in order to reach a thermal equilibrium. In general, extended diffusion periods are utilized and transition-metal surface layers will quickly be transformed to stable transition-metal silicides. Most of the transition metals form silicides which can change in their compositions (metal-rich and silicon-rich) when formed at different temperatures. It has been demonstrated that the concentration of iron in the bulk of a silicon sample covered with metallic iron is much higher in the starting phase than in thermal equilibirum, after some time when iron silicide has formed on the sample surface [3.5]. This initial "solubility" represents the initial iron concentration in the sample when the silicide is not yet completely formed at the surface and the sample had not yet reached thermal equilibrium. Recent experiments utlizing rapid thermal annealing on silicon samples contaminated with metallic iron showed that the iron concentrations exceeded the solubilities by more than onc order of magnitude at the same temperatures [3.6]. Because of the short annealing duration of 10 or 20 s, a very steep depth profile of iron concentration was obtained, which interfered with an inverse concentration profile due to the outdiffusion of iron during the cooling period of the sample (about 10s). Therefore, the exact maximum iron concentrations have not been determined so far. Corresponding experiments can hardly be performed in furnace tubes since the period of time required for the sample to reach the furnace temperature (about two minutes) by far exceeds the diffusion duration applied in these experiments.

In conclusion, the measurement of solubility requires complete formation of that metal silicide which is situated near silicon in the respective phase diagram. For the 3d transition metals the composition of these silicides is of the type MSi_2 (M = Ti...Ni) with the exception of copper, forming Cu_3Si. The Arrhenius equation for the solubility, see (2.2), then holds for temperatures below the eutectic temperature.

Transition metals which exhibit various stable silicides at different temperatures should exhibit discontinuities in the slopes of their solubility curves as a function of temperature if the formation enthalpies of the silicides are different. In general, these differences are expected with the exception of NiSi and $NiSi_2$ exhibiting equal formation enthalpies [3.7]. So far, measurements of the solubilities passing the temperature region where two stable silicides exist, have not been performed by the same researcher applying the same method to avoid large measurement errors.

Most solubility data were determined at high sample temperatures and, in general, there is only one stable high-temperature silicide modification. In this case a constant slope of the solubilitiy (in a logarithmic scale) versus the reciprocal temperature can be expected for temperatures below the eutectic temperature, as represented by the Arrhenius equation (2.2). The respective data for the solution entropies and enthalpies of the 3d transition metals as well as the temperature region in which they have been determined or are valid, are listed in Table 3.1 together with the respective references. In addition, the solubility at 1100°C was calculated with (2.2) to compare the results obtained for various metals with the exception of the solubilities for nickel and copper. These metals exhibit eutectic temperatures below 1100°C. Since an extrapolation beyond the eutectic temperature results in values which are too high, the respective data calculated for a sample temperature of 1100°C are indicated by parentheses. For the solubility of manganese two similar enthalpy and entropy data have been published by *Weber* [3.5] and *Gilles* [3.9]. Therefore, average values for both

Table 3.1. Solubilities of the 3d transition metals $[cm^{-3}]$ in intrisic silicon $S = 5 \cdot 10^{22} \exp(S_S - H_S/kT)$.
(No reliable data is available for Sc and V)

Metal	S_S	H_S	T region [°C]	S (1100°C)	Ref.
Ti	4.22	3.05	950÷1200	$2.1 \cdot 10^{13}$	3.8
Cr	4.7	2.79	900÷1300	$3.1 \cdot 10^{14}$	3.5
Mn	7.11	2.80	900÷1200	$3.2 \cdot 10^{15}$	3.5,9
Fe	8.2	2.94	900÷1200	$2.9 \cdot 10^{15}$	3.5
Co	7.6	2.83	700÷1200	$4.0 \cdot 10^{15}$	3.5
Ni	3.2	1.68	500÷ 950	(5.10^{17})	3.5
Cu	2.4	1.49	500÷ 800	(8.10^{17})	3.5
Zn	7.26	2.49	840÷1200	$5.2 \cdot 10^{16}$	3.10

parameters have been listed in Table 3.1. As inferred from this table, a difference of almost five orders of magnitude is obtained between the solubility of titanium and that of copper, both at $1100\,°C$.

From Table 3.1 it is obvious that the solution enthalpies H_s decrease with increasing atomic number from titanium to copper with the exception of zinc where this tendency is retrograde. However, up to zinc the dominant defect of all 3d transition metals is an interstitial defect (Sect. 3.3.1) although the tendency to form substitutional defects increases with increasing atomic number. Finally, zinc is the only one within this sequence where the concentration of substitutional defects exceeds by far the concentration of interstitial defects in thermal equilibrium [3.11]. This exception can be explained by the different defect structures. On the other hand, the solution entropies S_s increase at first with increasing atomic number from titanium up to iron and then they decrease again. Once more, zinc forms an exception. The results of the competing tendencies can be deduced from the calculated solubilities at a temperature of $1100\,°C$. In general, there is a remarkable increase of the solubility from titanium to copper, however, the values for manganese, iron, and cobalt are very close together and might be identical within an experimental error. This appears still more evident if we compare the slopes of all Arrhenius plots, as depicted in Fig. 3.1.

Figure 3.1 gives a comparison of the solubilities of the 3d transition metals as a function of the inverse temperature. The values were calculated according to (2.2) using the parameters listed in Table 3.1. From experience, it is assumed that the still missing solubility values of scandium are a little below those of titanium. The values for vanadium are expected between those of titanium and chromium since the chemical trend of the solubilities from Ti to Cu exhibits increasing values with increasing atomic numbers.

Due to the lack of theory for the solubilities of the transition metals in silicon, a reasonable explanation for the striking coincidence of the solubilities of manganese, iron, and cobalt within an experimental error cannot be given so far. As seen in Sect. 3.2, the diffusivities of manganese and iron are almost the same, those of cobalt are considerably higher and nearly agree with the values of nickel.

As well as the dependence of the solubility on surface conditions there is an additional dependence on the bulk properties of the silicon crystal. High concentrations of the doping elements (B, P, As) can considerably increase the solubilities of 3d transition metals, as proven by *Hall* and *Racette* for copper [3.12], and by *Gilles* for manganese, iron, and cobalt [3.13]. Usually the solubilities of the metals are determined in intrinsic silicon exhibiting low doping concentrations. Even the oxygen content in Czockralski (CZ) grown silicon can change the solubility under certain conditions

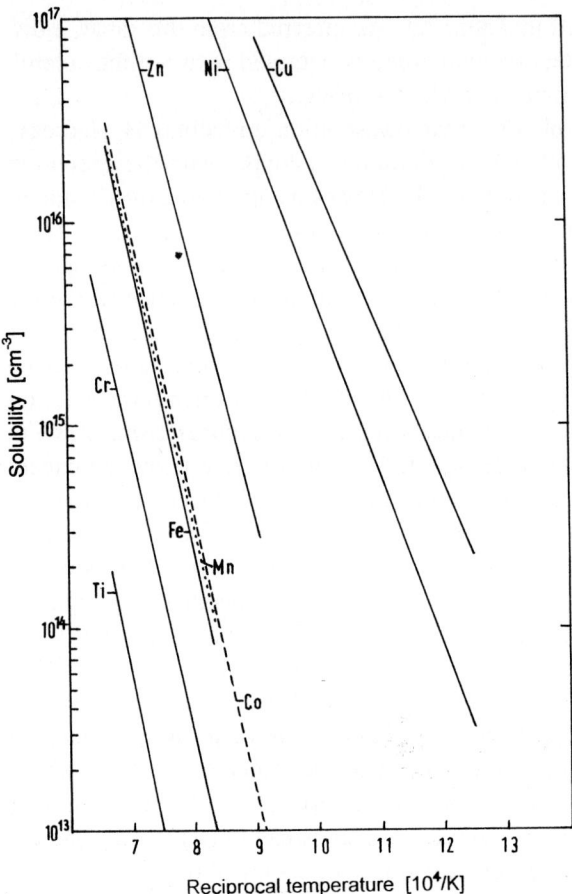

Fig. 3.1. Solubilities of the 3d transition metals as a function of the inverse sample temperature. Recalculated values using (2.2) and data from Table 3.1

[3.14]. Furthermore, the ratio between the amount of impurities situated on interstitial and on substitutional lattice sites can be affected [3.13].

In general, the 3d transition-metal impurities diffuse on interstitial sites [3.15] and mainly occupy interstitial sites in silicon [3.1]. However, there is an increasing tendency in the sequence of the 3d transition metals to occupy substitutional lattice sites with increasing atomic number of the impurity. Thus substitutional defects have been observed for the metals Mn, Co, Ni, Cu, and Zn. Applying common preparation conditions, manganese almost exclusively occupies interstitital lattice sites. Substitutional manganese can be formed by a co-diffusion of manganese and copper [3.16, 17] which is assumed to shift the ratio of vacancies and silicon self-interstitials. Substitutional defects of iron have never been observed. This is in accordance with theoretical predictions that substitutional iron is unstable in silicon [3.1]. Because of their high mobility, the interstitial defects of the heavier 3d transition metals Co, Ni, Cu, cannot be observed at Room Temperature (RT) since they diffuse to the sample surfaces during the cooling period where they precipitate or they precipitate in the bulk of the sample. The ratio between substitutional and interstitial defects seems to increase considerably with increasing atomic numbers within the series of the 3d transition metals. Thus, the existence of substitutional defects has been proven for cobalt, nickel, and copper. Zinc only forms stable substitutional defects. Little information is available to date on the ratio of the interstitial and substitutional solubilities. Only for zinc the temperature-dependent relation Zn_i/Zn_s has recently been deduced from diffusion profiles [3.11].

The common measurement method to determine the solubilities of impurities is the neutron-activation analysis. This technique determines the total impurity content and does not allow the lattice site where the impurity is located, to be distinguished. Techniques which enable this distinction via symmetry properties such as, for example, electron spin resonance have not yet been applied to samples at High Temperature (HT). Therefore, our knowledge of the interstitial or substitutional character of the solubility is limited to the most likely deductions from the general behavior of the respective metals. The HT solubilities of the lighter 3d transition metals from Sc to Fe belong to interstitial defects. The solubilities of the fast-diffusing metals Co, Ni and Cu may be affected by an increasing fraction of substitutional species. Finally, the retrograde solubility of Zn (Table 3.1) should mainly be governed by a substitutional defect.

All solution enthalpies compiled in Table 3.1 exhibit rather high values, and therefore the temperature dependence of the solubilities is clearly visible (Fig. 3.1). As a consequence, the extraplation to RT yields very low values (<1 atom/cm^3). Although extrapolations are dangerous, it is expected that the RT solubilities for all 3d transition metals are extremely low, and the interstitial defects, if detected at all, are metastable. This means that interstitial defects would not exist in thermal equilibrium but they can be observed if they were quenched-in. This, however, can only be achieved for those metals which exhibit rather low diffusivities (Sects. 3.2, 3).

3.2 Diffusivities

The solubilities of impurities in silicon determine the maximum impurity concentration in a sample after diffusion at a defined temperature for a sufficiently long time. The diffusivities of impurities in silicon fix the size of the contaminated area after the diffusion at a defined temperature for a defined period of time. In contrast to the solubility, the diffusivity is primarily a bulk property. The diffusivity also determines the period of time which is necessary for a sample of a given size to reach thermal equilibrium. Very low diffusivities prevent the impurity from reaching thermal equilibrium in a short time and enable the existence of quenched-in, metastable, electrically active defects at RT. High diffusivities prevent the impurities from forming such metastable, electrically active defects. Thus, the existence of electrically active point defects at RT is a criterion for a low diffusivity of the respective impurity metal in silicon.

For those transition metals which form interstitial and substitutional defects, we have to distinguish between an interstitital and a substitutional diffusion mechanism. In general, the interstitial diffusion will be faster than the substitutional diffusion which needs additional intrinsic defects such as vacancies or silicon self-interstitials. First attempts to calculate interstitial diffusivities of the 3d transition metals have been published recently by *Utzig* [3.4]. He investigated the influence upon the migration enthalpy of the elastic energy of the impurity atom in the silicon host lattice due to its atomic radius. It was found that this effect is important for the lighter 3d transition metals Sc, Ti, V, and, to a minor extent, for Cr. The calculated migration enthalpies agree fairly well with experimental data. To achieve better agreement a more sophisticated model must be developed.

Two different models are in competition for the substitutional diffusion mechanism:

(i) The *Frank-Turnbull* model or the **dissociative mechanism** [3.18] which assumes that interstitially diffusing impurity atoms occupy vacancies which are generated in thermal equilibrium. The missing vacancy is then replaced by a vacancy diffusing from the sample surface or from any other vacancy source within the bulk of the crystal (e.g., a dislocation). In this case the vacancy diffusion is the limiting factor since the interstitial impurity diffusion is, in general, much faster.

(ii) The **kick-out mechanism** [3.19, 20] which assumes that the interstitially diffusing impurity atom kicks a silicon atom from the lattice site into the host lattice and occupies this lattice location. In order to reach thermal equilibrium the silicon self-interstitial formed by this process must diffuse to the surface of the sample or any other sink for self-interstitials within the bulk of the sample (e.g., stacking faults or dislocations). The limiting factor in this mechanism is the diffusion of the silicon self-interstitial since the interstitial impury diffusion is, in general, much faster.

Several diffusion experiments demonstrated that the kick-out mechanism is predominant for certain impurities (e.g., Au, Pt, Zn) at least at higher diffusion temperatures [3.11, 19-21]. At lower diffusion temperatures the dissociative mechanism may become increasingly important. However, at lower temperatures the measurement error increases and, therefore, the experimental results and the conclusions become less reliable.

The experimental results for the diffusivity parameters of the 3d transition metals are listed in Table 3.2. Their calculation has been performed with (2.3). The temperature regions are listed in the table together with calculated values for a diffusion temperature of $1100\,^\circ C$. The same chemical trend which was observed for the solubility data (Table 3.1) can be found again in the diffusivity data showing increasing values for increasing atomic numbers within the series of the 3d transition metals. Again, the only exception is zinc which diffuses via the kick-out mechanism. Once again, deviations of up to about seven orders of magnitude and more are found between the slow and the fast diffusing 3d transition metals at the same temperature. Compared to the diffusion coefficients of the shallow doping elements such as boron or phosphorus, all these impurities, including titanium, diffuse extremely quickly and exhibit deviations to the dopant elements of more than four orders of magnitude at $1100\,^\circ C$.

The diffusivities of cobalt have been determined by *Gilles* [3.13], as well as by *Utzig* and *Gilles* [3.25]. The average of both sets of results are listed for simplification and clarity in Table 3.2. Although both sets of results agree fairly well the migration enthalpy of 0.37 eV published in [3.25] is even smaller than that of copper which in silicon is the fastest diffusing transition metal. Therefore, it does not fit in the trend of decreasing enthalpy values with increasing atomic numbers shown in Table 3.2. However, the migration enthalpy of 0.53 eV and also the preexponential factor resulting of an average of both measurements fit very well in this trend. Several experimental results for the diffusivities of the same metal (Cu and Ni) published by other researchers do not match well, although the temperature regions join or even overlap. Some of these values were determined from diffusion profiles of electrically active defects and therefore must be related to the substitutional diffusion coefficient. In order to avoid confusion, these results have not been listed in Table 3.2. For the 3d transition metals with the exception of zinc the interstitial diffusion mechanism is absolutely predominant. The values listed in Table 3.2 have also been selected by *Utzig* [3.4] for his comparison of experimental and calculated migration enthalpies.

In order to obtain a quick overview, the diffusivities of the 3d transition metals were calculated from (2.3) using the parameters listed in Table 3.2. The results are plotted in Fig.3.2 showing the diffusivities on a logarithmic scale versus the reciprocal temperature. From experience it is expected that the missing diffusivities of scandium are situated not too far below those of titanium. As already mentioned in Sect.3.1, the diffusivities of manganese and iron, on the one hand, and cobalt and nickel, on the other, nearly agree with one another. For the metals chromium, manganese,

Table 3.2. Interstitial diffusivities of the 3d transition metals in intrinsic silicon $D = D_0 \exp(-H_M /kT)$ [cm^2/s].
(No reliable data is available for Sc and Zn)

Metal	D_0 [cm^2/s]	H_M [eV]	T region [°C]	D (1.100°C) [cm^2/s]	Ref.
Ti	$1.2 \cdot 10^{-1}$	2.05	$600 \div 1150$	$3.6 \cdot 10^{-9}$	3.8,22
	$1.45 \cdot 10^{-2}$	1.79	$950 \div 1200$	$3.9 \cdot 10^{-9}$	3.8
V	$9.0 \cdot 10^{-3}$	1.55	$600 \div 1200$	$1.8 \cdot 10^{-8}$	3.22
Cr	$6.8 \cdot 10^{-4}$	0.79	$27 \div 400$	$(8.6 \cdot 10^{-7})$	3.22
	$1.0 \cdot 10^{-2}$	0.99	$900 \div 1250$	$2.3 \cdot 10^{-6}$	3.23
Mn[a]	$1.63 \cdot 10^{-3}$	0.71	$14 \div 1200$	$4.0 \cdot 10^{-6}$	3.9,22
	$6.9 \cdot 10^{-4}$	0.63	$900 \div 1200$	$3.6 \cdot 10^{-6}$	3.9
	$2.4 \cdot 10^{-3}$	0.72	$14 \div 90$	$(5.5 \cdot 10^{-6})$	3.22
Fe	$1.3 \cdot 10^{-3}$	0.68	$30 \div 1200$	$4.1 \cdot 10^{-6}$	3.5
Fe$^+$	$1.1 \cdot 10^{-3}$	0.66	$0 \div 1070$	$4.2 \cdot 10^{-6}$	3.22
Fe$^+$	$1.4 \cdot 10^{-3}$	0.69	$30 \div 400$	$(4.1 \cdot 10^{-6})$	3.24
Fe^0	$1.0 \cdot 10^{-2}$	0.84	$30 \div 400$	$(8.3 \cdot 10^{-6})$	3.24
Co[a]	$4.2 \cdot 10^{-3}$	0.53	$900 \div 1100$	$4.7 \cdot 10^{-5}$	3.13,25
	9.10^{-4}	0.37	$700 \div 1100$	$3.9 \cdot 10^{-5}$	3.25
Ni	$2.0 \cdot 10^{-3}$	0.47	$800 \div 1300$	$3.8 \cdot 10^{-5}$	3.26
Cu$^+$	$4.5 \cdot 10^{-3}$	0.39	$30 \div 970$	$(1.7 \cdot 10^{-4})$	3.27

[a] : Average of both references

(..) D values extrapolated to temperatures beyond the validity region (1.100°C)

iron, and copper diffusivity results for the RT region became available recently. For the low- and high-temperature regions of iron diffusivities *Weber* [3.5] presented already in 1983 a common fit of both which almost agrees with the recently published results of *Nakashima* [3.22] and even still better with those of *Heiser* et al.[3.24]. The low-temperature diffusivities for manganese [3.22] also fit very well with earlier high-temperature values measured by *Gilles* [3.9] as can be seen in Fig. 3.2. Hence, the fit of both is included in Table 3.2. *Mesli* et al. presented the diffusivities for copper in the whole temperature region, which also appear reliable. Finally the diffusivities for chromium at high temperatures [3.23] and at low temperatures [3.22] do not fit as well as the others although both extrapolated plots exhibit a cross over in the temperature region in between. In this case an aver-

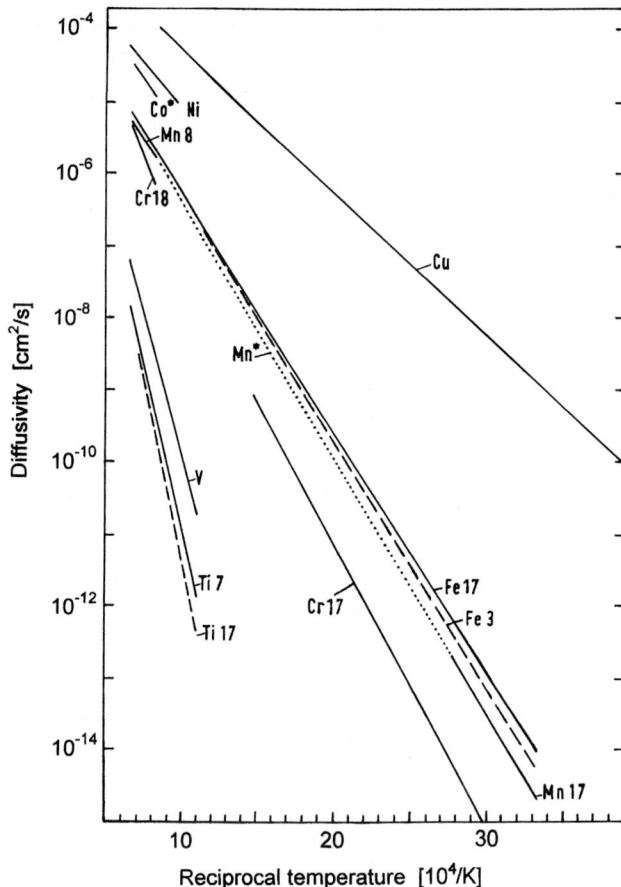

Fig. 3.2. Diffusivities of the 3d transition metals as a function of the inverse sample temperature. Recalculated values using (2.3) and data from Table 3.2

age value has not yet been calculated although it might be obvious from the general trend that the low-temperature slope could be steeper, whereas the high-temperature slope might be too steep.

As deduced from Tables 3.1 and 2, the migration enthalpies are much lower than the solution enthalpies of the respective metals. As a consequence, the solubilities extrapolated to RT exhibit extremely low values (<1 atom/cm^3) whereas the diffusivities of the sequence chromium to copper still exhibit reasonable values (Fig. 3.2). This agrees with the experience that these metals are still mobile at RT, it is proven by their RT reactions forming donor-acceptor pairs with the almost immobile shallow-doping elements boron, aluminum, gallium, and indium (Sect. 3.3.2).

An almost vanishing solubility means that the metal is unstable on interstitial sites in the host silicon lattice at RT. Consequently, there is a tendency for the supersaturated transition metals to leave the silicon lattice during the cooling period if the cooling rate is sufficiently low. This outdiffusion of transition metals to the sample surfaces is the physical basis for the formation of so-called **haze**, treated in more detail in Sect.3.4.2. For higher cooling rates or thicker samples, only those metal atoms which are situated near to a surface can reach it during the cooling period. The remaining atoms tend to precipitate within the bulk and form a new phase. Only a few 3d transition metals are able to precipitate by a homogeneous nucleation mechanism (Co, Ni, Cu) in the absence of heterogeneous nucleation centers such as other impurity precipitates or lattice defects. This is discussed in more detail in Sect.3.4.1.

Finally, transition metals which exhibit lower diffusivities remain supersaturated and dissolved on interstitial sites in the silicon lattice. Because of their instability they tend to form more stable complexes and compounds with other impurities even at RT. This will be treated in Sect.3.3. The wide temperature range for the diffusivities of Cr, Mn, Fe, and Cu listed in Table 3.2 was obtained from interpolations of HT measurements and RT values deduced from the formation of metal-boron pairs. Interpolations between high- and low-temperature diffusion data for manganese and chromium have been performed recently and yielded reasonable results [3.28], showing that the diffusivities can be calculated by applying only one Arrhenius equation in the entire temperature region. This was also demonstrates for copper by *Mesli* et al. [3.27] in a wide temperature range.

The same chemical trends observed for the solubilities and for the diffusivities in the sequence of the 3d transition metals cannot be generalized. The doping elements such as boron and phosphorus, for example, exhibit much higher solubilities but much lower diffusivities compared to the 3d transition metals. This discrepancy is not caused by the substitutional diffusion mechanism of the shallow doping elements in the silicon lattice. Interstitial oxygen in silicon exhibits similar relations: the solubility equals that of copper and nickel, but the diffusivity of oxygen in silicon is five to six orders of magnitude lower than that of copper or nickel.

3.3 Dissolved Impurities

The slowly diffusing light interstitial 3d-transition metals from scandium to iron (with respect to the fast diffusing heavy metals Co, Ni and Cu) can be kept dissolved even at RT where they are metastable due to their almost vanishing solubilities. Defect concentrations which nearly agree with the solubilities of the impurity at the respective diffusion temperature can be achieved by quenching the diffused samples to RT. In the sequence of the 3d transition metals, the concentrations of the quenched-in defects change abruptly with the transition from iron to cobalt. The concentrations of quenched-in defects for Co, Ni and Cu are of the order of magnitude between 10^{-2} and 10^{-4} of the respective solubility [3.29] depending on the metal, the temperature of the diffusion, and the sample preparation. The reason for this abrupt change is the increase of the diffusivity by about one order of magnitude at high temperature and two at RT between iron and cobalt. As a consequence, the three fast-diffusing metals Co, Ni and Cu precipitate almost quantitatively even during rapidly quenching the samples to RT.

However, as mentioned above, a small fraction of the fast diffusing metals remains dissolved and electrically active. They form deep energy levels in the band gap of silicon. In contrast to the slowly diffusing 3d transition metals which form defects on interstitial sites the isolated defects of the fast diffusing metals occupy substitutional sites in the host lattice of silicon. Beside these isolated point defects there are still other electrically active deep energy levels which could be correlated to pairs between the respective transition metal and other impurities such as shallow acceptors, hydrogen, or other transition metals. In general, these pairs are combinations of an acceptor state bound electrostatically to a donor state. Donor-acceptor pairs can even be formed between two atoms of the same metal where the acceptor state is due to a substitutional metal atom and the donor is due to an interstitial metal atom such as $Cu_s Cu_i$ which has been identified first [3.30].

As mentioned in Sect.3.1, the tendency to form substitutional defects increases with increasing atomic numbers within the sequence of the 3d transition metals but also in the sequences of the 4d and 5d transition metals. This means that the proportions of the solubilities of substitutional defects increase with respect to the solubilities of the interstitial defects. The solubility data listed in Table 3.1 are interstitial solubilities with the exception of those for zinc, which might mainly be a substitutional solubility since no interstitial defects of zinc are known. The diffusivities listed in Table 3.2, which can be fitted by an Arrhenius equation are likewise interstitial diffusivities. For a more detailed discussion see Sect.4.6.2. The substitutional solubilities of cobalt, nickel, and copper are very low which can be deduced from the extremely small concentrations of substitutional defects with respect to the high solubility values, although substitutional defects are more stable during cooling the sample in comparison to interstitial defects which tend to precipitate or diffuse to the surfaces because of their higher mobility.

Within the sequence of the 3d transition metals it is now well-known that the lighter metals from scandium to iron form interstitial defects, if the samples were quenched from high temperatures to RT. The heavier metals from cobalt to zinc, on the other hand, form substitutional defects although in low concentrations with respect to the much higher interstitial solubilities, with the exception of zinc. Although the boundary between interstitial and substitutional defects is not exactly known for the 4d and 5d transition metals it is tentatively assumed that this boundary will also be situated between the iron- and the cobalt group, following the supposition of *Lemke* [3.31]. The assumption is based on the simple model that the properties of the transition metals are mainly dependent on the sum N of the outer s and d electrons (Fig. 1.1) forming a regular sequence from 3 to 12. Then the groups with N from 3 to 8 form interstitial defects, and the groups from 9 to 12 would form substitutional defects. This, however, is proven so far only for the sequence of the 3d metals. Further peculiarities of the various groups will be dicussed in Sect. 3.3.1 to follow, where the electrical properties of the point defects will be treated in more detail.

3.3.1 Point Defects

The quenched-in interstitial defects of the slowly-diffusing 3d transition metals from scandium to iron are electrically active and form deep energy levels in the band gap of silicon. The concentrations of these defects were determined by electrical methods such as Deep Level Transient Spectroscopy (DLTS) [3.32], Electron Spin Resonance (ESR) [3.5], or Hall-effect measurements [3.33] (Sect. 6.2). In general, the application of these electrical methods results in $10 \div 30\%$ lower concentrations compared to the results obtained by means of a neutron-activation analysis, which has been widely used to determine the solubilities of the transition metals. The reason for this loss in concentration is not clear. *Weber* [3.5] speculated that it might be due to the loss of part of the interstitially dissolved metals during the quenching of the silicon sample to RT. However, so far it cannot be fully excluded that this discrepancy between the two different measurement methods may be at least partly due to the summing up of measurement errors. The methods applied are quite different and the measurement error for the determination of concentrations by taking advantage of electrical methods (in particular, DLTS) is estimated to be of the order of magnitude of 10%. This problem will be discussed once more in Sect. 8.2 where it will be shown that the existence of residual nucleation centers, even in as-received silicon wafers, seems to be the most likely cause of the measurement discrepancies in determining the defect concentrations.

In general, the electrically active centers of the quenched-in light 3d-transition metals are due to interstitial defects. This was proven by symmetry investigations applying ESR measurements [3.5, 15]. However, if special sample-preparation techniques are employed (e.g., co-diffusion of the respective metal with large amounts of copper [3.17]) substitutional defects can also be formed from some of these metals. The properties of the interstitial and the substitutional defects differ considerably. Common preparation techniques assumed the interstitial defects for the light 3d transition metals with $N \leq 8$ are always predominant and cause the technical importance of these impurities.

Because of the confusing multiplicity of different results for the activation energies of the deep levels of transition metals in the silicon band gap in the earlier literature [3.33, 34], on the one hand, and the missing data, on the other hand, systematic investigations were performed in the early 1980s to obtain reliable data [3.29, 35]. The resurgence in research was promoted by the application of the new measurement technique, DLTS, which was developed by *Lang* several years earlier [3.32]. For the chemical analysis of semiconductor materials the DLTS signals must be calibrated to correlate the temperature of the maximum DLTS signal to the known deep energy levels of definite impurities. Meanwhile, this new measurement technique proved to be useful and highly adaptable in solving many problems in device manufacturing. With the extremely high sensitivity for impurity detection by means of DLTS, new problems arose concerning a sufficiently clean sample-preparation technique. Since that date the importance of sample preparation has still increased. For a suitable preparation of defined transition-metal defects, all the knowledge about the behavior of the impurities in silicon during heat treatments and during the cooling period are required. It is one aim of this monograph to prescribe the conditions for a suitable sample preparation useful for detailed investigations of the properties of impurities in silicon.

Since the time when the first lists of activation energies for the deep levels of the 3d transition metals in the silicon band gap were published [3.29, 35], a series of particular investigations verified the new data within experimental errors. Thus average values can be presented today and it is not expected that there will be a significant modification of most of these data in future.

In general, the electrically active interstitial defects of the light 3d transition metals exhibit four different charge states: double positively charged (+ +), single positively charged (+), neutral (0) and, finally, single negatively charged (−). Consequently they can form double donors (+ +/+), donors (+/0), and acceptor states (0/−). Not all of the different charge states can be found for all defects within the band gap. Scandium exhibits a triple donor, and chromium as well as iron show only a single donor state. The experimental data for the respective activation energies determined in eV are summarized in Table 3.3. The activation energies are measured from the neighboring band edge with a negative sign indicating the distance

Table 3.3. Activation energies of the 3d, 4d, and 5d transition metals are listed with n being the sum of s+d electrons

n	Metal	i/s	ddd	dd	(E_∞)	d	(E_∞)	a	aa
	3d Sc	i	+0.20	−0.50		−0.21			
3	4d Y	i							
	5d La								
	3d Ti	i		+0.28!	(0.038)	−0.27!	(0.004)	−0.08	
4	4d Zr	i		+0.32!		−0.42		−0.13	
	5d Hf	i		+0.32!		−0.40		−0.10	
	3d V	i		+0.32!	(0.14)	−0.45		−0.18	
5	4d Nb	i		+0.18!		−0.62		−0.28!	
	5d Ta	i?		+0.19!		−0.58		−0.22!	
	3d Cr	i				−0.22			
6	4d Mo	i				+0.28!			
	5d W	i?				+0.40!			
	3d Mn	i		+0.27!	(0.07)	−0.42		−0.12	
7	4d Tc								
	5d Re	i?				−0.35!		−0.07	
	3d Fe	i				+0.39!	(0.043)		
8	4d Ru	i				+0.26!		−0.14!	
	5d Os	i?				+0.30		−0.22!	
	3d Co	s				+0.41!		−0.41	
9	4d Rh	s				−0.58		−0.32	
	5d Ir	s				−0.62		−0.24	
	3d Ni	s				+0.17		−0.41	−0.07!
10	4d Pd	s		+0.12!		+0.30!		−0.21	
	5d Pt	s		+0.08!		+0.32		−0.23	
	3d Cu	s				+0.22		+0.46	−0.16!
11	4d Ag	s				+0.38!		−0.55!	
	5d Au	s				+0.34		−0.55!	
	3d Zn	s						+0.32	−0.53
12	4d Cd	s						+0.5	−0.45
	5d Hg								

from the conduction-band edge (E_c−0.xx eV) and a positive sign giving the distance from the valence-band edge (E_v +0.xx eV).

Furthermore, Table 3.3 lists correction values (E_∞) for activation energies which have been determined by Arrhenius plots. The Arrhenius plot represents the temperature-corrected thermal emission coefficients $e_n \times (300/T)^2$ on a logarithmic scale as a function of the inverse absolute temperature $1/T$ (Chap.6). If the activation energies have been determined from measured capture and emission rates, the results must not be corrected. However, if the activation energies were determined by Arrhenius plots, which is a wide-spread technique, the results must be modified by adding the correction term E_∞. Corrections must be performed only if the majority-carrier capture cross-section is temperature dependent. E_∞ is the gradient of the carrier capture cross-section $\sigma_{n/p}$ as a function of the inverse temperature $1/T$ which is an Arrhenius plot for the carrier capture cross-section [3.36]:

$$\sigma_{n/p} = \sigma_{n/p\,\infty} \exp\!\left(\frac{-E_\infty}{kT}\right). \tag{3.1}$$

E_∞ increases with an increasing temperature dependence of the carrier capture cross-section. If the majority charge-carrier capture cross-section $\sigma_{n/p}$ of an energy level (σ_n for electrons if the energy level is situated in the upper half of the band gap, and σ_p for holes if it is in its lower half) is independent of the sample temperature, E_∞ disappears. Otherwise, E_∞ must be added to correct the result obtained by the Arrhenius plot since this technique assumes a carrier capture cross-section which is independent of the sample temperature [3.36]

$$e_{n/p} = \sigma_{n/p}\,v_{th}\,\frac{N}{g}\exp\!\left(\frac{-E_T}{kT}\right). \tag{3.2}$$

Inserting (3.1) into (3.2) and taking the logarithm results in

$$\ln(e_{n/p}) = A - \frac{E_\infty + E_T}{kT}. \tag{3.3}$$

This shows that the gradient of the Arrhenius plot for the emission rates leads to the sum ($E_T + E_\infty$) and must be corrected.

A common method to measure the carrier capture cross-section by DLTS is the pulse-width-variation technique (Sect. 6.2.1). Applying this method, the temperature dependence of the carrier capture cross-section can be determined and yields E_∞. The activation energies presented in Table 3.3 are correct values if E_∞ is known. The correction terms E_∞ are indicated in the table in order to enable the identification of unknown defects if the activation energy was only determined by means of an Arrhenius plot. The activation energy for interstitial iron in p-type samples determined by means of an Arrhenius plot, for example, would result in $E_T + E_\infty = 0.39 + 0.04 = 0.43$ eV. In Table 3.3 these enthalpy values are marked by an exclamation mark.

Table 3.3 is organized following the assumption of *Lemke* [3.31] that the properties of the transition metal dissolved in silicon depend mainly on its respective valence n which is the sum of the outer s and d electrons.

Therefore, the groups of equal n are listed together in the sequence 3d, 4d, and 5d transition metals. Since many of the so far unknown rare transition metals have been investigated recently by *Lemke* [3.31] most of the former blanks could be eliminated. As a consequence, the following chemical trends can be deduced from Table 3.3:

(i) The most conspicuous trend is the shift from the single triple donor state of Sc to the double acceptor states of Zn and Cd with increasing n from 3 to 12, although there are several exceptions from a strict regularity (Mn, Pd, Pt). All interstially dissolved transition metals exhibit at least one donor state. All substitutional transition metals exhibit at least one acceptor state. It can be expected that the blanks will fit in this system.

(ii) A further conspicuous property is observed in the group $n = 6$ where all three metals (Cr, Mo, W) exhibit only one donor state.

(iii) Almost all metals in one group exhibit very similar activation energies in the sequence from the 4d to the 5d transition metals: Zr-Hf, Nb-Ta, Ru-Os, Rh-Ir, Pd-Pt, and Ag-Au, (3 groups exhibit blanks for the 4d, 5d or both transition metals). Even in the so far only exception ($n=6$) the difference is still rather small. Exceptions from a similar behaviour within one group are only the 3d transition metals Fe, Ni, Cu, and eventually Mn which, however, is doubtful because of the blank for Tc. It appears strange that these three exceptions form the main impurities in silicon-device fabrication.

Because of the mentioned similarities of the properties of metals situated within one group the boundary between interstitial and substitutional defects is tentatively placed between $n = 8$ and $n = 9$. It is proven by experiment that iron forms an interstitial defect and the theory predicts that the substitiutional defect is not stable [3.1]. So far interstitial cobalt could not yet be detected. Because of its high diffusivity it diffuses to sample surfaces or precipitates even during quenching the sample from high temperatures to RT. On the other hand, signals observed in Mössbauer spectra on cobalt contaminated and highly boron-doped silicon samples were interpreted as CoB pairs [3.37]. However, the existence of these pairs formed during cooling the sample does not prove the existence of an interstitial cobalt at RT. After the dissociation of the pair the interstitial cobalt will diffuse immediately to the next neighbouring precipitate.

The data listed in Table 3.3 are often mean values averaged from those published results which were believed to be most reliable. For rare metals often only one result was available. Averaging has been performed to decrease measurement errors and to avoid the pain of the selection. In general, the deviation between the mean value and the published results is less than 0.2 eV, often only 0.1 eV which does not exceed the expected measurement error. The references are cited in Chaps.4 and 5 where the vari-

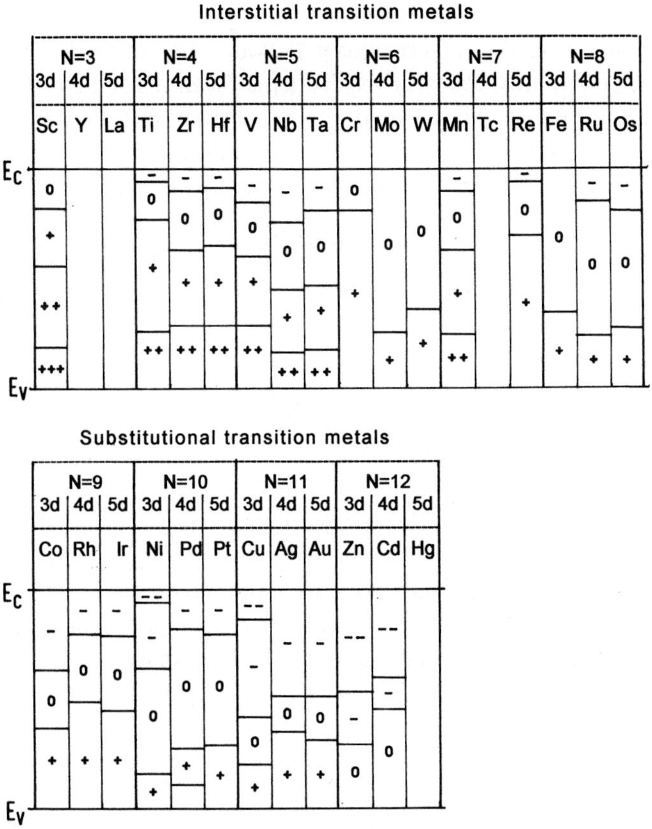

Interstitial transition metals

N=3			N=4			N=5			N=6			N=7			N=8		
3d	4d	5d	3d	4d	5d	3d	4d	5d	3d	4d	5d	3d	4d	5d	3d	4d	5d
Sc	Y	La	Ti	Zr	Hf	V	Nb	Ta	Cr	Mo	W	Mn	Tc	Re	Fe	Ru	Os

E_c (top) ... E_v (bottom). Charge states, read from E_c down to E_v:

- Sc: 0, +, ++, +++
- Y: —
- La: —
- Ti: −, 0, +, ++
- Zr: −, 0, +, ++
- Hf: −, 0, +, ++
- V: −, 0, +, ++
- Nb: −, 0, +, ++
- Ta: −, 0, +, ++
- Cr: 0, +
- Mo: 0, +
- W: 0, +
- Mn: −, 0, +, ++
- Tc: —
- Re: −, 0, +
- Fe: −, 0, +
- Ru: −, 0, +
- Os: −, 0, +

Substitutional transition metals

N=9			N=10			N=11			N=12		
3d	4d	5d	3d	4d	5d	3d	4d	5d	3d	4d	5d
Co	Rh	Ir	Ni	Pd	Pt	Cu	Ag	Au	Zn	Cd	Hg

E_c (top) ... E_v (bottom). Charge states, read from E_c down to E_v:

- Co: −, 0, +
- Rh: 0, +
- Ir: 0, +
- Ni: − −, −, 0, +
- Pd: −, 0, +
- Pt: − −, −, 0, +
- Cu: −, 0, +
- Ag: −, 0, +
- Au: − −, −, 0, +
- Zn: − −, −, 0
- Cd: −, 0
- Hg: 0

Fig. 3.3. Activation energies of the defects of interstitial and substitutional transition metals. (N: sum of s+d electrons, charge states are indicated)

ous metals are discussed seperately and in more detail. All activation-energy values compiled in Table 3.3 are also plotted in Fig. 3.3.

Soon after the publication of the reliable activation energies of the 3d transition metals in silicon [3.29] several groups started to develop models which allow the calculation of deep acceptor and donor levels of interstitial and substitutional 3d transition metals in silicon [3.1-3, 38-40]. Whereas the first cluster-model calculations [3.38, 39] exhibited substantial discrepancies from the experimental findings, the Spin-unrestricted Density-Functional (SDF) calculations for single transition-metal ions in an otherwise perfect, infinite crystal succeeded in predicting more realistic results. They agree fairly well with the experiments if low-spin ground states are assumed for the light interstitial, and the heavy substitutional 3d transition metals [3.1]. This assumption broke with the so far generally accepted Ludwig-Woodbury model which generally assumed high-spin configurations [3.17].

The SDF model predicted that interstitial iron forms only one donor level within the forbidden band and substitutional iron is not stable in silicon. As a consequence the unsuccessful experimental search for substitutional iron could be terminated. Thus far, the only substantial discrepancy between theoretical and experimental results for the interstitial 3d transition metals from Ti to Fe is the prediction of the existence of a chromium double donor which has never been observed in experiments. A further discrepancy is the triple donor of Sc discovered experimentally. Although there are quantitative differences between theoretical and experimental values which exceed the expected measurement error, the chemical trends agree well.

Further parameters which characterize the properties of deep levels are the capture cross-sections for electrons and holes. They can be measured by means of the pulse-width-variation technique using the DLTS equipment (Sect.6.2). Although the procedure is not too complicated the experimental results are not satisfactory, and theoretical predictions are not yet available. The reason for the rather large measurement errors is the requirement of extremely short pulses to determine high carrier capture cross-sections. Although short pulses can be generated by modern electronic equipment, they must be transferred to the sample which is mounted in a cryostat, avoiding pulse deformations and elongations of the decay times. However, this requires additional expense.

Thus, the data available in the literature scatter sometimes by orders of magnitude; and new, more reliable experiments are needed. Primarily, only majority-carrier capture cross-sections have been determined. This means that the values for electrons were measured for traps with energy levels situated in the upper half of the silicon band gap. Those for holes have been determined for traps with energy levels in the lower half of the band gap. In addition, the capture cross-sections can depend on the sample temperature. The respective activation energies are characterized by the known values E_∞, which are listed in Table 3.4 if they are known. Not all researchers who determined capture cross-sections of deep levels measured their temperature dependence. Therefore the data are not complete.

The experimental results for the majority-carrier capture cross-sections of the transition metals are compiled in Table 3.4. The type of the respective majority carrier (electrons or holes) is indicated by e and h, respectively. The few data which were available for minority-carrier capture cross-sections are not listed in this table. The carrier capture cross-section at a defined temperature was calculated by applying [3.41]

$$\sigma_{n/p} = \sigma_0 \exp\left(\frac{-E_\infty}{kT}\right) \tag{3.4}$$

Table 3.4. Majority-carrier capture cross-sections σ of the transition metals listed with N being the sum of s and d electrons

n		Metal	i/s	ddd	dd	d	a	aa
3	3d	Sc	i	h $1.1\cdot10^{-19}$	e $2\cdot10^{-14}$	e $3\cdot10^{-14}$		
	4d	Y	i					
	5d	La						
4	3d	Ti	i		h $1.9\cdot10^{-16}$ *	e $1.3\cdot10^{-14}$	e $3.5\cdot10^{-14}$	
	4d	Zr	i		h $1.3\cdot10^{-17}$	e $>10^{-14}$	e $>10^{-14}$	
	5d	Hf	i		h $>5.0\cdot10^{-18}$	e $>2\cdot10^{-14}$	e $>2\cdot10^{-14}$	
5	3d	V	i		h $2.2\cdot10^{-18}$ *	e $2\cdot10^{-15}$	e $1.6\cdot10^{-16}$	
	4d	Nb	i		h $3.8\cdot10^{-16}$	e $>5.5\cdot10^{-16}$	e $7.5\cdot10^{-18}$	
	5d	Ta	i		h $6.0\cdot10^{-17}$	e $>3.9\cdot10^{-15}$	e $2.2\cdot10^{-17}$	
6	3d	Cr	i			e $7.3\cdot10^{-15}$		
	4d	Mo	i			h $6.0\cdot10^{-16}$		
	5d	W	i			h $5.0\cdot10^{-16}$		
7	3d	Mn	i		h $2.0\cdot10^{-18}$ *	e $3.1\cdot10^{-15}$	e $3.1\cdot10^{-15}$	
	4d	Tc						
	5d	Re	i			e $5.1\cdot10^{-16}$	e $8.7\cdot10^{-16}$	
8	3d	Fe	i			h $3.8\,10^{-17}$ *		
	4d	Ru	i			h $9.2\cdot10^{-16}$	e $1.1\cdot10^{-16}$	
	5d	Os	i			h $8.0\cdot10^{-16}$	e $4.6\cdot10^{-17}$	
9	3d	Co	s			h $5.0\cdot10^{-18}$	e $2.2\cdot10^{-15}$	
	4d	Rh	s			e $2.0\cdot10^{-14}$	e $5.6\cdot10^{-15}$	
	5d	Ir	s			e $7.2\cdot10^{-14}$	e $9.1\cdot10^{-15}$	
10	3d	Ni	s			h $5.4\cdot10^{-15}$	e $1.2\cdot10^{-16}$	e $5.4\cdot10^{-18}$ *
	4d	Pd	s		h $6.5\cdot10^{-17}$	h $5.6\cdot10^{-16}$	e $1.6\cdot10^{-15}$	
	5d	Pt	s		h $3.5\cdot10^{-17}$	h $8.4\cdot10^{-15}$ *	e $2.9\cdot10^{-14}$	
11	3d	Cu	s			h $3.0\cdot10^{-14}$	h $1.5\cdot10^{-14}$	e $1.9\cdot10^{-17}$ *
	4d	Ag	s			h $7.0\cdot10^{-16}$	e $7.2\cdot10^{-17}$	
	5d	Au	s			h $2.5\cdot10^{-15}$	e $1.4\cdot10^{-16}$	
12	3d	Zn	s				h ...	h $2.5\cdot10^{-15}$
	4d	Cd	s				h ...	e ...
	5d	Hg						

where σ_0 is the value listed. Capture cross-sections which are known to depend on the temperature are marked by an asterisk (*) in Table 3.4. The value corresponds to the temperature of the DLTS maximum. Most of these temperatures are listed in the detailed tables of the respective metal in Chaps. 4 and 5.

In addition to interstitial defects, substitutional defects of the 3d transition metals are formed in silicon by applying the special preparation techniques mentioned before in Sect. 3.1. From theory it was predicted that titanium and iron do not form substitutional defects [3.1]. Within the sequence from scandium to iron discussed in this chapter, substitutional defects have only been detected and investigated for manganese [3.16]. For

Table 3.5. Experimental (exp) [3.16] and theoretical (theor) [3.1] energy levels of substitutional manganese [eV]

Charge state	ΔE_A (exp) [eV]	ΔE_A (theor) [eV]	Concentr. (exp) [cm^{-3}]
d	+0.34	+0.5	$10^{13} \div 10^{14}$
a	− 0.43	− 0.4	$10^{13} \div 10^{14}$
aa	...	− 0.22	...

the formation of the substitutional manganese *Lemke* reproduced the preparation technique proposed by *Ludwig* and *Woodbury* in the early 1960s [3.17] by applying co-diffusion of copper and manganese. From the five levels observed, three (one donor and two acceptors) were attributed to copper, and the remaining two levels were attributed to substitutional manganese. The results are listed in Table 3.5 together with the respective theoretical predictions [3.1] and the estimated defect concentrations. The deviations between theory and experiment equal those which were found for the interstitial defects. The existence of a double acceptor state was predicted by theory but could not be observed in the experiment. The concentrations of the supposed substitutional manganese defects are about one order of magnitude lower than the respective solubilities of manganese at the diffusion temperatures employed (950 ÷ 1050°C). The lower concentration of dissolved manganese can be explained by the addition of copper which forms precipitates during quenching the sample from HT. These copper precipitates can also act as sinks for interstitial manganese thus reducing the concentration of electrically active defects. The substitutional as well as the interstitial manganese is without any importance for device manufacturing to date. It should be mentioned that the stable form of manganese is the interstitial defect. The substitutional defect is only formed in the presence of self-interstitials in concentrations well exceeding those in thermal equilibrium, which was generated by copper co-diffusion.

The fast-diffusing 3d transition-metals cobalt, nickel, and copper also exhibit electrically active, deep levels which show concentrations usually well below 1% of the respective solubilities. Because of their high diffusivities even at low temperatures it is expected that the interstitial defects of these metals are not stable at RT because of outdiffusion and precipitation. On the other hand, substitutional defects are known to diffuse much slower and there is a chemical trend of increasing substitutional defects with increasing atomic numbers, as mentioned in Sect. 3.1. For zinc, for example, which is the heaviest 3d transition metal, only a substitutional defect is

Table 3.6. Experimental (exp) [3.42-45] and theoretical (theor) [3.1] energy levels of substitutional cobalt [eV]. (The number in parentheses gives the number of results used for averaging)

Charge state	ΔE_A (exp) [eV]	ΔE_A (theor) [eV]
d	+0.40 (4)	− 0.48
a	− 0.41 (3)	− 0.35
aa	...	− 0.25

known. Therefore, it could be useful to compare the experimental results published in the literature with predictions for the substitutional defects. Cobalt, however, is the heaviest metal for which calculations have been published [3.1]. In Table 3.6 the results for the activation energies of substitutional cobalt obtained from theory are compared with the mean values from experiments [3.42-45]. The numbers in brackets indicate the number of experimental results found in the literature which fairly well agree with the predictions of the theory only for the acceptor. So far, the predicted double acceptor of cobalt was not observed. Although an acceptor level at $E_c − 0.21$ eV has been determined experimentally in cobalt contaminated samples [3.45, 46] this could be correlated recently to a cobalt-hydrogen complex, formed by etching the sample before evaporating Schottky contacts [3.45].

More reliable results were reported meanwhile for the substitutional defects of nickel, copper, and zinc, which are compiled in Table 3.3 and plotted in Fig. 3.3. Their reliability is enhanced due to the obvious similarities of results within one group of equal valences. In contrast to nickel and copper which belong to the main impurities in silicon during device fabrication zinc is usually not found in the bulk of wafers, although it is widespread as an impurity and even can be detected on the surfaces of as-received wafers. Due to its high vapor pressure, zinc completely evaporates during heat treatments and condenses mostly as an oxide at places of lower temperature in the furnace tube (Sect. 5.7). Electrically active zinc defects can be formed only by an encapsulated diffusion process excluding the presence of oxygen.

For all these impurities many identified and unidentified levels were reported showing that the large amount of highly mobile interstitial defects react with one another and with any other impurity present in the crystal to overcome their supersaturation during cooling the sample.

3.3.2 Complexes

The electrically active interstitial point defects of the lighter 3d transition metals are metastable and, at least, some of them exhibit a mobility that is not too low at RT. They tend to form complexes and pairs to achieve a more stable form in the silicon host lattice. Of technological importance is the formation of donor-acceptor pairs which, for example, must be taken into account to determine the total impurity concentration of chromium, manganese, iron, and gold in a sample.

Another kind of complexes can be formed in the surface region of a sample by a chemical etching process, as reported in Sect.2.2.2. Etching is a common preparation step for the evaporation of Schottky barriers required for DLTS measurements. During the etching process hydrogen is formed, which penetrates the surface region of the sample even at room temperature. The driving force seems to be a drift of the positively charged atomic hydrogen in the space-charge region of the silicon surface since the penetration depth of hydrogen is independent of the temperature, of the etching duration, and of the selection of etch solution [3.47]. Furthermore, the penetration depth is changed with the application of a reverse bias to the Schottky barrier [3.45]. The hydrogen can then react with the doping elements mainly boron or phosphorus and with metal impurities forming electrically active or inactive hydrogen complexes.

Finally there is a further kind of complexes which might be a special form of donor-acceptor pairs with both charge states formed by the same metal: $Me_i Me_s$. These pairs can also be electrically active or neutral.

All three types of complexes, the donor-acceptor pairs, the hydrogen complexes, and the transition metal pairs are discussed subsequently as follows:

Donor-acceptor pairs are formed between the positively-charged donor state of the interstitial metal impurities and the negatively-charged acceptor state of the doping element which is usually boron in a p-type sample. Since iron belongs to the main impurities in device manufacturing, the iron-boron pairs are common defects in p-type samples which were cooled sufficiently quickly to Room Temperature (RT) after a diffusion process. For common doping concentrations exceeding about 10^{14} cm^{-3}, the Fermi level in p-type silicon at RT is located between the energy level of the iron donor ($E_v + 0.39$eV) and the energy level of the boron acceptor ($E_v + 0.044$eV). Therefore, the donor state of the iron is positively charged and the acceptor state of the boron is negatively charged. While the substitutional boron atom is immobile at RT, the interstitial iron atom is able to diffuse through the silicon lattice and will occasionally approach a boron atom. Opposite-charge states attract one another and form an iron-boron pair. The reaction is diffusion limited and highly dependent on the boron concentration in the sample, since this determines the mean distance between iron and boron atoms. Usually the iron concentration is orders of magnitude lower than the boron concentration and therefore does not noticeably influence the reaction time. The following example should give a rough estimate of the reaction times: in common p-type samples with resistivities between 5 and 10 $\Omega \cdot$cm about one half of the common iron concentrations is paired after a period of about one hour has elapsed after cooling the sample to RT. This time is usually needed to prepare the Schottky contacts for the DLTS measurements. High iron contents on the order of 10^{14} cm^{-3} are almost completely paired within one day following the iron diffusion and cooling the sample to RT. In samples of higher resistivities the pairing reaction needs considerably longer times, whereas the pairing is much faster in samples of lower resistivities.

The iron-boron pair is electrically active and forms a donor state in the lower half of the silicon band gap ($E_v +0.1$ eV) [3.48] and an additional acceptor state in the upper half of the band gap ($E_c -0.27$ eV) [3.49]. The pairing reaction can easily be observed by recording the donor state in the DLTS spectrum. The reaction takes place in the near-RT region up to about 200°C where iron starts to precipitate and reduces the interstitial iron content with time [3.48]. In this temperature region a thermal equilibrium between interstitial iron and substitutional boron, on the one hand, and iron-boron pairs, on the other hand, is formed:

$$[Fe^+] + [B^-] \longleftrightarrow [Fe^+B^-] \, . \tag{3.5}$$

The equilibrium is shifted towards the iron-boron pairs at low temperatures (0°C) and towards the dissociated pairs at about 200°C. At higher temperatures the iron concentration will be reduced with time due to the beginning of precipitation [3.48].

The equilibrium between the pairs and the dissociated defects can be shifted versus the isolated point defects by illuminating the sample with intense white light [3.48]. During illumination excess electron-hole pairs are generated in the sample, which shift the quasi-Fermi level toward the mid gap. This is completely analogous to an increase in the sample temperature. As a consequence, the interstitial iron atoms can become discharged and do not form pairs since the binding energy of the pairs is mainly determined by the electrostatic attraction. This was proven by the experimental finding that iron-boron pairs do not form in highly boron compensated n-type silicon even after extended storage of the samples at RT. In n-type samples the boron acceptor is still negatively charged but the iron donor is neutral, and a pairing reaction does not take place [3.50].

Dissociation of the iron-boron pairs can also be achieved by minority-carrier injection. This requires a pn junction and the application of a forward current [3.51]. The dissociation exhibits a linear dependence on the injection level and enables the determination of the electrostratic binding energy to 0.45 eV. Only a weak temperature dependence of 0.1 eV was measured [3.51]. Pair dissociation is assumed to be caused by discharging the iron donor state thus shifting the equilibrium (3.5) towards the dissociated defects.

The formation of iron-boron pairs affects the change of two important parameters:

(i) The resistivity of the p-type silicon sample increases if the iron concentration approaches the boron concentration by discharging the boron acceptor and changing it to a donor state. The acceptor state of the iron-boron pairs is situated in the upper half of the silicon band gap and is electrically neutral.

(ii) The rather high minority-carrier capture cross-section of interstitial iron in p-type silicon changes to lower values of the iron-boron pairs. As a consequence the pair formation increases the minority-carrier lifetime of the sample at low injection levels [3.52]. More details may be found in Sect. 4.1.4.

Both phenomena can be used to measure the iron concentration if other interfering impurities can be excluded (Sect.6.2). For a quantitative determination of the iron concentration in p-type wafers by means of DLTS, the concentrations of the interstitial iron and the iron-boron pairs must be summed up since both belong to different and independent defects. Spectra recorded at different stages of iron-boron pair formation are displayed in Fig.3.4 showing temperature scans soon after quenching the sample from HT (*1*), after two days (*2*), after 100 days storage time elapsed after quenching the sample to RT (*3*), and finally after sample illumination with bright light for one hour (*4*). For clarity the spectra are displaced in depth in form of a perspective graph. Detection of the FeB pairs by DLTS requires temperature scans down to about $50 \div 60$ K. Liquid-nitrogen cryostates cannot be used since they do not reach these low temperatures. If they are part of a commercial equipment they must be replaced by a closed-cycle helium refrigerator to enable the quantitative detection of iron in samples, including iron-boron pairs. This, of course, is more expensive and requires more measurement time (Sect.6.2.1).

Besides FeB pairs, other donor-acceptor pairs may be formed by RT reactions of quenched-in light 3d transition metals with shallow acceptors. In boron-doped p-type wafers, a small amount of unintentional aluminum doping is frequently observed. Like boron, the aluminum atoms form acceptor states which are positively charged. Hence, pairing with interstitial iron atoms is possible. In contrast to the single-defect structure of FeB pairs

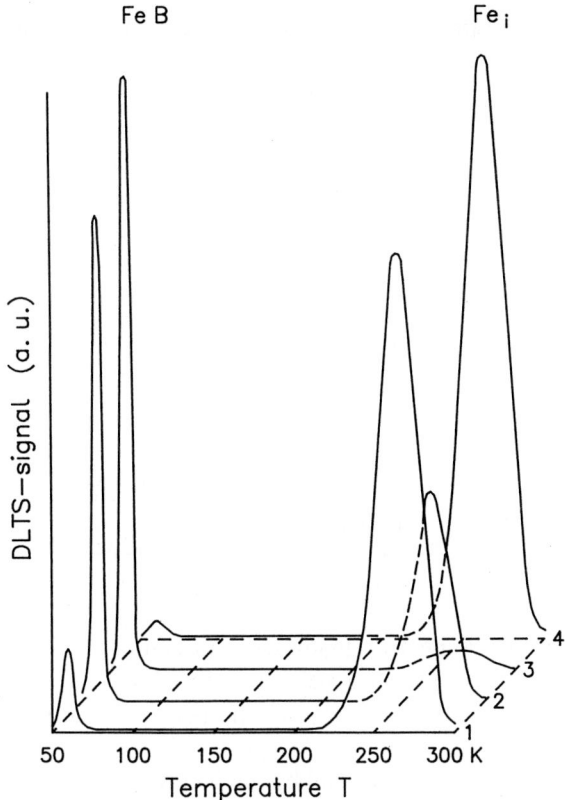

Fe B   Fe $_i$

DLTS–signal (a. u.)

50   100   150   200   250   300 K

Temperature T

Fig. 3.4. DLTS temperature scans of a quenched, iron-doped p-type sample. The displaced spectra were recorded at different times elapsed after quenching the sample and show stages of the iron-boron pair formation. Interstitial iron at 260 K, iron-boron pairs at 60 K. (1: soon after quenching the sample 2: after 2 days 3: after 100 days 4: after illumination for 1 hour)

with the boron atom on a substitutional lattice site bound to an iron atom on the next neighboring interstitial site in the $\langle 111 \rangle$ direction, FeAl pairs form two different defect structures with bindings in the $\langle 111 \rangle$ and $\langle 100 \rangle$ directions. Both defects are metastable and change from one configuration into the other by cooling the sample with and without an applied reverse voltage, respectively. Thus iron-aluminum pairs form two different donor states at $E_v + 0.2$ eV with pairing in the $\langle 111 \rangle$ direction, and $E_v + 0.13$ eV in the $\langle 100 \rangle$ direction [3.53].

As boron and aluminum, gallium and indium form shallow acceptor states in silicon on substitutional lattice sites, and they form as well iron-acceptor pairs in a similar way. Since these acceptors are less important im-

Table 3.7. Properties of donor-shallow acceptor pairs, the pair axis being in the $\langle 111 \rangle$ (1) or $\langle 100 \rangle$ (2) direction

Metal	Cr ΔE_A	d/a	E_B	Mn ΔE_A	d/a	E_B	Fe ΔE_A	d/a	E_B	
B	+0.29	d	0.4	−0.53	d	0.5	+0.10	d	0.6	(1)
							− 0.27	a		(1)
							<250K		metastable	(2)
Al	+0.45	d	0.8	−0.45	d	0.46	+0.20	d	0.7	(1)
							+0.13	d	metastable	(2)
Ga	+0.48	d	0.35	−0.42	d	0.29	+0.25	d	0.47	(1)
							+0.14	d	metastable	(2)
In							+0.27	d	metastable	(1)
							+0.15	d		(2)

purities in device fabrication, the respective donor-acceptor pairs are rarely found unless these acceptors have been intentionally introduced into the silicon crystal, for example, during gallium diffusion. The respective activation energies are listed in Table 3.7.

In addition, donor-acceptor pairs are formed by the interstitial 3d transition-metals chromium and manganese which likewise exhibit a sufficient mobility at RT. Donor-acceptor pairs of vanadium have been observed by EPR [3.17] but their activation energies have not yet been determined. The existence of cobalt- and nickel-acceptor pairs is expected. Signals in Mössbauer spectra on cobalt-doped samples have been interpreted as CoB pairs [3.37] but their electrical activity was not yet correlated to one of the unidentified deep energy levels observed by DLTS after intentional cobalt doping. The activation energies measured in eV of the various pairs between chromium, manganese, and iron as donors and boron, aluminum, gallium, and indium as acceptors are listed in Table 3.7 (Relevant references can be found in Chaps. 4 and 5).

By comparing the various activation energies of the same pair configuration (type 1 means the axis is in the $\langle 111 \rangle$ direction) as listed in the table, *Feichtinger* et al. [3.54] observed a chemical trend in the activation energies of pairs which follow those of the shallow acceptors, on the one hand, and those of the *double* donors of the transition metals, on the other. The binding energies of the pairs indicated in Table 3.7 amount to values between 0.3 and 0.8 eV and were taken from [3.49] by *Lemke*.

Besides the transition-metal donor-*shallow* acceptor pairs discussed above, there are additional transition-metal donor-*deep* acceptors pairs which can be of limited technological importance. These pairs are formed by the donors of the same interstitial 3d transition-metals chromium, manganese, and iron extended to vanadium, and the acceptor state of a substitutional transition metal such as zinc or gold [3.55]. The acceptor state of substitutional platinum should also form donor-acceptor pairs, as observed by EPR [3.17], but their activation energies have not yet been determined. The activation energies of the known donor-deep acceptor pairs are compiled in Table 3.8. Especially the iron-gold pairs can be of technological importance if gold is an intentional or an unintentional impurity. These pairs do not form at RT but at slightly elevated temperatures between 200° and 300°C [3.56] when slowly cooling the sample after a diffusion process or during a subsequent annealing process. It is assumed that the pair formation requires a positively charged donor state and a negatively charged acceptor state. Gold exhibits an acceptor state at $E_c-0.54$ eV and a donor state at $E_v+0.35$ eV. In n-type silicon at RT the gold acceptor is negatively charged but the iron donor is neutral. On the other hand, in p-type silicon the iron donor is positively charged but the gold acceptor is neutral. Thus, pairing takes place only at elevated temperatures where the Fermi level is shifted towards the midgap. At about 250 K the gold acceptor coincides with the Fermi level and the iron donor is slightly below the Fermi level. But at elevated temperatures the occupation statistics of electrons at deep energy levels do not exhibit an abrupt transition at the Fermi level. So the gold acceptor is partly negative and neutral whereas the iron donor is partly positive and neutral and pairing reactions can take place in n- and p-type wafers.

Since the binding energy of the iron-gold pairs with 1.22 eV [3.56] exceeds the highest value of the transition-metal shallow-acceptor pairs, the thermal dissociation of the iron-gold pairs starts at higher temperatures. The pairs formed at the elevated temperatures are stable at lower tempera-

Table 3.8. Energy levels of donor-deep acceptor pairs [eV]

Metal	d/a	V	Cr	Mn	Fe	Ni	Cu	Remark
Zn	d	+0.29		+0.18				
	a		−0.1		−0.47			
Pd					−0.32			
Au						+0.35	+0.32	bistable
						+0.48	+0.42	bistable
	d	+0.42	+0.35	+0.57	+0.43			
	a	−0.20		−0.24	−0.35			

tures. However, the pairing reaction is not quantitative, as found for iron-boron pairs. As a consequence, iron-gold pairs and substitutional gold acceptor and donor states are observed simultaneously in the sample. Since the iron-gold pair exhibits a donor state in the lower half of the silicon band gap and an acceptor state in the upper half, it can be observed by DLTS in n-type and in p-type samples. In order to get the total gold and the iron concentrations, the substitutional gold and the iron-gold pairs must be summed up as well as the interstitial iron, the iron-boron pairs and the gold-iron pairs, respectively.

In contrast to the formation of iron-boron pairs which increase the carrier lifetime of the specimen at least for low injection levels, the formation of iron-gold pairs decreases the carrier lifetime. A reduction by a factor of five compared to the substitutional gold of equal concentration has been observed [3.56]. This is a remarkable phenomenon since gold is known as a strong recombination center. Therefore it is important to avoid the unintentional formation of iron-gold pairs in electronic devices. However, this is only possible by avoiding the gold contamination since iron is always present in processed wafers at least in minor concentrations, even in the as-grown material before processing (Chap. 9).

The iron-zinc pairs are technologically unimportant. Although zinc is a wide-spread impurity which can be found, for example, on the surfaces of polished wafers, it does not usually diffuse into wafers because it evaporates immediately at higher temperatures due to its rather high vapor pressure (Sect. 5.7).

Hydrogen complexes can be formed in the surface region of a silicon sample if atomic hydrogen is introduced into this region. This can be performed by a chemical etching of the silicon surfaces using well-known acid or alkaline etch solutions. These may be various combinations of hydrofluoric acid, and nitric acid with or without an addition of acetic acid at room temperature, or potassium hydroxide at slightly elevated temperatures ($>60°C$). Another possibility to introduce atomic hydrogen into the silicon surface region is the application of a hydrogen plasma at voltages between 200 and 300 V and sample temperatures between 100° and 300°C. Other techniques such as ion implantation or hydrogen diffusion at high temperatures ($>1100°C$) are less convenient because of a possible introduction of additional and unwanted defects.

Due to early investigations of *Van Wieringen* and *Warmholtz* [3.57] hydrogen exhibits a very high diffusivity in silicon with a migration enthalpy of 0.48 eV probably valid for the positively charged ion H^+. In recent experiments performed by chemically etching silicon samples with different doping concentrations *Jost* [3.58] could demonstrate the high diffusivities. From his published penetration depths of hydrogen in silicon a rough estimation of the diffusivity could be deduced exhibiting a value in the order of

10^{-9} cm^2/s for H$^+$ at room temperature, which agrees fairly well with the results obtained by *Van Wieringen* and *Warmholtz*.

Atomic hydrogen can be found in silicon in three different charge states: neutral, single negatively charged, and single positively charged. As a consequence it forms one acceptor and one donor state within the band gap of silicon. The positions of the two levels within the silicon band gap is not yet well understood. *Johnson* et al. [3.59] proposed a negativ U-character of hydrogen in silicon exhibiting a donor level at $E_c - 0.16$ eV and an acceptor level at $E_c - 0.52$ eV. Assuming this case, hydrogen would change its charge state from a positive to a negative state by shifting the Fermi level in n-Si between both levels. In between both levels there should be an unstable neutral state.

As mentioned in the beginning of this section, the driving force for the introduction of hydrogen into silicon is probably a drift of the hydrogen ion in the space charge region at the sample surface. Therefore, the penetration depth is, in general, independent of the etch conditions, but it depends on the charge-carrier concentration in the sample.

As a first consequence of the high chemical reactivity of atomic hydrogen and its high mobility even at room temperature the formation of complexes with all common doping elements such as the shallow acceptors (B, Al, Ga, In, Tl) and the shallow donors (P, As, Sb) can be observed resulting in an effective passivation of the dopants at room temperature [3.60, 61]. These hydrogen-dopant complexes are electrically inactive. The electrical activity of the dopants can be restored by a thermal annealing of the samples at temperatures between 100° and 300°C, where the complexes dissociate. The dissociation energy depends on the chemical nature of the dopant. A strongly marked charge-carrier concentration profile at the sample surface, however, may yield an erroneous determination of the impurity concentration applying DLTS since the signal height depends on the changing local charge-carrier concentration. The DLTS measurement is performed within the space-charge region of a Schottky barrier which usually is found in a depth of a few μm beneath the surface. This is also the depth in which hydrogen can be introduced by chemically etching of the sample.

As a second consequence of the high reactivity and mobility of hydrogen in silicon the formation of complexes between hydrogen and deep-energy level impurities mainly due to dissolved transition metals is observed. These hydrogen complexes can be electrically neutral, which results in a more or less effective passivation of the respective deep-energy level or they can be electrically active forming new deep-energy acceptor or donor states. Once more the complexes dissociate at elevated temperatures but the thermal stability, in general, seems to be higher compared to hydrogen-shallow acceptor or donor states and their dissociation temperature is usually beyond 400°C.

The investigation of hydrogen-transition metal complexes is still in its beginning. Nevertheless a multitude of results have been published recently since often various complexes for one metal were observed, which differ in composition and structure. Differences in composition can be caused by the local hydrogen concentration. Complexes with one and with two hydrogen atoms can be formed depending on the respective hydrogen concentration. The depth profiles of the complexes can be complicated if a reverse voltage is applied or not to the Schottky barrier during annealing the sample, since dissociated hydrogen ions can drift in the electric field. Although there are some proposals for the atomic structure of particular hydrogen-transition metal complexes, most of the structures are still unknown and it can be assumed that many hydrogen complexes are still not yet detected. So far results were published on the following transition metals: Ti [3.58, 65], V [3.62], Cr [3.60, 62], Co [3.45, 58], Pd, Ag [3.60, 64], Pt [3.58, 60, 63], and Au [3.60, 66]. These will be discussed in more detail in the respective sections of Chaps. 4 and 5.

During former investigations sometimes problems arose to correlate particular deep energy levels detected by DLTS to the transition metal which has been intentionally diffused into the silicon sample (e.g., Ti, Co, Pd, Pt). Some of these problems could be solved recently by correlating levels to hydrogen complexes of the respective impurity. On the other hand deep levels were completely passivated by hydrogen and therefore not detected in earlier investigations (Pd donor and double donor) or they were partly passivated and problem arose to correlate them to the impurity because of lower concentrations with respect to the other energy level of the same impurity (Pt donor).

Although the donor-acceptor pairs are the most common complexes in wafers after technological processes, there are still *other complexes* which may form during the cooling period. This is because of the supersaturated state of the interstitial impurities in silicon at low temperatures. Only a few defects could be correlated to definite complex configurations, for example, the Mn_4 complex. A special type of donor-acceptor pairs where both atoms are formed by the same impurity, one on a substitutional site, the other one on an interstitial site, was reported recently to be more common than believed earlier. The first example was the $Cu_s Cu_i$ pair which was detected electrically by DLTS in 1981 [3.29] and correlated to the mentioned structure by photoluminescence measurements [3.30]. Meanwhile *Lemke* reported many other examples which are electrically inactive and therefore passivating the other levels (Co, Ni, Rh, Pd), or form new deep energy levels (Cu, Ag, Ir, Pt, Au) [3.67]. All these metals exhibit a stable substitutional defect at room temperature but their interstitial defect concentration at high temperatures is still high enough to form those pairs. Most of them form also haze (Co, Ni, Rh, Pd, Cu) or precipitate in the bulk of the sample (Ag) during the cooling period. The remaining metals Pt and Au are known to diffuse both, very fast on interstitial sites and slowlier by the kick-out mechanism. Many other defects described in the literature are only *related* to a metal impurity. Silicon solid-state chemistry, although making progress, is still in its infancy and further rapid development can be expected. However, higher concentrations of still unknown defects are rarely found in DLTS investigations of samples after technological processes. With decreasing impurity content in wafers in the last decade the importance of such defects likewise decreased. On the other hand, new techniques may require defects with special properties, for instance, for lifetime doping. For those applications the properties of complex defects may be of interest.

3.4 Precipitated Metals

All transition metals are unstable at RT due to their extremely low solubilities, as mentioned before. The easiest way to overcome this instability is to outdiffuse the impurities to the specimen surfaces or to precipitate them in the bulk of the sample in the cooling process so that they form a new phase.

For the outdiffusion, the diffusivity of the respective impurity must be sufficiently high with respect to the duration of the cooling period. Furthermore, the diffused impurity must be able to precipitate at the surface, or to evaporate into or to react with the ambient atmosphere in order to avoid back-diffusion. In general, quantitative outdiffusion of an impurity from samples with thicknesses of common wafers is rarely observed. Only *cobalt* tends to a quantitiative outdiffusion from thin samples forming a high density of small precipitates at the specimen's surfaces.

However, there are a few metals which do not form either precipitates or deep energy levels:

Zinc, as mentioned in Sect.3.3, evaporates at the sample surface and therefore avoids back-diffusion. Zinc introduced into a silicon sample, for example, during a preceding diffusion in a closed ampoule or by ion implantation will quantitatively diffuse out of the sample during the HT processes.
Mercury is another metal exhibiting a very high vapor pressure and therefore reacts in the same manner as zinc.

Only fast-diffusing transition metals will diffuse to the sample surfaces where they form precipitates. Because of this outdiffusion, denuded zones remain beneath both surfaces which correspond to the denuded zones obtained by internal gettering pre-annealing where the oxygen has been diffused out of the sample before precipitation of the oxygen takes place. The depth of these precipitation-free zones depends on the diffusivity of the respective impurity metal and the duration of the cooling period applied at the end of the diffusion process. Phenomena related to the surface precipitation will be discussed in Sect. 3.4.2. The conditions for precipitation in the bulk of the specimen will, however, be discussed now.

3.4.1 Volume Precipitates

For the precipitation of transition metals within the bulk of a sample, the following conditions must be fulfilled concerning the properties of the respective impurity, the process parameters, and the conditions in the bulk of the sample:

- High supersaturation of the impurities in the sample.
- High diffusivity of the impurities in silicon.
- Sufficiently-low cooling rates.
- For metals which are not able to precipitate by a homogeneous nucleation mechanism, the presence of nucleation centers in the form of lattice defects or in the form of other precipitates already present in the sample.
- Excess vacancy or excess self-interstitial concentration in the silicon crystal, depending on the lattice constant of the respective silicide formed during precipitation. In general, the crystal is not in thermal equilibrium. During precipitation silicon atoms will be emitted or absorbed. The precipitation is favoured if thermal equilibrium is approached.

Because of the variety of parameters which determine the precipitation of impurities the morphology of the precipitates can differ considerably: rapid quenching of samples containing fast-diffusing impurities will result in precipitates of high density, small size, and non-characteristic morphology of the precipitates as well as their etch pits formed by preferential etching of the sample. The same sample after applying low cooling rates will exhibit quite different precipitates of low density, extended size and a more characteristic structure of the etch pits. Typical morphologies of etch pits can be rod-like or cross-like structures, extended star-like structures with 90° or 60° angles on ⟨100⟩ and ⟨111⟩ silicon surfaces, respectively, defects which resemble snow flakes, or just irregularly shaped accumulations of etch pits.

The outdiffusion of the impurities to the sample surfaces is affected by the cooling rate. In the case of quenched wafers the distribution of defects in the cross-section of the wafer will be homogeneous. After applying low cooling rates to the sample, the cross-section of the wafer will show an accumulation of bulk precipitates in the middle region of the wafer between the two denuded zones beneath both surfaces. Examples for inhomogeneously distributed impurity precipitates after applying moderately high cooling rates to samples contaminated with two different impurities are displayed in Fig.3.5. The figure presents two X-ray section topographies on samples cut from 7.5 cm CZ-grown wafers doped with cobalt and palladium, respectively. Equal cooling rates were applied for the preparation of both samples. In the wafer doped with cobalt no indication for the formation of bulk precipitates can be observed. The "Pendelloesung" fringes parallel to the sample surfaces observed in this topography are typical for almost defect-free crystals. During the cooling period the cobalt atoms diffused nearly quantitatively to the sample surfaces where they precipitated by forming high densities of small particles. As a consequence, the surfaces are highly stressed and appear dark in the section topography. In the palladium-doped wafer, however, precipitates are observed at both surfaces and in the central zone of the cross-section. The regions beneath both surfaces exhibit denuded zones. Similar denuded zones are well known from oxygen precipitation after performing three-step internal-gettering pre-annealing processes. The dark spots visible in the topographies do not represent the precipitates themselves, they only decorate the stress areas surrounding the precipitates, which originate from the lattice misfit between the silicon lattice and the silicide phases. To a certain extent the size of this diffraction image is related to the size of the respective precipitates.

Systematic investigations of the structure, morphology and the composition of several transition-metal precipitates in silicon have recently been performed by means of TEM [3.68-71]. It was found that iron, cobalt, nickel, copper and palladium precipitate in the form of crystalline silicides epitaxially grown on defined planes of the silicon host lattice. Because of the

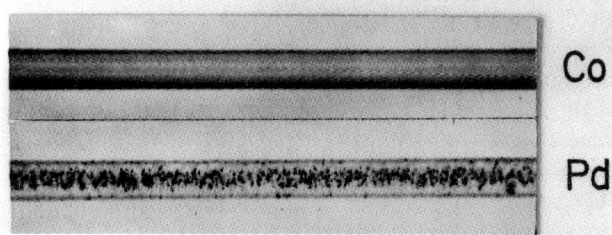

Fig. 3.5. X-ray section topographies of CZ-grown wafers after diffusion of cobalt and palladium, respectively. The samples were moderately fast cooled to RT. All cobalt atoms diffused to the sample surfaces. Palladium exhibits strong precipitation on both surfaces and in the bulk of the sample, showing denuded zones beneath both surfaces

different misfit of the various silicides with respect to the silicon lattice, secondary defects in the form of dislocations can be formed inside the precipitate, or punched-out into the surrounding host lattice. The size of the precipitates, the size of precipitate colonies and especially the size of punched-out dislocation loops strongly depend on the cooling rate applied to the sample or on the annealing processes subsequently performed at lower temperatures (Figs.4.1-3, 5).

Table 3.9 summarizes the properties of the silicide precipitates formed by iron, the fast-diffusing 3d transition-metals cobalt, nickel, copper and the fast-diffusing 4d transition-metal palladium. These metals are known to form haze, as will be discussed in Sect.3.4.2. The table lists the composition of the respective silicides, their morphology and their mean size after application of moderately fast cooling rates (haze program, cooling rate about 4 K/s). Secondary defects observed within the precipitates or in the surrounding host lattice are reported in addition (Figs.4.1-3, 5).

As may be inferred from Table 3.9, iron, cobalt and nickel precipitates consist of the respective stable HT silicon-rich metal (Me) silicide modification $MeSi_2$. Copper and palladium form metal-rich silicide modifications with Cu_3Si being likewise an equilibrium phase. Within this group of haze-forming metals, iron is assumed to be the only impurity which cannot form precipitates by a homogeneous nucleation mechanism at least during moderately fast cooling of the sample from HT ($> 900°C$). Thus, iron needs foreign nucleation centers to form silicide precipitates (heterogeneous nuclea-

Table 3.9. Properties of precipitates after moderately fast cooling (extr. disl.: Extrinsic dislocations)

Metal	Precipitate	Morphology	Dimension $[\mu m/\mu m^2]$	Secondary defects
Fe	α/β-$FeSi_2$	rod-like defects	0.1	none
Co	$CoSi_2$	platelets $\| \{111\}$	(0.005×0.2)	none
Ni	$NiSi_2$	tetrahedra platelets $\| \{111\}$ agglomerates	≤ 5 $\leq (0.05 \times 100)$ ≈ 10	dislocations in precipitates and in Si matrix
Cu	Cu_3Si	star-like colonies precipitates	$0.5 \div 80$ $0.007 \div 0.02$	extr. disl. loops $\| \{110\}$
Pd	Pd_2Si	star-like colonies precipitates	≈ 20 $0.015 \div 0.05$	extr. disl. dipoles $\| \{111\}$ and $[110]$

tion), whereas all other impurities listed in this table are able to form pre-cipitates via a homogeneous nucleation mechanism without the help of fore-ign nuclei.

When quenching the sample from HT, interstitial nickel and cobalt im-purity atoms arrange to form small and very thin silicide platelets parallel to the {111} silicon lattice planes. Platelets existing of only two neighboring nickel silicide monolayers have been observed [3.70-72]. In this case each nickel atom belongs to the surface of the precipitate. This might be one of the simplest morphologies of precipitates, and hence it forms very quickly even when quenching the sample from HT. In addition, these silicides are easily formed since the lattice parameters of $NiSi_2$ are almost equal to those of the silicon lattice, especially at HT. In Fig.3.6 the lattice parameter of silicon, $CoSi_2$, and $NiSi_2$ are plotted as functions of the sample temperature [3.7, 73]. At about $500°C$ the lattice constant of $NiSi_2$ agrees with that of silicon. At this temperature the silicide formation does not absorb self-inter-stitials or emit silicon atoms into interstitial sites of the surrounding silicon host lattice. At this temperature there is no volume change during precipi-

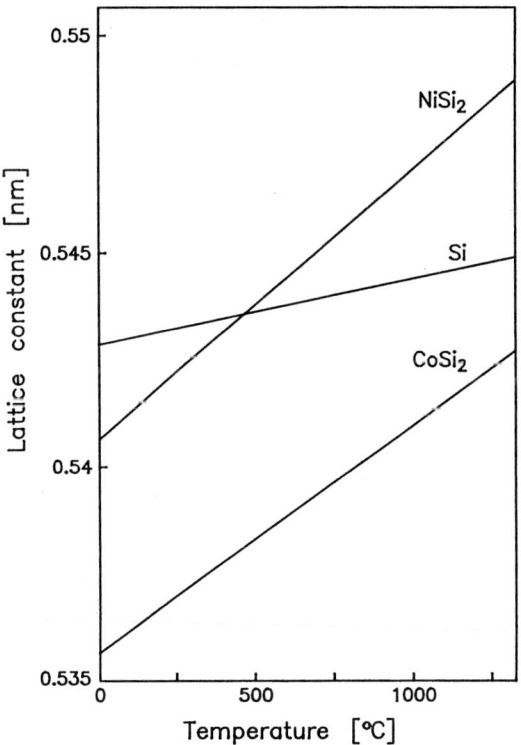

Fig. 3.6. Lattice constants of Si, $CoSi_2$, and $NiSi_2$ as a function of the sample temperature [3.73]

tation and hence no stress in the host lattice. As a consequence, no dislocations should be formed or punched-out.

A ripening process of the nickel precipitates takes place at reduced cooling rates or during subsequent annealing processes at a lower temperature. During this ripening process large precipitates grow at the expense of neighboring smaller precipitates. This, however, leads to the formation of dislocations within the nickel silicide and in the surrounding silicon matrix.

Cobalt silicides do not grow in the same way. Cobalt exhibits the smallest precipitates of all the impurities listed in Table 3.9, which may mainly be caused by its lower solubility in silicon exhibiting values about three orders of magnitude lower than that of nickel. During the cooling at a low rate, the cobalt atoms tend to diffuse to the nearest sample surface where they precipitate (Sect. 3.2). If the cooling period is too short for a quantitative outdiffusion of cobalt, the precipitate platelets grow in thickness. The same effect was observed during the subsequent annealing of the sample at lower temperatures.

All other impurity precipitates compiled in Table 3.9 exhibit a larger misfit between their lattice parameters and the respective silicon-lattice parameters. As a consequence the precipitation causes a more-or-less severe change in volume. The volume is contracted or expanded corresponding to the relation of the respective silicide lattice constant with respect to the silicon lattice constant. Table 3.10 lists the change in volume due to precipi-

Table 3.10. Change in volume due to precipitation. $[\pm N_{Si}$: Number of emitted $(+)/$ absorbed $(-)$ Si atoms per precipitated metal atom]

Metal	Precipitate	N_{Si}	Precipitate temp. [°C]	Stable configuration at T
Fe	α-FeSi$_2$ β-FeSi$_2$	-0.11	25	$T > 915°C$, tetragonal $T < 915°C$, orthorhombic
Co	CoSi$_2$	-0.08 -0.03	25 1000	
Ni	NiSi$_2$	-0.025 0 0.036	25 450 1000	
Cu	Cu$_3$Si	≈ 0.5		
Pd	Pd$_2$Si	0.55		

tation for the impurities discussed above [3.70-72]. The number N_{Si} denotes the quantity of emitted silicon atoms per precipitated metal atom. Negative signs signify an absorption of silicon atoms and positive signs the emission of silicon atoms into interstitial sites in the surrounding silicon lattice. As may be inferred from the table, the precipitation of iron, cobalt and nickel (at low sample temperatures) cause an absorption of silicon atoms whereas silicon atoms are emitted during precipitation of copper and palladium (as well as nickel at HT). Because of the high quantity of emitted silicon atoms into the surrounding silicon lattice, copper and palladium precipitates differ in their morphology from the other precipitates. Their N_{Si} values corresponds to that of oxygen precipitation forming SiO_2 (N_{Si} = 0.5) [3.74]. They form colonies of accumulated precipitates coupled with the appearance of extrinsic dislocation loops inside the colonies and punched-out into the surrounding. These dislocations grow parallel to definite silicon-lattice planes and directions. Due to these punched-out dislocations in the surrounding silicon lattice, the colonies exhibit a star-like morphology in TEM and in visual inspection after preferentially etching the sample, as listed in Table 3.9 or depicted in Figs.3.14,17-19 of Sect.3.4.2.

During a simultaneous precipitation of copper and iron, which both belong to the main impurities in device manufacturing, the emission and absorption of silicon atoms can compensate, at least partly. This may enhance the precipitation velocity since the number of silicon self-interstitials or vacancies is drastically reduced compared to the formation of precipitates of a single species. Otherwise, the excess intrinsic defects beyond the thermal equilibrium, being silicon self-interstitials or vacancies, must diffuse to sinks where they can recombine. In an undistorted silicon single crystal such sinks are limited to the sample surfaces which may be far away, taking into account the rather low diffusion velocity of intrinsic defects compared to the diffusion velocity of nickel or copper. In fact, iron precipitates in combination with copper precipitates have been observed by TEM [3.70, 73]. It is assumed that the copper precipitate formed the nucleus for the heterogeneous precipitation of iron.

Because of the emission or absorption of silicon atoms due to the formation of silicides the precipitation process also depends on the properties of the silicon host crystal which can exhibit excess vacancies or excess self-interstitials. The vacancies recombine with emitted silicon atoms, and self-interstitials act as sources for absorbed silicon atoms. In general, commercial crystals are not in thermal equilibrium. Usually excess vacancies are found predominantly at the seed and the tang ends of the crystal and in a smaller peripheral region near the edge of the crystal rod. In the bulk of the crystal excess self-interstitials prevail. However, this distribution can be changed since it depends on the parameters of the crystal growth.

Recently, *Lemke* [3.75] observed that the distribution of excess vacancies and excess self-interstitials changes drastically the concentration of substitutional transition metals in contaminated crystals grown by the floating-zone technique in an argon atmosphere. Still more impressive are recent results due to *Falster* and co-workers [3.76]. By applying a sophisticated rapid thermal annealing technique they succeeded in tailoring a specified profile of the excess-vacancy concentration [Magic Denuded Zone (MDZ) wafers] which enables one to produce efficient internal gettering preannealed in a single short temperature process. The enhanced vacancy concentration beneath the denuded zone causes a drastically amplified precipitaion rate of SiO_2 which is known to emit self-interstitials. Finally, in the early haze experiments [3.77] applying Pd contamination a striking difference in the precipitation behaviour between the peripheral wafer zone and the middle region was observed which, at that time, could not be interpreted reasonably. In order to get more detailed information on the influence of vacacies and self-interstitials on the precipitation behaviour of various transition metals it would be interesting to emply the new MDZ wafers.

3.4.2 Surface Precipitates - Haze

In the previous subsection we have seen that the outdiffusion of impurities during the cooling period of the sample is one possibility to overcome the supersaturation of the impurities at low temperatures. However, a continuous outdiffusion requires the precipitation of the impurities at the sample surfaces to avoid higher local supersaturations compared to the bulk of the wafer. During the cooling of the sample, its surface exhibits the lowest temperature which locally increases the supersaturation of the impurities and reduces the back-diffusion of the impurity atoms into the bulk [3.78]. On the other hand, an enhanced precipitation of impurities at the sample surface could be caused by a lowering of the nucleation barrier. *Seibt* [3.73] assumed that such a lowering of the nucleation barrier can be expected for two reasons:

(i) The sample surface acts as a sink or source for silicon self-interstitials and vacancies, respectively, which are generated or annihilated during impurity precipitation, as discussed above (Table 3.10).

(ii) It has been observed by means of TEM that precipitates of haze-forming transition metals (Co, Ni, Cu, Pd) in the sample's surface region cross the original surface and penetrate, for instance, into the surface oxide layer [3.68]. This could be interpreted as an energy lowering at the silicide-oxide boundary on the top of the sample compared to the silicide-silicon boundary within the bulk of the sample.

All of the different effects may, in addition, overlap and render the silicon surface an effective site for forming nuclei for those impurities which are able to precipitate by a homogeneous nucleation mechanism. Several years ago outdiffusion to the sample surface and surface precipitation was also observed for iron after a moderately fast cooling of the sample from HTs. However, in recent investigations applying an equal treatment to wafers grown by modern techniques, the outdiffusion of iron and the formation of haze could no longer be observed. Now, iron precipitation turns out to be limited to a heterogeneous nucleation mechanism only. The reason for the difference in former and recent investigations must be due to differences in the pulling parameters during the single-crystal growth, and is not yet known in detail. There may be a change in the surface treatment of the wafers but there may also be a change in the purity of the bulk crystal with regard to its content of grown-in intrinsic defect clusters (Sect. 8.2.2). In conclusion, surface precipitation and outdiffusion to uncoated silicon surfaces are, in general, limited to impurities which precipitate by a homogeneous nucleation mechanism.

Precipitates of impurities at the surface of a wafer can be revealed by preferential etching of the wafer. Each precipitate causes the formation of a shallow etch pit. Its size depends on the impurity, the respective etchant applied, its temperature, the duration of the etching process, and on the orientation of the surface. Most of the etch solutions developed to reveal lattice defects in the bulk of a sample are also suitable etchants to reveal impurity precipitates at the surface. However, there are significant differences between the various etch solutions concerning the surface orientation as well as the species of impurity precipitates. Thus, it is well known that the *Sirtl* etch [3.79] is suitable to reveal surface precipitates on (111)-oriented surfaces but it does not reveal the same precipitates on (100)-oriented wafers. The *Yang* etch [3.80], on the other hand, exhibits most impurity precipitates on both kinds of surface orientations although it is simply a diluted *Sirtl* etch. Only cobalt precipitated on (100)-oriented wafers cannot be revealed by the *Yang* etch which is, however, possible by applying solutions of other compositions [3.81].

A systematic study of the application of the various preferential etch solutions on wafers exhibiting surface precipitates is not available. To date, the preferential etch solutions were, in general, developed to reveal lattice defects such as stacking faults or dislocations in the bulk of samples. Some of them were especially developed to suppress the so-called haze effects [3.82] which in most cases might have been surface precipitates. Several well known etchants are listed below, together with their composition and specific properties:

Sirtl Etch [3.79]: A mixture of 500 g chromium trioxide CrO_3 dissolved in 1ℓ water and the same quantity of hydrofluoric acid (49%). This solution must be continuously agitated during etching. It affects only wafers with (111) orientation.

Secco Etch [3.83]: A mixture of 44 g potassium bichromate $K_2Cr_2O_7$ dissolved in 1 ℓ water and the same quantity of hydrofluoric acid. It has especially been developed for preferential etching of (100)-oriented wafers. The solution must not be agitated during etching but it exhibits low etching rates.

Yang Etch [3.80]: A mixture of 150 g chromium trioxide dissolved in 1 ℓ water and an equal volume of hydrofluoric acid. It is reactive on all common wafer orientations (100), and (111), (110). It consumes less chromium and the solution must not be continuously agitated during etching.

Wright Etch [3.82] has been developed to reveal lattice defects by simultaneous suppression of shallow etch pits and therefore it is not really suitable to reveal haze. It is a composition of nitric acid, hydrofluoric acid, acetic acid, chromium trioxide, copper nitrate and water.

Schimmel Etch [3.84]: A mixture of 75 g chromium trioxide dissolved in 1 ℓ water and double the volume with hydrofluoric acid. For samples with resistivities less than 0.2 $\Omega \cdot cm$, additional 1.5 parts per volume water is added. The etchant should be continuously agitated during etching.

MEMC Etch [3.85]: A chromium trioxide-free etchant for delineating dislocations and slips in wafers, which has been developed recently to reduce pollution. The etchant is a mixture of 36 mℓ HF, 25 mℓ HNO_3, 18 mℓ acetic acid, 21 mℓ De-Ionized (DI) water, and 1 g $Cu(NO_3)_2:3H_2O$ dissolved in 100 mℓ of the solution. Although suitable for revealing lattice defects, this etchant is not suitable for revealing surface precipitates because of the formation of more-or-less strong staining films which interfere with the hazy appearance of surface precipitates.

New Etch [3.81] has especially been developed to reveal haze with a chromium trioxide-free etch. It is a mixture of HF, HNO_3, and acetic acid of various compositions. For any fraction HNO_3/HF a suitable volume content of acetic acid in the mixture to reveal the precipitates of all haze-forming metals can be calculated according to

$$CH_3COOH = (1 \pm A/10)\,B\exp(C \cdot HNO_3/HF) \qquad (3.6)$$

with

$$A \approx 1 \quad \text{for} \quad HNO_3/HF \leq 6 \,,$$
$$A \approx 2 \quad \text{for} \quad HNO_3/HF > 6 \,,$$
$$B = 58 \quad \text{and} \quad C = -0.06 \,.$$

A suitable composition is, for instance, 10 mℓ HF (40%), 40 mℓ HNO$_3$ (65%), 50 mℓ acetic acid. Lower concentrations of acetic acid will favor the etching of copper haze, higher concentrations will favor nickel haze. A suitable composition for revealing only copper is HF:HNO$_3$ = 1:50 without the addition of any acetic acid. This is the first etch where copper and nickel haze can be identified separately. The etch works on all orientations of the wafer surface.

All etchings must be performed in plastic beakers for $1 \div 20$ minutes. To finish etching, the beaker is filled up with DI water; the samples are rinsed several times and dried by means of compressed air or nitrogen. The procedure has been reported in the DIN as well as the Americal Society of Testing and Materials (ASTM) Standards [3.86].

The accumulation of shallow etch pits on the surface of a wafer is generally called **haze** or **fog**. Haze can be revealed in a spot light where it appears bright in non-reflecting directions because of the scattering of the incident light at the shallow etch pits. Surface areas without haze appear dark in non-reflecting directions. In this meaning haze is always caused by an accumulation of etch pits on the wafer surface and not by separate single etch pits or lines of etch pits which may be due to lattice defects or dislocations and slip lines. The microscopic structure of haze is the accumulation of shallow etch pits which are also called **saucer pits** in the literature. In contrast to precipitates, line defects crossing the wafer surface, such as dislocations, cause triangular or rectangular etch pits (depending on the wafer orientation) exhibiting a defined top of the etch pit pyramid.

Besides impurity precipitates on the surface of a wafer after being subjected to a heat treatment, the following defects can cause a similar appearance which may interfere with haze in its stronger definition:

(i) Surface damage due to sawing or lapping, which has not been removed completely by etching and polishing, can cause haze after preferentially etching the sample. In contrast to haze caused by impurity precipitates at the wafer surface, this haze due to damage appears even without any heat treatment. In addition, this kind of haze is not gettered, for instance, by scratches on the front or on the reverse side of the wafer.

(ii) Oxygen precipitates in the bulk of a wafer can be revealed at the surface by preferential etching. Due to the striated distribution of the oxygen in the crystal the appearance of this haze is striated as well. A striated haze pattern is an almost unequivocal peculiarity of oxygen precipitation. Oxygen precipitates can be decorated by metal impurities which are then hardly detected if all of the haze is gettered. On the other hand, partly gettered haze can reveal oxygen precipitates in the form of striations if homogeneous haze is gettered partly by striations of high-density oxygen precipitates.

(iii) Haze can also appear on the surface of highly doped wafers after preferential etching. The respective etch pits are very small and the light-scattering effect remains low if detectable at all. This haze again appears without any heat treatment. It is assumed to be due to precipitates of the doping element within the bulk of the wafer.

(iv) Very strong haze is caused by HT processes performed in broken furnace tubes. This kind of haze can be observed without the aid of a spot light. It often exhibits a characteristic stream-line-shaped structure, and its strength exceeds by far the strongest haze due to precipitates of palladium. This kind of haze cannot be reduced by cleaning the furnace tube, but only by exchanging it. The orign of this haze is not yet known. It is assumed that it may be due to silicon-carbide precipitates originating from the CO_2 content in the atmosphere.

In order to exclude haze due to the defects listed above, a stronger definition of haze caused by impurity precipitates at the surface of a wafer is required. The main difference is the appearance of haze with and without a preceding heat treatment. So the following definition for haze in this restricted meaning is proposed: *Haze is the accumulation of shallow etch pits at the surface of a wafer due to metal-impurity precipitates formed during the cooling period of a preceding short HT heat treatment of several minutes and a subsequent preferential etching of the wafer. Without this heat treatment the wafer should be free of haze. As far as it is known at present, this haze is only caused by cobalt, nickel, copper, rhodium and palladium impurities. Iron haze can only be formed during subsequent low-temperature annealing processes. But iron can strengthen haze in the presence of other haze-forming impurities.*

Haze in this restricted meaning can be used to analyze wafers before and after technological processes (Sects. 6.3 and 7.1). Because of the limited number of impurity metals which form haze, and the limited number of main impurities in modern technology, this haze investigation is a suitable method to detect nickel and copper on the top of wafers or in their bulk.

The following **haze test** can be carried out quickly, only requires a furnace tube, an etching facility and a spot light. The waver to be investigated is heat treated, for instance, at 1050°C for 7 minutes in an inert gas atmosphere (Ar or N_2) without any preceding cleaning or etching process. This annealing duration is adapted to 7.5 and 10 cm wafers with standard thicknesses and enables iron atoms to cross the whole wafer thickness by diffusion. The cooling period at the end of the heat treatment should be standardized for a better comparison of the impurity distribution on different wafers. We usually apply a definite, moderately-fast pull-out velocity (for instance, 220 mm/min, or a faster pull-out velocity with stops for one minute each at definite positions within the region of the temperature gradient of the furnace tube. Both procedures result in cooling rates of about 4K/s). After a preferential etching of the sample, preferably by the New etch, the haze pattern, as inspected under spot-light illumination, represents the distribution of the impurities in the bulk of the wafer or (before performing the diffusion) on the top of its surfaces. Regarding the annealing conditions, the strength of the haze is a measure of the respective impurity which, however, differs with the impurity copper or nickel. For a quantification of the haze, the relative fraction of scattered light can be determined after calibrating the measurement facility by means of a polished wafer without haze used as a reference sample. Setting the scattered light of the polished wafer to 100 arbitrary units (a.u.) haze resulting in 101 or 102 a.u. corresponds almost to the limit of visibility in the spot light. Haze of unintentional copper or nickel impurity contaminations will be in the region between 101 and about 170 a.u., whereas haze of intentional contamination, for instance by palladium, may result in several hundred a.u.

The diameter of the haze pattern originating from a spotlike impurity source depends on the diffusion length of the respective impurity. This, in turn, is a function of its chemical nature, the diffusion temperature and its duration including the cooling period. In order to clearly reveal spotlike impurity sources, the temperature of 1050°C and the annealing duration of 7 minutes including a moderately-fast cooling period is suitable, as reported above. In order to exactly delineate the finger prints of the impurity contaminations, a shorter heat treatment combined with quenching of the sample to RT is recommended. This provides useful information about the contamination source, and often the respective impurity source can immediately be identified because of its characteristic structure.

3.4.3 Haze Phenomena

As mentioned above, haze is only formed by the transition metals cobalt, nickel, copper, rhodium, and palladium, which all precipitate via a homogeneous nucleation mechanism. To give an idea of the appearance of haze, several examples are illustrated in the following pictures. Figure 3.7 represents a 7.5 cm wafer (111)-oriented, which has been spotwise contamined by metallic nickel in the center of the reverse side after removing unintentional metal contamination and the native oxide by short etching of the wafer in a mixture of HNO_3 and HF. Subsequently, the wafer was heat treated at 1050°C for 20 min in a nitorgen atmosphere and quenched to RT. The haze was revealed by a preferential etch applying a Sirtl etch [3.79]. Finally the wafer was photographed under spot light. As can be seen, a circular, bright haze pattern was formed by the outdiffusion of nickel in all directions. It covers a large part of the wafer surface. When

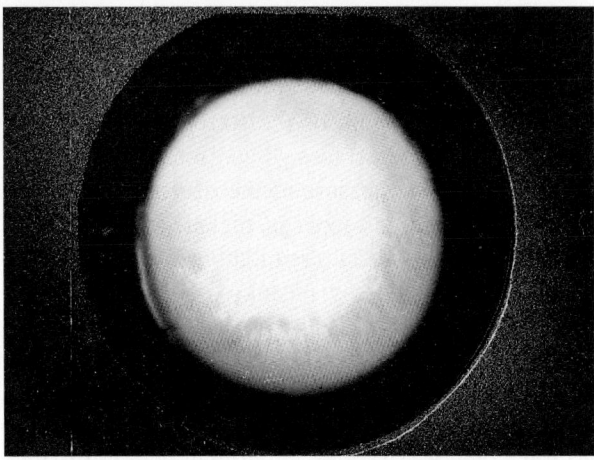

Fig. 3.7. 75 mm wafer showing a circular haze pattern formed by outdiffusion of an intentional spot-like nickel contamination in the center of the wafer. Diffusion at 1050°C for 20 min in a nitrogen atmosphere and Sirtl etched

quenching the wafer from diffusion temperatures to RT, the nickel atoms diffused to the sample surface where they precipitated. The preferential etching formed shallow etch pits which scatter the spot light in all directions and therefore the hazy area appears bright, whereas the polished surface near the edge of the wafer appears dark, because of the lack of etch pits and light scattering. Since the size of the haze pattern is determined by the diffusion of the respective metal, a prolonged diffusion time and an increased diffusion temperature will increase the size of the hazy area independently of the respective form of the contamination. But there is also a dependence on the respective metal with its characteristic solubility, diffusivity and precipitation behavior.

Figure 3.8 exhibits a comparison of the haze patterns formed by nickel, palladium, cobalt and copper after applying preparation conditions as for Fig. 3.7. Only for revealing the iron haze (old wafer) the cooling period was considerably prolonged. In spite of this extended cooling period the diameter of the iron haze pattern is by far the smallest one. Nickel exhibits the largest haze area followed by palladium, cobalt and copper. If we compare the solubility and diffusivity data of these metals (Figs. 3.1, 2), this sequence is not self-evident. It is expected that the haze pattern of iron is the smallest one since both values are low in comparison with the data of the other haze-forming metals. Furthermore, iron requires heterogeneous nucleation centers which might not have been present in high densities. Nickel, on the other hand, forms the most extended haze area, although its solubilities and diffusivities do not exhibit the highest values. As discussed in Sect. 3.4.1, the HT modification of nickel silicide fits the silicon host lattice

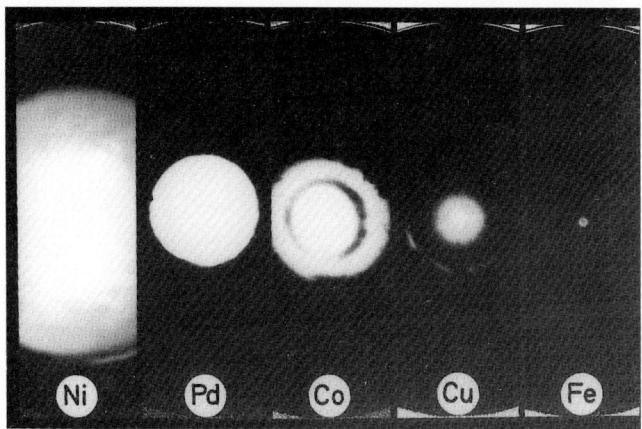

Fig. 3.8. Comparison of the haze patterns due to intentional spot-like contaminations with haze-forming metals in the center of the wafers. Diffusion at 1050°C for 20 min in nitrogen, quenching to RT, and preferential etching. The iron-contaminated wafer was moderately fast cooled to RT

well and may therefore require a low formation energy. Consequently, a high density of nuclei is expected. Although this also holds for cobalt silicide, the hazy area of cobalt is considerably smaller. However, this can be explained with the solubility of cobalt, which is orders of magnitude smaller than that of nickel, and agrees with that of iron. As a consequence, the cobalt concentrations are about three orders of magnitude lower than those of nickel for an equal distance from the contamination source and under equal diffusion conditions. Therefore, it appears reasonable that the cobalt haze pattern is less extended than that of nickel.

In contrast to nickel, copper and palladium silicides cause significant changes in volume due to precipitation (Table 3.10). Therefore, their precipitation is coupled with the emission of silicon self-interstitials. The structures of the respective precipitates are quite different (spherical compared to platelets). The ripening process, especially of copper precipitates, is strongly marked. So extended precipitate colonies still increase in size while smaller precipitates in their neighborhood disappear during subsequent annealing at lower temperatures or during slow cooling of the sample. However, this ripening reduces the density of etch pits on the surface of a wafer and therefore reduces the light scattering effect and the brightness of the haze, too. This will be demonstrated in the next illustration.

Figure 3.9 displays a haze pattern in the form of a face on the polished surface of a 7.5 cm wafer contaminated spot-wise (Pd, Co), and line-shaped (Fe, Cu, Ni) by the same metals as in Fig. 3.8. The diffusion was performed at 1050°C for six minutes in an argon atmosphere followed by a moderately-fast cooling period within about three minutes (the sample was with-

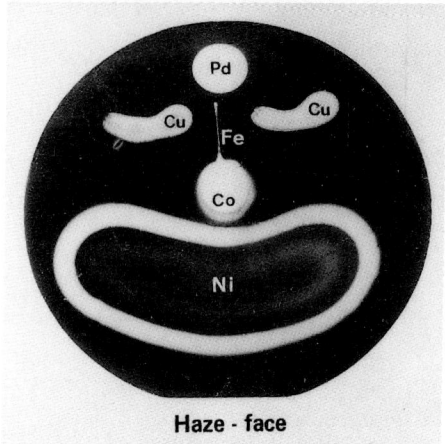

Fig. 3.9. Haze-face formed by spot-like and line-shaped contaminations with metals as indicated. Diffusion at 1050°C for 6 min in argon, cooled to RT within 3 min, and preferential etching

drawn from the furnace tube with a pull velocity of 230 mm/min). The haze patterns of palladium and copper are similar to those exhibited in Fig. 3.8. The line-shaped haze pattern of iron shows broadening at both ends, which indicates the effective formation of nucleation centers by small amounts of cobalt on one side, and palladium on the other side, even in regions where palladium and cobalt haze is not revealed without the presence of iron. This demonstrates the strengthening of haze by iron. A similar appearance is observed for cobalt haze in the vicinity of nickel, showing that the extension of the outdiffused nickel is much larger than the visible haze area. Thus, cobalt and palladium form nuclei for the precipitation of iron, nickel forms nuclei for the precipitation of cobalt, but cobalt obviously does not form nuclei for the precipitation of nickel.

The nickel haze is due to a line-shaped contamination. In the region of high nickel concentrations near to the contamination source, almost no haze is revealed. In this region the density of the precipitates at the wafer surface has strongly been reduced by ripening, as reported above. This process must be taken into account when haze is analyzed quantitatively. The cooling rate must be adapted to the respective impurity concentration to maintain a linear dependence between the contrast of haze and the respective impurity concentration. In addition to the cooling rate, the ripening process depends on the chemical nature of the impurity, as shown in the next series of pictures.

Figure 3.10 depicts line-shaped haze patterns of the same metal and in the same sequence as in Fig. 3.8. In contrast to Fig. 3.8, the samples were cooled to RT within five minutes after applying a pull velocity of 900 mm/min with four interim stops for one minute each at decreasing temperatures (for the haze program, see Sect. 7.1). Especially the haze patterns of palladium and copper exhibit strongly marked central regions where the for-

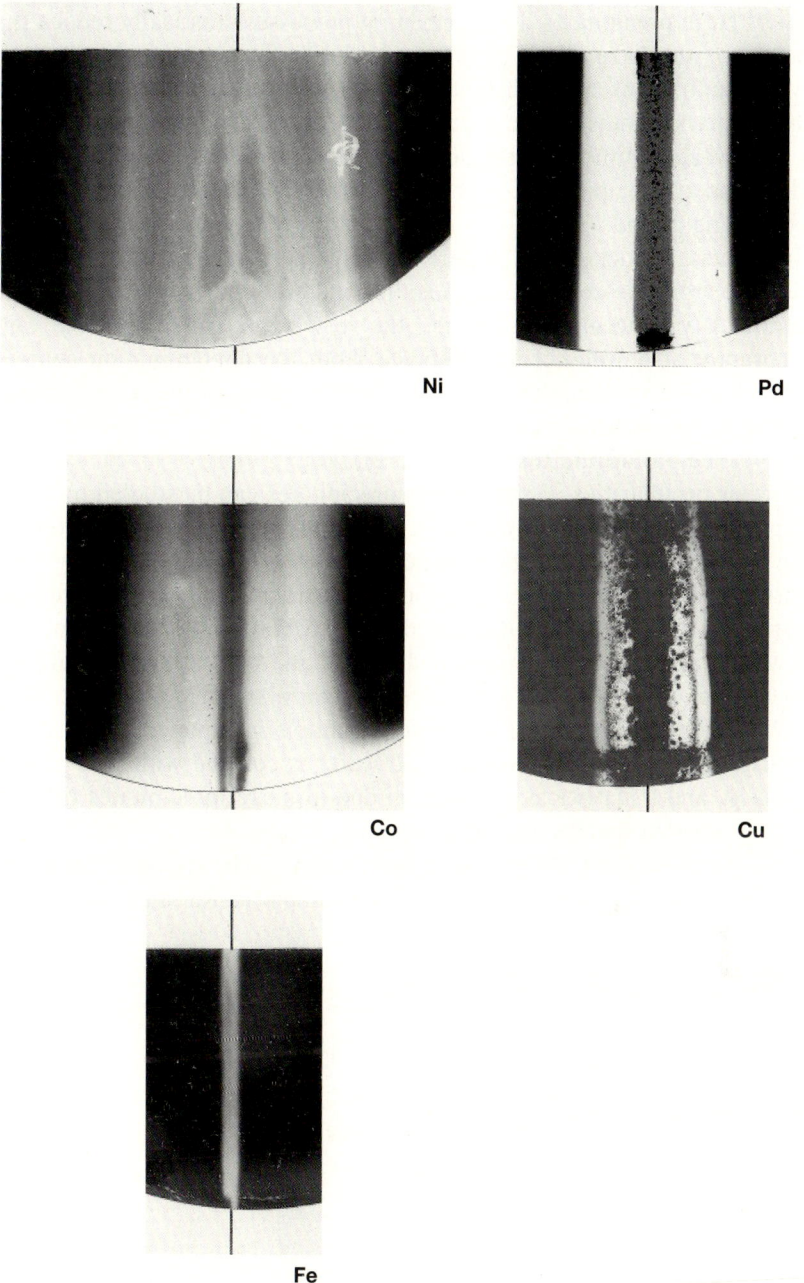

Ni

Pd

Co

Cu

Fe

Fig. 3.10. Comparison of haze patterns of different metals after line-shaped contamination. Diffusion at 1050°C for 7 min, moderately fast cooling within 5 min, and preferential etching. The contamination lines are marked

mation of large precipitates due to ripening processes drastically reduce the brightness of the haze. Haze is only revealed in the regions of lower impurity concentrations more distant from the original contamination lines which are marked outside the picture. In the case of copper, many dark spots are revealed in the bright haze area showing denuded zones which surround extended precipitates formed by ripening processes during the extended cooling period.

In contrast to nickel haze, where the brightness of the haze drops with the distance from the contamination source in agreement with the expected lateral impurity concentration profile, the very bright palladium haze vanishes abruptly at a well-defined distance from the contamination source. This enables one to accurately measure the diameter of the haze pattern. This appearance, which so far is not yet well understood, has been applied in the so-called **palladium test** (Sect. 8.2.2).

Further peculiarities of the various precipitates can be studied by a microscopic inspection of the etch pits at the surfaces of the sample after revealing haze. In addition to the chemical nature of the impurities, the surface orientation of the sample influences the structure and morphology of the respective precipitates, as illustrated by the following pictures. Contaminated wafers after diffusion, quenched to RT, and preferential etching exhibit a haze of high density, which does not show specific peculiarities of the etch pits on a microscopic scale. One example is depicted in Fig. 3.11. The polished surface of a (100)-oriented wafer is covered with etch pits of high density and a statistical distribution due to nickel precipitates of high concentrations. No specific structures can be distinguished.

The morphology of the etch pits due to nickel haze changes considerable after a moderately-fast cooling of the sample to RT and subsequent

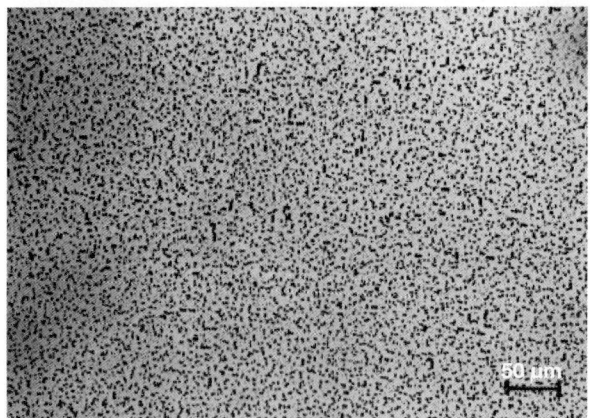

Fig. 3.11. Etch pits on a (100)-oriented wafer due to intentional nickel haze after quenching the sample to RT and preferential etching. Microscopic investigation, scale: 50 μm

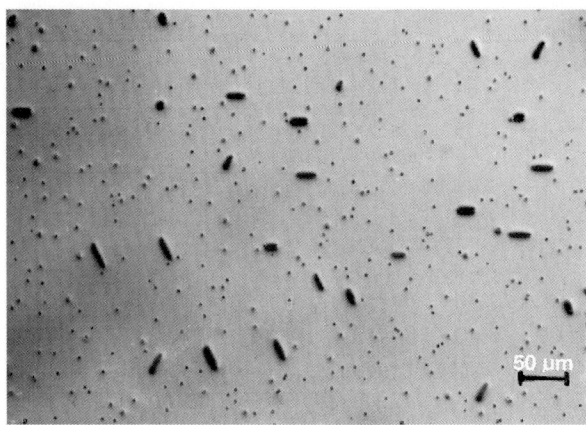

Fig. 3.12. Etch pits on a (111)-oriented wafer due to intentional nickel haze after moderately fast cooling and preferential etching. Microscopic investigation (scale: 50 μm). The section has a high nickel content

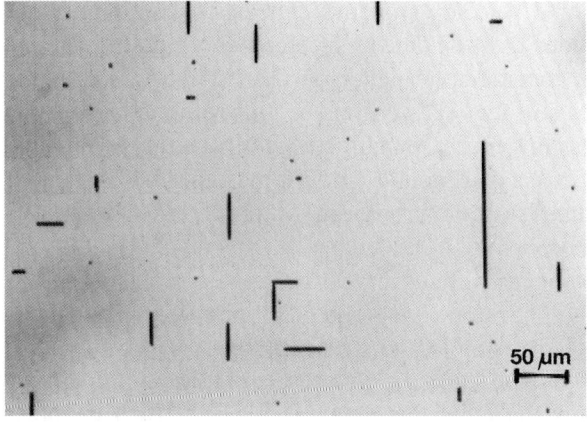

Fig. 3.13. Etch pits on a (100)-oriented wafer due to intentional nickel haze after moderately fast cooling and preferential etching. Microscopic investigation (scale: 50 μm). The section has a high nickel content

preferential etching. Differences between (111)- and (100)-oriented wafers are evident, as shown in the Figs. 3.12 and 13, respectively. Differences are observed in the appearance of 60° and 90° angles between elongated etch pits due to punched-out dislocation loops on the surface of (111)- and (100)-oriented wafers, respectively. In addition, differences in the length of the etch pits and in the density of the small pits situated between the larger ones are strongly marked. The characteristic shape of the etch pits may

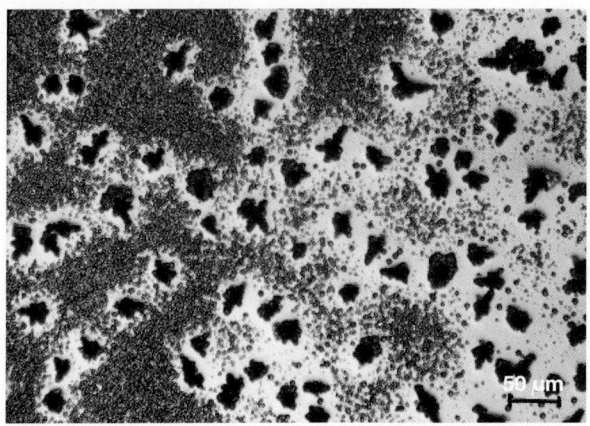

Fig. 3.14. Etch pits on a (111)-oriented wafer due to intentional palladium haze after moderately fast cooling and preferential etching. Microscopic investigation (scale: 50 μm). The section has a high palladium content

easily lead to a confusion with surface-stacking faults exhibiting quite similar structures, and this could lead to erroneous conclusions. In comparing both figures it is conspicuous that the density of etch pits which determines the brightness of haze is considerable higher on the (111)-oriented wafer than on the (100)-oriented surface. This correlates well with the experience that (111)-oriented wafers, in general, exhibit stronger haze than (100)-oriented wafers. However, this experimental finding must not be correlated with different impurity concentrations but can simply be due to the different size, density and morphology of the etch pits.

Less significant differences were observed in the microscopic structure of etch pits due to palladium haze on (111)- and (100)-oriented wafers after moderately fast cooling of the sample and subsequent preferential etching. Examples are shown in Figs. 3.14 and 15, respectively. The etch pits exhibit crystal-like morphologies surrounded by high-density haze with denuded zones in the neighborhood of larger precipitates. The mean size of the crystal-like precipitates seems to be much larger on (111)-oriented wafers compared to the higher density of the smaller quadratic precipitates on the (100) wafer surface. Both pictures were taken near the contamination zone where high palladium concentrations are expected. The ripening process is very evident on one half of the picture while the other half still shows extended regions covered with overlapping small etch pits. Thus, the darkness on the photograph correlates with the brightness of the respective haze.

Quite different behavior was observed for cobalt haze on (111)- and (100)-oriented wafers. Whereas cobalt haze is strongly marked on (111)-oriented surfaces even after moderately fast cooling, as shown in Figs. 3.8-10 and by the microscopic observation in Fig.3.16, (100)-oriented

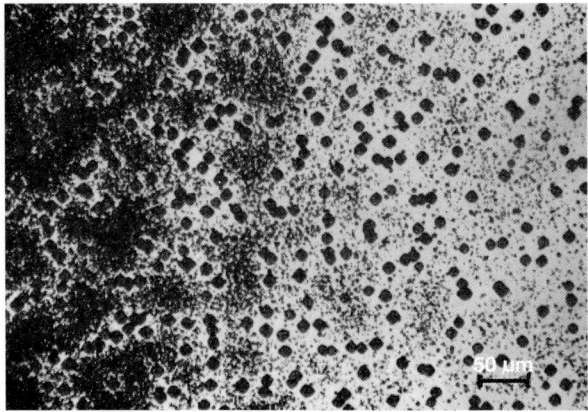

Fig. 3.15. Etch pits on a (100)-oriented wafer due to intentional palladium haze after moderately fast cooling and preferential etching. Microscopic investigation (scale: 50 μm). The section has a high palladium content

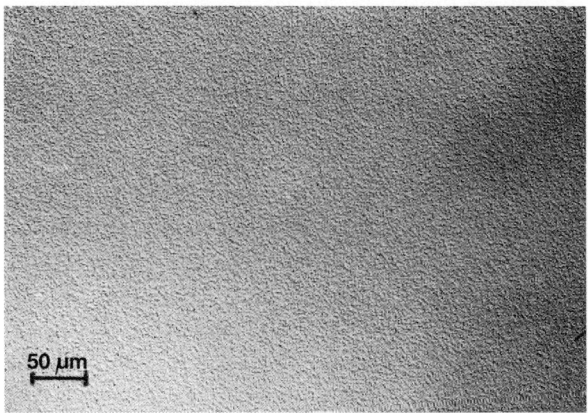

Fig. 3.16. Etch pits on a (111)-oriented wafer due to intentional cobalt haze after moderately fast cooling and preferential etching. Microscopic investigation (scale: 50 μm). The section has a high cobalt content

wafers intentionally doped with cobalt do not exhibit any haze after applying a *Yang* etch (no figure). As mentioned in Sect. 3.4.1, cobalt forms small platelets of $CoSi_2$ precipitates parallel to {111} silicon lattice planes. After moderately-fast cooling the thickness of these platelets amounts to about 3 nm whereas their diameter amounts to 0.1 to 0.6 μm [3.73]. In (111)-oriented wafers an accumulation of cobalt-silicide precipitates are formed at the wafer surface with orientations parallel to the surface. These surface precipitates exhibit a homogeneous distribution on the four different {111}

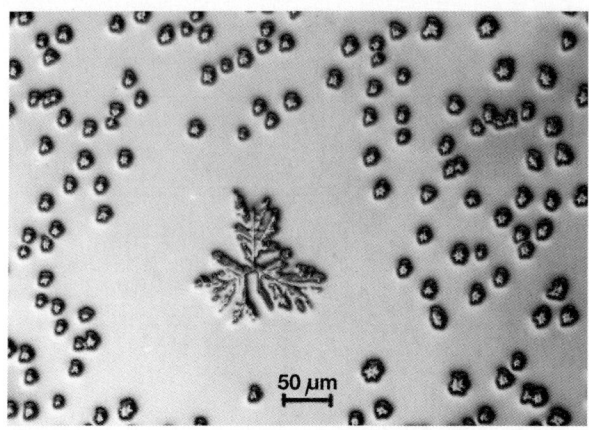

Fig. 3.17. Etch pits on a (111)-oriented wafer due to intentional copper haze after moderately fast cooling and preferential etching. Microscopic investigation (scale: 50 μm). The section has a high copper content

silicon-lattice planes. They can be preferentially etched and then form the high-density haze showing a microscopic structure of overlapping small etch pits, as represented in Fig.3.16. In (100)-oriented wafers, however, the silicide platelets on {111} lattice planes form an angle of about 60° with the silicon surface and the distorted surface area is almost equal to the cross-section of the platelet ($<0.6\times 0.03\mu m^2$) and consequently much smaller than the surface of the platelet itself ($0.6\times0.6\mu m^2$). Since the etch pits of cobalt haze on (111)-oriented wafers are already very small, the etch pits on (100)-oriented wafers should be at least 20 times smaller. Therefore, they may not be resolved and do not scatter the visible light. In spite of these small areas of crystal distortion at the wafer surface, other preferential etch solutions developed recently reveal the cobalt haze on (100)-oriented surfaces [3.81]. This demonstrated the existence of cobalt haze on (100)-oriented wafers.

Finally, copper forms again different microscopic structures of the etch pits on (111)- and (100)-oriented wafers after moderately fast cooling and preferential etching of the sample. The microscopic structures of the etch pits resemble the structures of the respective precipitates [3.69]. In addition, they resemble those of palladium precipitates when applying short etching durations [3.69]. Examples are shown in Figs.3.17 and 18. On (111)-oriented surfaces (Fig.3.17) rather large etch pits, which are similar to dendritic structures or snow-flakes, are observed. They are surrounded by small etch pits with irregular shapes and denuded zones in the neighborhood of the large precipitates. On (100)-oriented surfaces (Fig.3.18) few larger crosses are mixed with accumulated rectangular small crosses. Once more, denuded zones are formed in the vicinity of the large pits.

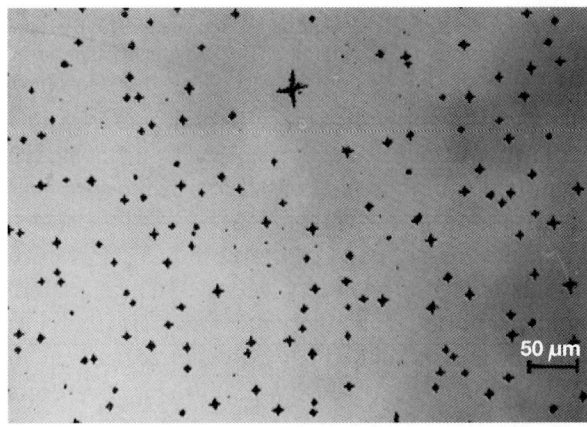

Fig. 3.18. Etch pits on a (100)-oriented wafer due to intentional copper haze after moderately fast cooling and preferential etching. Microscopic investigation (scale: 50 μm). The section has a high copper content

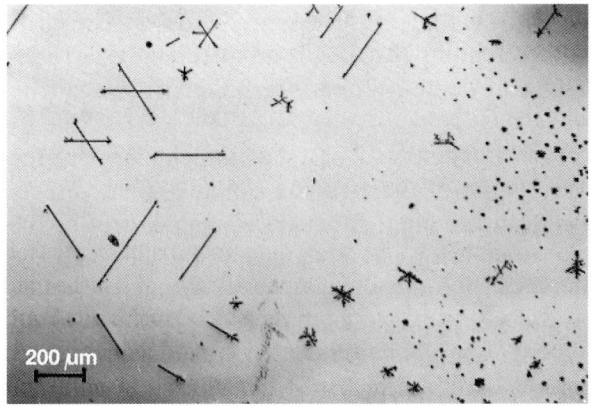

Fig. 3.19. A microscopic structure of etch pits on a (111)-oriented wafer due to intentional copper haze after slowly cooling and preferential etching. Microscopic investigation (scale: 200 μm)

If high copper concentrations coincide with low cooling rates at the end of the copper diffusion, the precipitates and consequently the etch pits of copper haze can further grow to sizes up to about 100 μm. They then form extended crosses in crystallographic orientations parallel to {110} planes due to the directions of dislocations, which are associated with the copper silicide colonies. The microscopic structure of the etch pits on (111)-oriented wafers is depicted in Fig. 3.19. It is evident that these large precipitation colonies are formed during a ripening process since there are

73

Table 3.11. Etch pit structures of haze-forming metals after a Yang etch on (111)- and (100)-surface planes

Metal	(111) surface	(100) surface
Fe	no haze visible	no haze visible
Co	overlap. spots, high density	no haze visible
Ni	small lines 60 deg + spots	extended lines 90 deg few spots
Cu	dendrites, snow-flakes, small and large	crosses, 90 deg small and large
Pd	crystal-like spots + dense irregular spots	squares and dense spots

no small etch pits left in the surrounding of the extended crosses. The characteristic morphologies of the etch pits revealed from different haze-forming metals on (111)- and (100)-oriented wafers are summarized again in Table 3.11.

For the identification of unintentional haze-forming impurities, the characteristic microscopic structure of the etch pits can help if the respective impurity concentration is rather high. Otherwise the structures of the etch pits do not exhibit enough information to determine the chemical nature of the respective impurity. Furthermore, different impurities can be simultaneously present in the sample exhibiting different concentrations and distributions, which complicates their identification. However, in general, only copper or nickel, or both, are expected as unwanted metal impurities in wafers, forming haze. Thus, the haze test can easily detect their distribution on the surface of the wafer or in its bulk. Their concentration can be estimated from the strength of the scattered light. The trace of the contamination source is better revealed by applying short diffusion times and high cooling rates. It can help to identify the source of the contamination which can then be eliminated. For this purpose the chemical nature being copper or nickel must not be known exactly since both impurities are detrimental for device performance and should be avoided.

A typical example for haze observed on as-received wafers is depicted in Fig. 3.20. One half of a (111)-oriented wafer has been heat treated at 1050°C for 20 min and moderately-fast cooled to RT. After a preferential etch, haze was observed in several areas at the edge of the wafer. The overlapping semicircular haze patterns indicate that the contamination source

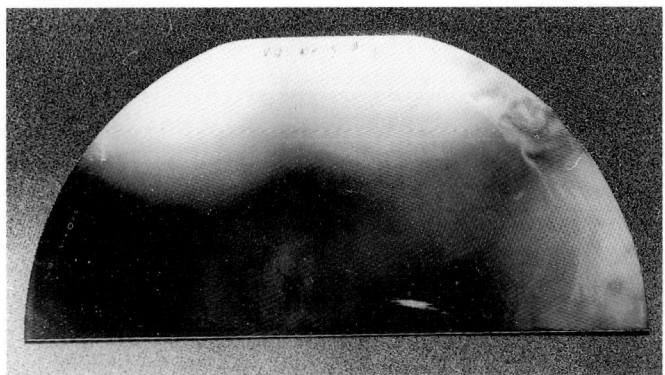

Fig. 3.20. Bright haze patterns at the edge of a half (111)-oriented wafer after application of the haze test. Detection of unintentional contamination with Ni or Cu by handling systems

was located near to the edge of the wafer. It might be due to any handling or carriage of the respective wafer. Because of its brightness it can be assumed that this haze may be due to nickel-contaminated tweezers or carriers. Further details will be presented in Sect. 7.1.

4. Properties of the Main Impurities

In the past decades significant progress has been made in the techniques for the detection, and the investigation of deep energy levels and precipitates of transition-metal impurities in silicon. Therefore, our knowledge of the impurity contamination and of its distribution on wafer surfaces has considerably increased. As a consequence, a number of contamination sources in device manufacturing could be eliminated. The purity of the wafers before and after processing has been improved between one and three orders of magnitude.

From the variety of transition metals only a few are wide-spread and form the main impurities of device production. These are iron, nickel, and copper. Several other transition metals such as molybdenum, platinum, and gold can be found occasionally after certain processes. Their respective contamination sources are mostly known or can easily be determined. Finally, a few other impurities are rarely found in minor concentrations after processing such as titanium, vanadium, chromium, and cobalt. Zinc, silver, and mercury have been detected on wafer surfaces but not yet in the bulk after HT processes. Because of their different behavior during processing, not all of these impurities are detrimental to devices.

In the following we shall provide a survey of specific properties for main and rare impurities, respectively. Each section reports on the properties of a single metal impurity, its behavior during technological processing and its influence on the final device, as far as it is known today. Since these sections are intended to summarize the properties of definite impurities, some of the previously treated data are repeated if they are needed for calculations or estimations. For each impurity the review section starts out with a general overview and is then build up: (1) Solubilities as a function of temperature, (2) diffusivities as a function of temperature, (3) behaviour during heat treatment, (4) electrical activity, (5) properties of the precipitates, (6) known impurity sources and common concentrations, and (7) avoidance of contamination.

4.1 Iron

Iron is one of the main impurities in device fabrication. Although iron is found only in minor concentrations of less than 10^{12} cm^{-3} in the bulk of a CZ-grown crystal near to the tang or poly end of the rod, its concentration may increase during wafer fabrication by one order of magnitude or even more. Due to the position of iron in the middle region of the sequence of 3d transition metals, it does not belong to the fast-diffusing metals (Co, Ni, Cu) nor to the slowly-diffusing ones (Sc, Ti, V). Together with manganese and chromium, interstitial iron forms a group of moderately fast-diffusing 3d transition metals (Fig.3.2). As a consequence, iron forms electrically active defects during quenching or moderately-fast cooling the iron-diffused sample from HT; it precipiates during slow cooling of the sample from HT or during subsequent annealing performed at lower temperatures. It can even appear simultaneously as electrically active iron dissolved on interstitial sites and as α-FeSi$_2$ precipitates in the same sample.

As deduced from theoretical considerations [4.1], substitutional iron is not stable in silicon. The interstitial, electrically active iron exhibits only one donor level in the lower half of the band gap, as substantiated by theory [4.1]. In p-type wafers the interstitially dissolved iron reacts with the dopant at RT by forming donor-acceptor pairs such as FeB, FeAl, FeGa and FeIn (Table 3.7). All these pairs exhibit donor levels with activation energies between 0.1 and 0.2 eV from the valence-band edge. Furthermore, iron forms donor-acceptor pairs with gold in n- and p-type wafer forming two energy levels, a donor and an acceptor state. In addition, a donor-acceptor pair with zinc has been observed (Table 3.8).

4.1.1 Solubility

Within the temperature range indicated the solubility of iron in silicon follows the Arrhenius equation (Table 3.1 with references and Fig.3.1)

$$S = 5 \cdot 10^{22} \, e^{(8.2 - 2.94/kT)} \quad [\text{cm}^{-3}] \quad (900\,^\circ\text{C} < T < 1200\,^\circ\text{C}) . \quad (4.1)$$

The solubility was determined by neutron-activation analysis probably at equilibrium with the HT modification of the iron silicide FeSi$_2$ as a diffusion source on the sample surface. Solubilities measured by electrical methods after quenching the samples to RT generally result in lower concentrations. It is assumed that this discrepancy is due to a gettering effect by intrinsic defect clusters (Sect.8.2.3). Iron exhibits a very low RT solubility beyond the detection limit. As a consequence the interstitially dissolved iron is not stable at RT.

Wafers contaminated with metallic iron and diffused for short times, for instance by applying rapid thermal annealing, exhibit iron concentrations which by far exceed the solubilities calculated from the Arrhenius equation (4.1) [4.2, 3]. The reason for the different "solubilities" is due to the different boundary conditions at the sample surface, formed by iron silicide and metallic iron, respectively. The enhanced solubility put in quotation marks suggests that the high iron concentration obtained after short annealing times cannot be maintained until the sample reaches the thermal equilibrium. After a rather short annealing time the metallic iron has completely reacted with the silicon substrate by forming iron silicide. After completing this chemical reaction the maximum soluble iron concentration decreases considerably, and the now supersaturated iron in the bulk diffuses back to the silicon-silicide boundary which acts as a getter for the iron. Therefore, the depth profile for iron after short annealing times exhibits a maximum iron concentration which may more-or-less exceed the solubility limit of iron according to the Arrhenius equation. This maximum iron concentration is reduced to the solubility value after applying a longer annealing period. In conclusion, iron concentrations exceeding the cited solubility can be achieved by annealing samples contaminated with metallic iron (for a short period of time); but the high concentrations cannot be maintained during extended annealing times. Therefore, enhanced iron concentrations are not real solubilities since the sample has not yet reached thermal equilibrium.

4.1.2 Diffusivity

$$D_{Fe} = 1.3 \cdot 10^{-3} \, e^{-0.68/kT} \quad [cm^2/s] \quad (30\,°C < T < 1200\,°C) \quad [4.2] \quad (4.2)$$

$$D_{Fe} = 1.1 \cdot 10^{-3} \, e^{-0.66/kT} \quad [cm^2/s] \quad (0\,°C < \ T < 1070\,°C) \quad [4.4] \quad (4.3)$$

$$D_{Fe^+} = 1.4 \cdot 10^{-3} \, e^{-0.69/kT} \quad [cm^2/s] \quad (30\,°C < T < 400\,°C) \quad [4.5] \quad (4.4)$$

$$D_{Fe^0} = 1.0 \cdot 10^{-2} \, e^{-0.84/kT} \quad [cm^2/s] \quad (30\,°C < T < 400\,°C) \quad [4.5] \quad (4.5)$$

For the diffusivity of iron in silicon four different results were reported in the literature (Table 3.2). Three of them [4.2, 4, 5] are correlated to Fe^+ at least in the RT region where diffusion has been determined from the reaction velocity of FeB pair formation in p-type silicon [4.2, 4]. The extended temperature region in which the Arrhenius equation should be valid results from interpolations between the diffusivities measured at HT, on the one hand, and the calculated values at room temperature, on the other hand. These three diffusivity values agree well within an experimental error. The most extreme two [4.2, 4] are depicted in Fig. 3.2. It is obvious that they coincide at high temperatures. But even at RT the differences are unimportant.

However, a large difference, far beyond an experimental error, has been found for the results correlated to neutral Fe^0, the charge state of iron in n-type silicon. It was only determined in the low-temperature region between 30 and 400 °C [4.5]. Whereas the difference to the other three results at HT is only about a factor of 2 (Table 3.2), it amounts to more than a factor of 50 at RT. Both diffusivities for Fe^+ and Fe^0 have been determined from outdiffusion experiments in the low-temperature region [4.5] which takes place at the sample surface within the space-charge region. Here ions drift in the electric field toward the surface or the bulk of the sample depending on the charge state of the ion and the direction of the electric field. The kinetics differ considerably and can be understood in terms of a carrier-emission-limited iron drift. So far results from bulk measurements in n-type silicon have not yet been published. Therefore, this diffusivity should be considered with caution unless it has been confirmed by additional experiments

4.1.3 Behavior During Heat Treatment

Due to its high vapor pressure at HT, metallic iron evaporates from surfaces in the starting period of a diffusion process. In this way wafers contaminated with iron may, in turn, contaminate other clean wafers on neighboring positions within the same furnace tube (Fig. 2.1), and they may contaminate the wall of the furnace tube itself. However, at HT, metallic iron reacts rather quickly with silicon by forming the HT iron-silicide modification α-$FeSi_2$. It is assumed that the silicide does not evaporate to the same extent as iron. Therefore, the iron evaporation is limited to the starting period of the diffusion process where metallic iron is still present on the wafer surface. When the reaction is completed, the silicide on the surface acts as a diffusion source during extended HT processing until it is exhausted, or the iron concentration in the sample has reached the respective solubility limit. Thereafter, the thermal equilibrium is reached and the indiffusion of iron equals its outdiffusion.

During the cooling period at the end of a HT process, the interstitially dissolved iron atoms become supersaturated. The iron atoms tend to diffuse out of the sample and form precipitates at the surfaces of the sample and in its bulk. Since iron precipitates only via a heterogeneous nucleation mechanism, the outdiffusion and the precipitation depend on the presence of nucleation centers formed by lattice defects or other (haze-forming) impurities. If a silicide layer is still present at the sample surface, it acts as a sink for outdiffusing iron atoms. By this process the iron concentration in the bulk beneath the iron-contaminated surface can become lower than the iron concentration beneath the opposite non-contaminated surface [4.3]. Since fast-diffusing transition metals which precipitate via a homogeneous nucleation mechanism (Co, Ni, Cu) can act as nucleation centers, iron forms haze if these impurities are present even in low concentrations. In addition, iron haze is formed during extended annealing processes of the iron-diffused sample at lower temperatures (Sect.6.3.1). In this case the supersaturation of iron increases and a homogeneous nucleation might be possible. Another explanation is based on the extended period of time for the iron atoms to diffuse to nucleation centers which may be present at the surface of the sample in lower concentrations.

In a recent paper *Heiser* and *Mesli* [4.5] postulated a difference in the outdiffusion behaviour of iron in n-type and in p-type silicon due to the neutral and positive charge states of iron, respectively. This leads to two diffusivities (4.4 and 5) listed in Sect.4.1.2. In contrast to the outdiffusion and precipitation of neutral iron (n-Si) at the sample surface *Heiser* and *Mesli* could not observe any outdiffusion and surface precipitation of positively charged iron (p-Si). Since these experiments were performed on Schottky barriers with and without applied bias we wanted to control these findings on bulk material [4.3] since there is still the contrast between iron haze observed on former silicon samples (Figs.2.1, 3.8 and 9) and the lack of iron haze in modern wafers contaminated and treated in the same way, which is not yet well understood. It cannot be reliably excluded that our earlier experiments were performed on n-type wafers and more recent investigations on p-type wafers. However, a direct comparison of the behavior of n- and p-type samples substantiated the former results: no difference could be found between n- and p-type wafers, both appeared haze-free also independent of the wafer orientation [4.3]. Although the explanation given by *Heiser* and *Mesli* [4.5] for the different behavior of neutral and positively charged interstitial iron due to the repulsive force of equal electrostatic charges which prevent the ionized iron from forming precipitates appears plausible it cannot explain the reported haze phenemena. Once more it was approved that iron enhances haze if foreign nucleation centres are present even if their haze is scarcely perceivable.

4.1.4 Electrical Activity

Substitutional iron in silicon is unstable [4.1]. Interstitial iron forms two charge states: a neutral state and a positively-charged state which results in one donor state. It is situated in the lower half of the silicon band gap with a mean activation energy [4.6] of

$$E_v - E_T = 0.39 \pm 0.01 \text{ eV} \qquad\qquad (4.6)$$

and

$$E_\infty = 0.043 \text{ eV} . \qquad\qquad (4.7)$$

If the activation energy is determined by an Arrhenius plot, the sum of both energies $0.39 \text{eV} + 0.04 \text{eV} = 0.43 \text{eV}$ is measured since the majority-carrier capture cross-section is temperature dependent (Sect. 3.3.1). The mean of the activation energy of this temperature dependence, E_∞, amounts to 0.043 eV.

Since the iron donor is situated in the lower half of the silicon band gap, its detection by means of DLTS using Schottky barriers is limited to p-type samples. However, in p-type wafers the interstitial iron reacts with the respective doping element (boron) by forming FeB pairs at RT. The reaction velocity increases strongly with increasing boron concentration. A rough estimate of the reaction velocity yields for boron-doped p-type wafers of about 7 $\Omega \cdot$cm ($2 \cdot 10^{15} \text{ cm}^{-3}$) that almost one half of the total iron content is paired to FeB after one hour of storage time at RT after quenching the sample. In addition, the reaction time depends on the storage temperature because the pairing reaction is diffusion limited. So the pairing can be avoided by storing the sample at liquid nitrogen, which freezes the mobility of iron. At elevated temperatures above RT the pairing reaction is accelerated but the thermal equilibrium between the pairs and the isolated ions is shifted versus that for isolated ions. At about 100°C, for example, the paired fraction is only 60% of the total iron amount [4.7]. A dissociation of the iron-boron pairs is also induced by minority carrier injection [4.8] and by a strong illumination with white light [4.7]. In both cases the quasi-Fermi levels are shifted versus the mid gap. As a consequence, the negative charge of the interstitial iron is at least partly compensated and the pairing reaction by forming electrostatic bindings between the iron and boron atoms is reduced.

The formation of iron-acceptor pairs is limited to p-type wafers where the interstitial iron is positively charged and the substitutional acceptor is negatively charged. In partly boron-compensated n-type wafers pairing cannot be observed. The reaction does not take place if the iron atom is electrically neutral. Then the interstitial iron can be detected even after years have passed since quenching the sample from HT.

In the presence of other doping elements (Al, Ga, In), the corresponding iron-acceptor pairs can be formed simultaneously. Two different structures of iron-acceptor pairs were observed that exhibit binding in the $\langle 111 \rangle$-direction (type 1 in Table 3.7) and in the $\langle 100 \rangle$-direction (type 2 in Table 3.7), respectively. These bistable configurations have been found for FeAl, FeGa and FeIn [4.9]. For FeB also other configurations were dete-

Table 4.1. Properties of iron shallow-acceptor pairs. The majority-carrier capture cross-sections were measured at the following temperatures of the DLTS signal (FeB: 50K, FeAl: 100K, FeGa: 120K)

Pair	ΔE_A [eV]	d/a	δ_M [cm^2]	Axis orient.-tion	E_b [eV]	Remarks	Ref.
FeB	-0.27	a	$1.6 \cdot 10^{-15}$	$\langle 111 \rangle$	0.6	anneal > 400K	4.10
	$+0.10$	d	$6.4 \cdot 10^{-14}$	$\langle 111 \rangle$		anneal > 400K	4.10
FeB	-0.43			2. Td	0.75	metast. < 250K	4.11
	-0.46			?		metast. < 250K	4.11
	-0.52			?		metast. < 250K	4.11
	-0.54			4. Td	0.73	metast. < 250K	4.11
	$+0.53$			3. Td	0.68		4.11
FeAl	$+0.20$	d	$1.1 \cdot 10^{-15}$	$\langle 111 \rangle$	0.7		4.12
	$+0.13$			$\langle 100 \rangle$		metast. at RT	4.12
FeGa	$+0.25$	d	$4.0 \cdot 10^{-15}$	$\langle 111 \rangle$	4.7		4.13
	$+0.14$			$\langle 100 \rangle$		metast. at RT	4.13
FeIn	$+0.27$	d		$\langle 111 \rangle$		metast. at RT	4.9,14
	$+0.15$			$\langle 100 \rangle$			4.9,14

tected but they dissociate below RT [4.11]. The formation of the respective pair configuration is charge-state-controlled and can be changed during sample cooling with and without the application of a reverse voltage at the rectifying contact in the sample, e.g., during recording a DLTS spectrum. The activation energies of the various deep energy levels are compiled in Table 4.1.

Besides pairing of iron with *shallow* acceptor atoms, pairing with *deep* acceptors was observed, too. Iron-gold pairs [4.15] are technologically most important since both impurities can be found occasionally in device production. Furthermore, gold is employed for carrier-lifetime tailoring. In contrast to the pairing of iron with boron, the formation of FeAu pairs is enhanced at elevated temperatures ($200 \div 250\,^{\circ}$C) where the FeB pairs had already dissociated. The dissociation of the FeAu pairs takes place beyond $250\,^{\circ}$C [4.15]. Thus, a DLTS spectrum of a p-type sample contaminated with iron and gold can exhibit signals of interstitial iron, FeB pairs, substitutional gold, and FeAu pairs. To determine the total iron content, the concentrations of the interstitial iron, the FeB pairs, and the FeAu pairs must be summed up, whereas the total gold concentration results from the sum of the substitutional gold and the FeAu pairs.

Besides pairing iron with gold, pairing iron with zinc has also been observed [4.16]. Although zinc is a wide-spread impurity and has been detected, for instance, on top of as-received polished wafers, it is usually not found in the bulk of wafers. Due to its high vapor pressure, zinc evaporates from the surface during the first heat treatment. So the diffusion of zinc into a sample is avoided. Even zinc-implanted samples do not exhibit these impurities in the bulk after a diffusion process. For a successful introduction into a sample, zinc must be diffused in a closed ampoule to avoid the disappearance of its vapor with the ambient atmosphere. As a consequence, iron-zinc pairs are technologically unimportant. The activation energies of the iron deep acceptor pairs are listed in Table 4.2.

Table 4.2 Properties of iron deep-acceptor pairs

Pair	ΔE_A [eV]	d/a	δ_M [cm^2]	E_L [eV]	Ref.
FeAu	-0.35	a	$2.1 \cdot 10^{-15}$	0.8	4.15
	$+0.43$	d	$2.6 \cdot 10^{-15}$		4.15
FePd	-0.32				4.17
FeZn	-0.47	a			4.16

In p-type silicon interstitial iron acts as a lifetime killer. The ratio between the capture cross-section for electrons and that for holes was estimated to 5000 at a temperature of 100 K [4.15, 18] and to 200 at RT [4.10]. The absolute values of the majority-carrier capture cross-section in p-type silicon published to date vary by about a factor of 50 with a most probable mean value of

$$\sigma_h (Fe_i) = 2.8 \cdot 10^{-16} e^{-0.043/kT} \quad [cm^2] . \tag{4.8}$$

The temperature dependence of this hole capture cross-section has an activation energy of 0.043 eV, as indicated in the temperature-dependent exponent. For 300 K a value of $5.3 \cdot 10^{-17}$ cm^2 is calculated, which agrees fairly well with $1.3 \cdot 10^{-16}$ cm^2 [4.18] and with results obtained by carrier-lifetime measurements in n-type samples correlated to DLTS results for p-type wafers prepared simultaneously [4.3]. This correlation enabled the determination of the capture cross-section for holes as an upper limit for carrier lifetimes in iron-doped samples. This procedure performed for hundreds of measurements at RT resulted in a capture cross-section for holes at RT of about $5 \cdot 10^{-17}$ cm^{-2}, in good agreement with the reported values.

For the minority-carrier capture cross-section σ_e only two results were presented in the literature, the RT value (4.9) by *Zoth* [4.18] and the temperature-dependent value (4.10) by *Lemke* [4.6]. At RT both values differ from one another by about 30%, which is considered a rather good agreement for capture cross-sections.

$$\sigma_e(\text{Fe}_i) = 2.6 \cdot 10^{-14} \quad [\text{cm}^2] \quad \text{at } 300 \text{ K} . \tag{4.9}$$

$$\sigma_e(\text{Fe}_i) = 10^{-10} T^{-1.5} \quad [\text{cm}^2] . \tag{4.10}$$

Minority-carrier lifetimes, or more properly diffusion lengths, in p-type wafers have been used to detect iron and to determine its concentration [4.19]. Since the electron capture cross-section of interstitial iron is rather high, the respective diffusion length is rather short. Due to the pairing reaction in boron-doped wafers, the interstitial iron disappears with the time elapsed after the HT treatment, and it is replaced by FeB pairs. Because of the lower activation energy of the FeB pairs their capture cross-sections are considerably reduced and the low-level minority-carrier diffusion length at low injection increases during the pairing reaction. As already mentioned, the pairs can be dissociated by a thermal treatment (about 200°C for several minutes). So the low-level diffusion length can be determined before and after this thermal treatment. If iron is the only effective recombination center in the sample, the low-level diffusion length L will be enhanced after the pairing by a constant factor equal to the square root of the relation between the minority-carrier capture cross-sections of interstitial iron $\sigma_e(\text{Fe}_i)$ and of the iron-boron pairs $\sigma_e(\text{FeB})$:

$$\frac{L(\text{FeB})}{L(\text{Fe}_i)} = \sqrt{\frac{\sigma_e(\text{Fe}_i)}{\sigma_e(\text{FeB})}} . \tag{4.11}$$

The equation implies that the interstitial iron was quantitatively transformed into iron-boron pairs before the heat treatment and that the precipitation of iron during the heat treatment is negligible. This diffusion-length enhancement, which amounts to about a factor of three [4.19], can serve for a qualitative detection of iron in the sample. However, the calculation of $\sigma_e(\text{FeB})$ according to (4.11) becomes more difficult since FeB exhibits two energy levels in the band gap of silicon: a donor level in the lower half and an acceptor level in the upper half. *Zoth* et al. [4.18] published values for the various capture coefficients α at RT, which are the product of capture cross-section σ and thermal velocity v of the respective carriers

$$\alpha_{e/h} = v_{e/h} \cdot \sigma_{e/h} . \tag{4.12}$$

For RT $v_e = 1.93 \cdot 10^7$ cm/s and $v_h = 1.56 \cdot 10^7$ cm/s. The recalculated values are compiled in Table 4.3. The origin of these values is not known, and they have not been verified by other researchers. The advantage of this method is the short measurement time needed to determine the diffusion length. The disadvantage is the poor accuracy of the resulting iron concentration which can considerably be deteriorated by the presence of additional impurities. For instance, a small amount of gold may cause the formation of FeAu pairs. Gold is well known as a strong recombination center, and FeAu has been found to act as a recombination center which is still stronger by a factor of five [4.15]. Measured majority-carrier capture cross-sections have been published for the acceptor and donor levels of the iron-gold pairs [4.15] and are listed in Table 4.2.

During the thermal treatment of the sample at 200°C to dissociate the FeB pairs, additional FeAu pairs can be formed which may even decrease the diffusion length instead of increasing it. In this case a DLTS spectrum may explain the unusual behavior of the diffusion length in this sample.

A tremendous complication for carrier-lifetime measurements to determine the iron content of a sample is the strong dependence of the lifetime due to interstitial iron on the injection level $\Delta N/N_0$. The injection level varies with the illumination intensity, since a higher light intensity generates more free carriers ΔN. On the other hand, the injection level depends also on the charge-carrier concentration of the sample N_0 (Sect.6.2.2). The injection dependence for defects exhibiting only one deep energy level within the bandgap of silicon is determined by the relation between the capture cross-section for electrons and that for holes. For interstitial iron this relation is about 200 at RT, as mentioned above. As a consequence, the carrier lifetime of iron exhibits a strong injection dependence. The calculation of the injection dependence of the lifetime due to FeB pairs is still more complicated, since the pairs exhibit two levels within the bandgap of silicon, a donor and an acceptor state. Both values are characterized by a capture cross-section for electrons and another one for holes which, in general, can be quite different. In the special case of FeB pairs these relations are almost equal and amount to about 6 and 1, respectively (Table 4.3). As a consequence, the injection dependence of the lifetime due to pairs is almost negligible, whereas that of interstitial iron steeply increases with the injection. As mentioned above, the low-level lifetime for interstitial iron is about a factor of 3 lower than that for pairs but it increases fast and for a medium-high injection level a cross-over of both lifetimes is found. In conclusion, lifetimes measured on different samples or by different equipments can be much easier compared if they were determined after complete pairing, since in most equipments lasers are used for carrier injection, generating medium-high injection levels. Diffusion-length measurements, however, are usually determined by applying low injection levels, therefore larger discre-

Table 4.3. Room-temperature capture cross-sections of FeB pairs [4.18]

Pair	d/a	α_e [cm^2]	α_h [cm^2]
FeB	d	$2.6 \cdot 10^{-13}$	$6.4 \cdot 10^{-14}$
FeB	a	$1.6 \cdot 10^{-15}$	$2.0 \cdot 10^{-15}$

pancies result by comparing lifetime measurements with those of the re-calculated diffusion-length measurements (Sect. 6.2.2).

4.1.5 Properties of the Precipitates

Iron precipitates via a heterogeneous nucleation mechanism, at least during cooling the sample from HT (Sects. 3.4.1, 2). Thus, lattice defects or other impurity precipitates are needed as nuclei to form iron precipitates. From TEM investigations it was deduced that iron precipitates as α-FeSi$_2$ in the form of rod-like defects [4.20] or as β-FeSi$_2$ during subsequent annealing processes [4.21]. One example for a rod-like defect is displayed in Fig. 4.1. The precipitate grew during a moderately-fast cooling of the sample from HT (1050°C applying the haze program for cooling the sample). In some cases, iron precipitates have exhibited additional small precipitates [4.21].

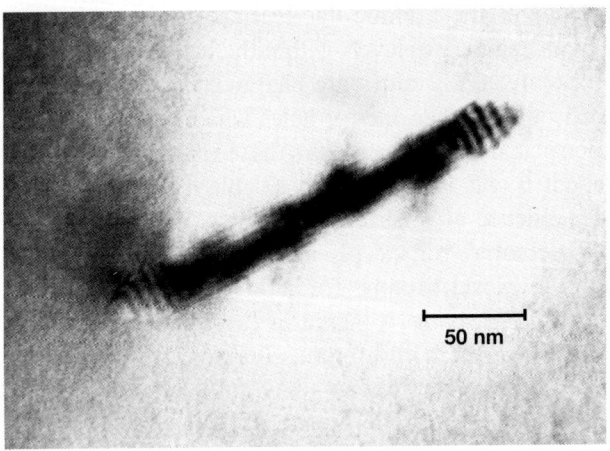

50 nm

Fig. 4.1. TEM micrograph of a rod-like α-FeSi$_2$ precipitate in silicon. The rod axis is parallel to Si$\langle 110 \rangle$

These adjacent, small crystalline particles resemble precipitates of copper silicide, which may have formed the nuclei for the iron precipitation. In addition, $FeSi_2$ precipitates were detected in decorated stacking faults [4.21]. Both HT modifications (α- and β-$FeSi_2$) have been found, which are the stable silicide phases above and below 915°C, respectively [4.20, 21].

Because of their accumulation at a silicon-silicon dioxide interface and their crossing of this interface, the iron precipitates cause the breakdown voltage of devices to deteriorate drastically [4.21, 22]. Furthermore, iron precipiates increase the leakage current [4.22] and cause soft reverse current-voltage characteristics. A weak spot due to α-$FeSi_2$ is formed since it behaves electrically like a metal.

If in modern device production low cooling rates are applied after the heat treatment, the iron impurities are mostly precipitated since the iron atoms have sufficient time to find suitable nuclei for their precipitation. Consequently, the iron impurity content must be kept as low as possible to avoid yield losses. In order to detect iron by means of DLTS in devices the sample must be heat-treated once more and quenched to RT to dissolve the iron atoms and keep them on the electrically active interstitial sites.

4.1.6 Known Impurity Sources, and Common Concentrations

Impurity sources for iron contamination during wafer processing are manifold. Any mechanical contact of the wafer surface with iron or stainless steel will produce a nearly unlimited source for a subsequent diffusion process. Although the electronegativity of iron resembles that of silicon, a liquid-phase contamination of wafer surfaces during wet processing is likewise possible. Strong contamination effects have been observed during the etching of wafers in alkaline solutions. But even acid cleaning solutions can contaminate wafers if iron impurities have accumulated in the chemicals after an extended use of the solutions.

Strong contamination has been observed during ultrasonic cleaning processes in vessels containing iron. The evaporation of iron during HT processes causes a vapor-phase contamination of other clean wafers and of the furnace tube. During processes at very high temperatures a large amount of iron is injected into the furnace tube from the resistor heating, regardless of whether the tube consists of quartz, polysilicon or silicon carbide. Boron-nitride diffusion sources can contain iron concentrations in the ppm region. Therefore, they cause a high iron contamination of the boron-diffused wafers of the order of $10^{13} \div 10^{14}$ cm^{-3}. Iron is easily sputtered onto wafer surface during ion implantation or during plasma processing if any metal containing iron is present in the vacuum chamber.

All these different impurity sources accumulate and yield an iron level which is characteristic for the cleanness of the whole process line. This iron level decreased considerably in the past decade from maximum amounts beyond 10^{14} cm^{-3} to average values in the region below 10^{12} cm^{-3} and much less in modern technology. Simultaneously, the iron contaminations on top of polished, as-received wafers decreased in the same way and by the same order of magnitude (Fig. 9.1).

4.1.7 Avoidance of Contamination

There are two general possibilities to reduce the impurity content in a wafer:

- Reduction of the impurity contamination, and
- gettering of impurities.

Both will be discussed in more detail in Chaps. 7 and 8, respectively. In principle, the reduction of iron contamination is relatively simple because the iron content can easily be controlled by DLTS measurements which yield quantitative results. However, routine applications of this method can consume extended measurement times to detect and eliminate the main sources of contamination.

If in a production line the final device is highly contaminated with iron, this line must, in general, be controlled at various stages starting with the unprocessed ingot. In order to reduce the quantity of control measurements to detect the process where the main iron contamination takes place, several technological processes performed in sequence can be checked together by a determination of the iron content in the final stage of this sequence. That sequence exhibiting a high concentration can then be controlled in more detail. The control measurement by means of DLTS can be performed on the original wafers if the starting material is p-type, with resistivities in the region of several $\Omega \cdot$cm. For other materials, monitor wafers with suitable properties are processed simultaneously. After sampling, the wafer is heat treated to diffuse possible surface contaminations into the bulk of the wafer. A heat treatment at $1050\,°$C for $10 \div 15$ min in an inert gas atmosphere is sufficient to achieve an almost homogeneous distribution of the iron content in wafers of common thickness. The wafer is then quenched to RT to keep the iron atoms electrically active on interstitial sites. After etching and evaporation of Schottky contacts, the DLTS measurements will yield the total iron concentration in the bulk by summing up the interstitial iron content and the concentration of iron-boron pairs or other iron-related defects.

After having detected the group of processes in which the iron contamination takes place, each single process is controlled in the same way. In general, it will be easy to determine the reason for iron contamination if the respective process has been discovered. Various possibilities known from experience in the past were listed above. In a production line, iron contamination of different sources can sum up to high iron levels. Therefore, all iron sources must be found and eliminated in order to reduce the total iron content to a sufficiently low level. The highest acceptable iron concentration depends on the device specifications and the gettering efficiencies of processes included in the specific production line.

As demonstrated with Fig.9.1, the requirements for a very low iron-impurity content in ULSI devices of high complexity will be more and more severe. This aim can only be achieved if all knowledge on impurity contamination and behavior is fully utilized and even extended further.

Much easier, faster, and more suitable for in-line control measuements is the application of fully automatic lifetime-measurement equipments which are now on the market. Since in most processed wafers the carrier lifetime is determined by the content of interstitial iron the only speciment preparation is a fast cooling process at the end of the last HT process, to keep the iron dissolved. The equipments enable lifetime plots of the whole wafer with the possibility to select a suitable local resolution. The colored plots enable the immediate detection of typical contamination marks (e.g., by handling, transportation, rinsing, droplets, replating) even from not quantitative measurements, for instance, due to high surface-recombination velocities. For quantitative measurements the wafer surfaces must be suitably prepared (Sect.6.2.2), then a direct calculation of the iron contents can automatically be performed by a computer. In general, the results agree well with control measurements performed by DLTS.

4.2 Nickel

Besides iron, nickel is one of the main impurities in device production. In analogy to iron the contamination with nickel takes place during production and processing of wafers. However, the concentration in the bulk of as-grown crystals, is low. Nickel can occasionally be found as a surface contaminant on as-received, polished wafers.

Nickel belongs to the fast-diffusing 3d transition metals and exhibits high solubilities in silicon at high sample temperatures. As a consequence nickel precipitates almost quantitatively even during quenching the sample from HT. The precipitation takes place via a homogeneous nucleation mechanism which does not require foreign nuclei. Only a residual quantity between 10^{-2} and 10^{-4} of the respective solubility is electrically active after quenching the sample, and forms various deep energy levels. Consequently, nickel cannot be detected quantitatively by means of DLTS. In contrast to iron, which forms only interstitial defects, nickel can be situated on interstitial and on substitutional sites in the host lattice. Since it precipitates almost quantitatively, the electrically active nickel defects do not play an important role in the deterioration of device performance.

Nickel belongs to the haze-forming transition metals [4.23]. This can be used for detection and examination of its distribution in contaminated wafers. In general, it is difficult to discriminate between the various haze-forming impurities. However, this is less important because of the small number of haze-forming impurities. From the main impurities in device production only nickel and copper form haze. The residual haze-forming metals such as cobalt, rhodium, and palladium are almost never found as impurities in silicon technology. The behavior of nickel and copper impurities is quite similar and the search for the contamination source and its elimination is therefore independent of the chemical nature of the impurity being copper or nickel. For special problems which require an accurate identification of nickel, modifications of the New etch (Sect.3.4.2) can be applied to distinguish haze formed by copper or by nickel impurities. Several alternative detection methods are available, which will be discussed in Chap.6.

4.2.1 Solubility

Within the temperature region indicated in (4.13) below the eutectic temperature (993°C) the solubility of nickel in silicon follows an Arrhenius equation [4.2] (Table 3.1 and Fig.3.1).

$$ S = 5 \cdot 10^{22} \, e^{(3.2 - 1.68/kT)} \quad cm^{-3} \quad (500°C < T < 950°C) . \quad (4.13) $$

The solubility was determined by a neutron-activation analysis after contaminating the wafer surface with metallic nickel [4.2]. However, at HT metallic nickel reacts with silicon by forming the silicides NiSi or $NiSi_2$, and this reaction is fast. Thus, it seems likely that (4.13) presents the solubilities with the stable silicide as a boundary at the sample surface. Like the other transition metals, nickel exhibits a very low RT solubility that is below the detection limit. Therefore, interstitially dissolved nickel is unstable at RT.

4.2.2 Diffusivity

Nickel diffuses mainly on interstitial sites but at RT only substitutional defects are detected. Therefore different diffusivities can be expected, D_i on intersitial sites (4.14) [4.24] (Table 3.2 and Fig.3.2), and D_S on substitutional sites (4.15) [4.25]:

$$D_i = 2.0 \cdot 10^{-3} e^{-0.47/kT} \quad [\text{cm}^2/\text{s}] \quad (800\,°\text{C} < T < 1300\,°\text{C}), \quad (4.14)$$

$$\left[D_S = 0.1 e^{-1.92/kT} \quad [\text{cm}^2/\text{s}] \quad (450\,°\text{C} < T < 800\,°\text{C}) \right]. \quad (4.15)$$

Besides the diffusivity data published by *Bakhadyrkhanov* et al. [4.24], additional values have been determined in lower or overlapping temperature regions. However, these data exhibit much higher activation energies for diffusivities such as 1.92 eV [4.25], and 1.52 eV [4.26]. At least the largest discrepancy is explained by that method which yields the diffusivity of substitutional Ni [4.25] in contrast to the much faster interstitial diffusivity evaluated in the data of [4.24]. Since the diffusion of nickel takes place essentially by an interstitial mechanism, these data are more important for understanding the diffusion of nickel during HT processing. Furthermore, the reported data agree well with recent diffusivity data for cobalt [4.27] and those for copper (Table 3.2), and therefore fit the expected chemical trend within the series of the 3d transition metals.

For nickel the substitutional diffusion is still unimportant since the solubility of interstitial nickel defects exceeds the solubility of substitutional nickel defects by about two orders of magnitude. In contrast to the interstitial diffusivity the substitutional diffusivity cannot be expressed in a simple equation because it still depends on the concentartion and on the presence of sinks for the annihilation of intrinsic defects such as vacancies or silicon self-interstitials. The substitutional diffusion of nickel in an oxidizing atmosphere, for instance, is drastically reduced because of the increased concentration of silicon self-interstitials that are generated by the oxidation and exceeding by far the thermal equilibrium [4.28]. As a consequence, a high nickel concentration is observed only below the surface which, however, exceeds the solubility. Therefore, the validity of (4.15) is limited and this is symbolized by the large parentheses. Substitutional diffusion will be treated in more detail in connection with the discussion of the properties of platinum, gold and zinc for which many diffusion experiments have been performed.

4.2.3 Behavior During Heat Treatment

Due to the high values of the solubilities and diffusivities of nickel at HT this impurity diffuses very quickly and results in high concentrations if the contamination source is not exhausted during the diffusion period. Metallic nickel exhibits a low hardness; therefore, a wafer can easily be contaminated by the pure metal. At elevated temperatures nickel forms the sequence of silicides presented in Table 4.4, showing increasing silicon content with increasing formation temperatures [4.29]. The HT modification $NiSi_2$ exhibits the same cubic lattice symmetry as silicon with an almost equal lattice constant (5.395Å compared to 5.406Å for silicon at RT [4.21, 30]). The silicide grows epitaxially on the silicon substrate. Both, the epitaxial growth and the equal lattice constant render the detection of $NiSi_2$ on top of a silicon substrate by means of X-ray diffraction extremely difficult.

Table 4.4. Formation temperature T of nickel silicides

Silicide	T [°C]	Remarks
Ni_2Si	$200 \div 350$	
$NiSi$	$350 \div 750$	
$NiSi_2$	≥ 750	stable HT modification

During heat treatment nickel silicide is formed very quickly and then acts as a diffusion source. Nickel diffuses so fast in silicon that a heat treatment at 1050°C for 20 min contaminates about 2/3 of the diameter of a 7.5 cm wafer with haze after a spot-wise contamination in the wafer center (Fig. 3.7). Considering the large number of diffusion processes needed for device production, it is evident that even a single nickel contamination at any point of a wafer is sufficient to contaminate the whole wafer. To achieve high yields it is important to avoid nickel contamination during the whole process or to limit this contamination to concentrations which can be gettered.

During the cooling period after a diffusion process the nickel atoms become supersaturated and diffuse to the surfaces of the wafer where they precipitate and form haze. Nickel atoms which cannot reach the surfaces during a short cooling period, form precipitates within the bulk. Due to outdiffusion to the surfaces, denuded zones beneath both surfaces are created and are free of precipitates, as shown for palladium in Fig. 3.5. The density of haze and the depth of the denuded zone depend on the cooling rate. Besides homogeneous nucleation of nickel it precipitates also via a heterogeneous nucleation mechanism if nucleation centers are present and can be reached by the nickel atoms during the cooling period. Therefore, nickel can be gettered by various methods, as known from the literature. On the other hand, nickel precipitates, formed by a homogeneous nucleation mechanism, can act as gettering centers for other impurities, e.g., for iron which is not able to precipitate without the presence of foreign nucleation centers.

4.2.4 Electrical Activity

For a better understanding of the behavior of nickel at RT the diffusion length of interstitial nickel atoms for a diffusion duration of 100 hours have been calculated and compared to those of interstitial iron and of substitutional nickel. If the HT diffusivities of iron and nickel in silicon are extrapolated to RT, the respective diffusion lengths can be determined. The results are listed in Table 4.5.

Table 4.5. Diffusivity D and calculated diffusion length L for iron and nickel at RT

Defect	D [cm^2/s]	L for 100 h [μm]
Ni$_i$	$2.5 \cdot 10^{-11}$	30
Fe$_i$	$4.8 \cdot 10^{-15}$	0.4
Ni$_s$	$5.4 \cdot 10^{-34}$	$4 \cdot 10^{-11}$

Although unstable at RT, interstitial iron may be quenched-in into samples and can be detected even after long periods, at least, in n-type silicon, as discussed above. In p-type silicon interstitial iron reacts with boron by forming iron-boron pairs. The reaction time is diffusion limited and depends on the mean distances between the substitutional boron atoms which are immobile at RT. The calculated diffusion length for interstitial iron atoms at RT for a diffusion duration of 100 hours is of the same order of the mean distance between boron atoms in a sample exhibiting boron concentrations of $2 \cdot 10^{15}$ cm^{-3}. This agrees fairly well with the experience of iron-boron pair formation.

However, interstitial nickel diffuses much faster at RT, and the resulting diffusion length is nearly two orders of magnitude higher compared to iron. Therefore, it is expected that unstable interstitial nickel is not present in wafers at RT because of its outdiffusion to the surfaces, precipitation, or reaction with other impurities.

On the other hand, the respective diffusion length for substitutional nickel is extremely short. Therefore, we would expect the existence of substitutional nickel at RT, although the fraction of substitutional nickel compared to the total nickel content may be small. Indeed, the fraction of the electrically active nickel is less than 1% of the respective solubility [4.23]. Furthermore, it is composed of various defects which have not all been identified. Some of the defects change their concentration during storage of the sample at RT. The mean activation energies of the identified levels, as observed by several researchers [4.31-34], are listed in Table 4.6.

From the multitude of nickel-related deep-energy levels (*Lemke* [4.35] detected up to 15 different DLTS signals) onyl recently a probably reliable correlation to substitutional nickel could be found. It exhibits a double acceptor situated very close to the conduction-band edge (only one researcher), an acceptor state (4 researchers), and a donor state (3 researchers). The DLTS maximum of the double acceptor appears at a temperature of 45 K, this may be the reason why it was not detected by other researchers so far. The donor level was observed to increase in concentration within a period of about 10 days, whereas a neighboring hole trap at $E_v +0.22$ eV decreases in the same time. *Nakashima* [4.32] studied the thermal stability of the donor and the acceptor levels, and found that they remain stable up to a temperature of 600°C which is in accordance with their correlation to a single substitutional defect. Many other levels could no yet be identified. To our knowledge nickel-hydrogen complexes have not yet been investigated. It is imaginable that several levels, especially those which decrease with time, could be explained by complexes of this type. However, the uncertainty of the defect identification is not a serious technological problem since all these defects appear only in low concentrations after quenching the sample to RT.

Table 4.6. Properties of substitutional nickel and nickel-related defects

Defect	ΔE_A energy [eV]	a/d	δ_M cross-section [cm^2]	T_{DLTS} [K]	Ref.
Ni_s	-0.07 ± 0.01	aa	$\sigma_e = 5.4\cdot10^{-18}$ *	45	4.31
Ni_s	-0.39 ± 0.02	a	$\sigma_e = 5.6\cdot10^{-17}$	200	4.32
Ni_s	$+0.17\pm0.01$	d	$\sigma_h = 1.1\cdot10^{-15}$	90	4.33
$Ni_s Ni_i$	electrically inactive				4.34

4.2.5 Properties of the Precipitates

All nickel precipitates investigated so far by TEM exhibit the configuration of the equilibrium HT silicide $NiSi_2$ epitaxially grown parallel to $\{111\}$ silicon lattice planes. One example for a $NiSi_2$ platelet is displayed in Fig.4.2 in a cross-sectional TEM micrograph. The nickel-doped sample was cooled moderately fast to RT applying the haze program. The precipitate touches the wafer surface S where its nucleus may have formed. The additional lattice defects indicated by the arrows are punched-out dislocations.

In several precipitates a twin orientation to the silicon matrix has been observed [4.21, 36-38]. In samples quenched from HT small (0.4 to $0.8\mu m$ in diameter) and very thin platelets have been observed, which partly consisted only of two neighboring silicide monolayers. Thus, each nickel atom is a boundary atom. In this case no dislocations were punched-out into the surrounding host lattice [4.38, 39].

Applying low cooling rates or subsequent annealing processes at lower temperatures, a ripening process of the precipitates was observed where the larger particles grow in size and the smaller ones shrink. In addition, the precipitates also increase in thickness. The following morphologies have been observed [4.37, 39]:

(i) Tetrahedral particles in regions of high nickel concentrations exhibit interfacial dislocations and dislocations within the precipitate.

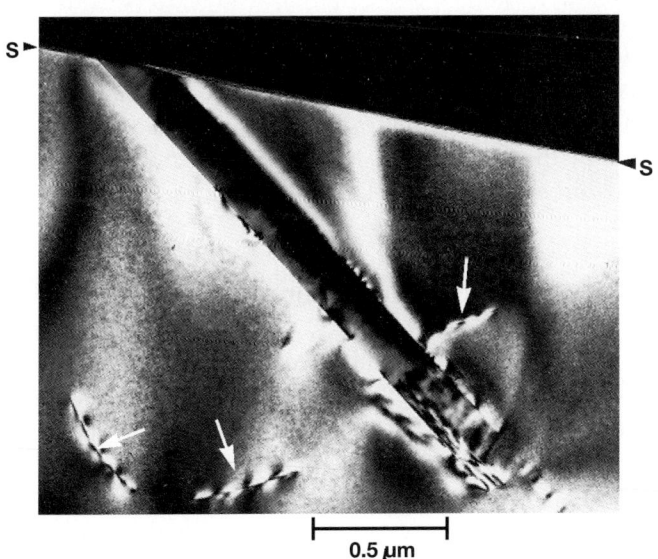

|— 0.5 μm —|

Fig.4.2. Cross-sectional TEM micrograph of a $NiSi_2$ platelet at the surface (S) of a wafer. The arrows indicate dislocations punched-out by the precipitate

(ii) Silicide platelets in regions of lower nickel concentrations show interfacial dislocations and dislocations punched-out into the silicon matrix from the edge of the platelet [4.37, 39]. Here, twin orientation was observed.

(iii) Agglomerates of both types of particles.

The different particles observed by TEM can be correlated to respective structures of etch pits, as revealed by preferential etching of the samples. In regions of higher nickel concentrations, irregularly shaped etch pits were revealed, whereas in regions of lower nickel concentrations rod-like etch pits were found (Figs. 3.12 and 13) [4.37]. A high density of small etch pits was obtained after preferential etching of samples which were quenched to RT (Fig. 3.13).

Moderately fast cooling of the samples from HT leads to an accumulation of precipitates at the sample surface, and the formation of haze and denuded zones beneath both sample surfaces. In this case the haze-forming nickel precipitates cross the original sample surface [4.38], as has been reported for iron precipitates. In oxidized wafers the precipitate penetrates the oxide and reduces the local oxide thickness. Since the conductivity of the silicide by far exceeds that of the oxide the breakdown voltage of a MOS capacitor is reduced, which may result in lower yield.

As shown in Fig. 3.6, the lattice parameter of $NiSi_2$ agrees with that of silicon for a temperature of about 450°C. Due to different temperature coefficients the lattice of $NiSi_2$ is larger than that of silicon at high temperatures [4.21]. Therefore, silicon self-interstitials are emitted during nickel precipitation at HT. On the other hand, silicon self-interstitials are absorbed if precipitation takes place at temperatures below 450°C. Primarily the precipitation will, at least, start at higher temperatures and silicon self-interstitials will be emitted, which then leads to the formation of extrinsic dislocations in the neighborhood of the precipitates. Indeed, extrinsic dislocation loops could be detected by *Augustus* applying TEM investigations [4.20].

4.2.6 Known Impurity Sources, and Common Concentrations

Nickel, as a component of steel, is widely employed in the production of equipment. Therefore, a mechanical contact of the wafer with a handling system or tweezers can easily take place and results in a strong contamination of the wafer. More than two decades ago the elimination of steel tweezers lead to the first success in reducing the up-till-then strong nickel and iron contaminations of wafers during device production. But even today nickel can be found, now and then, on the surface of as-received polished wafers. These wafers can contaminate the chemicals of the cleaning equipment, and consequently nickel can be replated on other clean wafers passing the same facility.

Another contamination source can be found in Reactive Ion Etching systems (RIE). Once contaminated a wafer can hardly be decontaminated by a common cleaning facility. Only repeated wet etching of a silicon surface removes most of the metal contamination. Investigations of wafers intentionally contaminated with iron, chromium and nickel have shown that iron and chromium can be removed from the wafer surfaces by a single wet etching; but nickel cannot because it replates on the etched silicon surface [4.40]. Therefore, nickel must be absolutely avoided inside the reaction chamber applying RIE. During a subsequent oxidation process iron and nickel cause the formation of stacking faults, but chromium does not. Even after removing 20 nm from a nickel-contaminated wafer surface by etching, the residual nickel impurities can still be detected and cause the formation of stacking faults during a subsequent oxidation process [4.40].

A further contamination source was found in the resistance heating of furnace tubes. Large amounts of nickel, copper and iron have been detected after a wet oxidation of wafers at $1000\,°C$ for 1 hour, exhibiting concentrations beyond 10^{14} cm^{-3}. The impurity concentrations increase exponentially with increasing oxidation temperatures [4.41].

4.2.7 Avoidance of Contamination

The mechanical contact of wafers to any metallic nickel must be avoided. It should be borne in mind that a nickel contamination can also be caused by handling wafers with Teflon tweezers if nickel-contaminated wafers have been handled before with the same tweezers. So all handling systems and tweezers must be cleaned periodically to avoid this kind of contamination.

Nickel has to be avoided inside vacuum chambers where sputtering takes place. This includes dry etching and ion-implantation facilities. Furthermore, nickel and its compounds must not be in chemicals used for wet processes because of the replating of nickel on uncovered silicon surfaces. In order to avoid nickel contamination in furnace tubes they have to be cleaned periodically by floating them with oxygen at HT with or without the addition of small amounts of HCl or trichlorethylene. The temperature during cleaning should be about $50\,°C$ higher than for the standard production processes. If cleaning is performed in the idle times of the tube, the subsequent processes will remain free of contamination within the detection limit of haze control. The corresponding nickel concentration is assumed to be in the region of less than 10^{12} cm^{-3}.

4.3 Copper

The third main impurity in device production besides iron and nickel is copper, which also belongs to the 3d transition metals. It can be found in higher concentrations in wafers after processing and on surfaces of as-received polished wafers. The copper concentrations in the bulk of the as-grown silicon crystal are very low, as already reported for iron and nickel. However, a copper contamination can start with the wafer production. Due to its electronegativity of 1.9 compared to that of silicon of 1.8, copper is easily replated on clean wafer surfaces during any wet chemical process such as cleaning, etching, polishing, or removing of oxide layers. Therefore, the copper impurity content in chemicals used in device-production lines should be as low as possible.

The properties of copper resemble those of nickel since both exhibit high interstitional solubilities and diffusivities at elevated temperatures. Copper exhibits the highest values of all transition metals. Consequently, interstitial copper precipitates almost quantitatively even during quenching samples from HT. Due to a homogeneous nucleation mechanism, copper diffuses to the sample surfaces during the cooling period after a heat treatment where it precipitates and forms haze. Simultaneously copper precipitates also in the bulk of the wafer beneath the denuded zones which are formed by outdiffusion of the copper atoms to the surfaces of the sample. In contrast to iron and nickel, the copper precipitates do not consist of the silicon-rich silicide $MeSi_2$ which is not stable for copper in thermal equilibrium, but they consist of a metal-rich silicide with the composition Cu_3Si [4.21, 42]. This copper silicide does not fit the lattice parameter of the silicon as well as $NiSi_2$. Because of its larger lattice parameter, more silicon self-interstitials are emitted per precipitated copper atom compared to nickel precipitation (Table 3.10). Therefore, the copper precipitates are decorated by a more-or-less extended extrinsic dislocation network which forms characteristic star-like etch pits in regions of high copper concentrations after slow cooling of the sample from HT or subsequent annealing at lower temperatures.

Because of its almost quantitative precipitation, only a small fraction of the respective copper solubility is electrically active and forms several deep energy levels. Interstitially dissolved copper atoms are not expected at RT because of their high mobility. The tendency to form substitutional defects may still be more obvious compared to nickel. Because of their low concentrations these electrically-active defects are not technically important, as already found for nickel.

4.3.1 Solubility

The interstitial solubility of copper follows the Arrhenius equation [4.2] within the temperature region indicated in (4.16) (Table 3.1 and Fig.3.1):

$$S = 5 \cdot 10^{22} \, e^{(2.4 - 1.49/kT)} \text{ cm}^{-3} \quad (500\,°C < T < 800\,°C). \qquad (4.16)$$

The solubility was determined by a neutron-activation analysis probably in equilibrium with the silicides at the sample surfaces. Like all other 3d transition metals, copper exhibits a very low (extrapolated) RT solubility which is below the detection limit. Therefore interstitially dissolved copper is unstable at RT.

4.3.2 Diffusivity

For the interstitial diffusivity of copper in silicon several slightly differing equations were advanced, which are reproduced below. Diffusivities exhibiting stronger deviations which could be explained by other researchers, are not included in (4.17-20). The first experimental results reported in (4.17) were obtained already in the sixties by *Hall* and *Racette* [4.43]. The measurements were performed on higher-doped material and therefore corrected for intrinsic silicon about three decades later with a most reliable result from *Mesli* et al. [4.44], reported in (4.18). Confusion resulted from the discovery that the so-called **X-defect**, which exhibits an extremely high mobility at room temperature [4.45] reported in (4.20) was ultimately identified as interstitial copper [4.46]. The X-defect was postulated to explain unexpected resistivity changes observed after having polished wafers in the presence of (contaminated) chemicals [4.47]. But even these diffusion problems could be solved by taking into account an additional drift of the Cu^+ ions in the electric field of the surface space-charge region and the influence of interstial oxygen dissolved in the material [4.44].

In p-type silicon Cu^{i} reacts with boron by forming electrically inactive CuB pairs. These pairs are stable at temperatures slightly below RT [4.44] and slightly above RT [4.49] RT and dissociate by forming silicide precipitates and reactivated boron. Finally, the results in (4.18 and 15) agree fairly well within a measurement error and enable the calculation of the diffusivity of copper in an extremely wide temperature range between $-90\,°C$ and $930\,°C$, see (4.19):

$$D(Cu_i) = 4.7 \cdot 10^{-3} \cdot e^{-0.43 eV/kT} \quad 400\,°C < T < 900\,°C \quad [4.43] \qquad (4.17)$$

$$D(Cu_i) = 5.0 \cdot 10^{-3} \cdot e^{-0.40 eV/kT} \quad 400\,°C < T < 900\,°C \qquad (4.18)$$
$$\text{(results of [4.43] have been corrected in [4.44])}$$

$$D(\text{Cu}_i{}^+) = 4.5 \cdot 10^{-3} e^{-0.39\text{eV}/kT} \quad -90°C < T < 930°C \tag{4.19}$$
$$(\text{FZ, CZ: } T > 30°C \text{ [4.46, 48]})$$

$$D(\text{X})_{\text{eff}} = 5.0 \cdot 10^4 e^{(-0.665\text{eV}/kT)} \quad -53°C < T < +7°C \tag{4.20}$$
$$(\text{uncorrected) [4.45]} .$$

With the best data given in (4.19) we can calculate the diffusivity at RT without an extrapolation of the measured HT values. Assigning to this diffusivity a diffusion length for copper at RT and a period of 100 hours results to about 200 μm, which is large compared to the respective values for nickel (30 μm) and iron (0.4 μm) under equal conditions, listed in Table 4.5.

Besides the interstitial diffusion of copper there is also a substitutional diffusion which yields substitutional defects forming deep energy levels which are stable at RT. Both diffusion mechanisms may run independently from one another. As mentioned in Sect. 4.2.2 for nickel, the substitutional diffusion of copper is less important because it is much slower and the solubility of interstitial copper still exceeds by far that of substitutional copper. This is deduced from the small fraction of electrically active defects in relation to the solubility of copper at the duffusion temperature. Similar to nickel, the substitutional diffusivity is drastically reduced by an additional oxidation of the sample during diffusion [4.28].

4.3.3 Behavior During Heat Treatment

Because of its low hardness, metallic copper can easily contaminate a wafer surface. So an intentional contamination can be performed by mechanical scratching of the sample with copper, especially on the rougher reverse side. However, the rubbed copper does not stick very tight to the silicon surface even after removing the native oxide, and it happens that particles fall off. Copper is also easily replated on wafers from any solution containing this impurity. The homogeneity of the replated copper is poor, and any variation of the surface will be decorated due to the formation of droplets during the preceding rinsing and drying.

During heat treatment, copper diffuses very fast into the bulk of the sample. After short annealing times a η'-Cu$_3$Si silicide [4.49] is formed at the surface and acts as a further diffusion source. During the cooling period copper diffuses to the sample surfaces and forms haze. Simultaneously it precipitates almost quantitatively in the bulk of the sample beneath the denuded zones formed by outdiffusion. Copper precipitates via a homogeneous nucleation mechanism at the sample surface and in its bulk as Cu$_3$Si [4.50]. In this way it may form nuclei for other impurities, for example, for iron. Applying lower cooling rates, large copper precipitates grow and smaller precipitates in their neighborhood shrink or even vanish, forming denuded zones surrounding the extended precipitation colonies. This ripening process works by a heterogeneous nucleation mechanism forming large precipitate colonies along an extended extrinsic dislocation network. In the bulk of the samples other precipitates in the form of platelets have been observed, they also appear after quenching the sample from HT [4.20, 38].

The ripening process of the precipitate colonies during extended cooling periods cause an increase of the individual colonies and a decreasing density of colonies. Therefore, the brightness of haze (the intensity of scattered light due to the etch pit density) is reduced in regions of high copper concentration (Fig. 3.10). The microscopic observation reveals star-like etch pits in this region, as shown in Fig. 3.19. For a detection of copper by applying the haze test the sample should be quenched from HT to RT in order to enhance the sensitivity of the haze inspection. The interference of nickel contamination for evaluating the haze test was discussed in Sect. 4.2 on the properties of nickel. In the case of large amounts of copper in the sample, a microscopic investigation of slowly cooled samples may help to identify copper due to the appearance of extended star-like etch pits exhibiting $60°$ angles on (111)-oriented surfaces (Fig. 3.19) and rectangular crosses on (100)-oriented wafer surfaces (Fig. 3.18). Copper haze can also be distinguished from nickel haze by varying the composition of the New etch (Sect. 3.4.2).

4.3.4 Electrical Activity

The fraction of the electrically-active copper, after quenching the samples from HT, is very small. At diffusion temperatures up to $800°C$ it amounts to about 10^{-3} of the respective solubility. At higher temperatures the absolute concentration of the electrically-active defects decreases by about an order of magnitude [4.23]. Due to the higher solubility of copper at higher temperatures the copper concentration increases drastically with the temperature and consequently the fraction of precipitated copper increases likewise. In the lower-temperature region only one donor state is formed, which is situated in the lower half of the silicon band gap (Table 4.7). This defect is correlated to a photoluminescence signal [4.51, 52] which is due to a copper atom on an interstitial site paired to one on a substitutional site. The DLTS signal shows a linear dependence on the luminescence signal and this depends on the squared intensity of the copper concentration.

Applying diffusion temperatures higher than $950°C$ up to $1050°C$, the concentrations of different electrically-active defects after quenching the sample to RT increase again. The concentrations are still enhanced by a subsequent annealing of the sample between $200°C$ and $500°C$ and can

Table 4.7. Properties of substitutional copper and copper-related pairs

Defect	ΔE_A [eV]	d/a	δ_M [cm²]	T_{DLTS} [K]	Ref.
Cu$_s$	-0.16 ± 0.01	aa	$\sigma_e = 1.9\cdot10^{-17}$ *	80	4.53
Cu$_s$	$+0.46\pm0.02$	a	$\sigma_h = 1.5\cdot10^{-14}$	230	4.54
Cu$_s$	$+0.22\pm0.01$	d	$\sigma_h = 3\cdot10^{-14}$	100	4.55
Cu$_s$ Cu$_i$	$+0.09\pm0.01$	d	$\sigma_h = 3.5\cdot10^{-15}$	80	4.56
Cu$_i$ Au$_s$	$+0.42/+0.32$		bistable		4.57
Cu$_i$ B$_s$	elect. inactive, diss. 200°C				4.49
Cu$_i$ In$_s$	elect. inactive			4.46	
M center	$+0.21$	a	metastable		4.44

then amount to about 10^{14} cm^{-3} [4.51, 53]. The mean activation energies of the deep levels are listed in Table 4.7. They have been observed by different researchers (more than those cited). These deep energy levels which are formed at higher diffusion temperatures, are assumed to represent different charge states of the same level [4.53]. Due to the reported experimental conditions for the sample preparation, which are needed to obtain the highest concentrations it is assumed that these deep levels are correlated to the substitutional copper.

Beside these defects a variety of deep energy levels were reported in the literature, which occupy almost the whole band gap of silicon with the exception of rather shallow levels and a gap between E_C-0.3 and 0.5 eV. Only a few are correlated to definite structures. The other energy levels are not presented since correlations to defined defect structures do not exist. Furthermore, the concentrations of these defects are small and it cannot be excluded that several of these levels may be due to any other impurity which happened to be present in the same sample. In conclusion, the electrically-active copper-related defects play only a minor role because of their low concentrations compared to the total copper content in a sample. The main deterioration of the device performance due to copper impurities is caused by the precipitates and not by the electrically-active defects.

4.3.5 Properties of the Precipitates

The structures of copper precipitates as well as the corresponding etch pits can be different in appearance, as shown in Figs. 3.17-19. They depend on the copper concentrations and the cooling rate applied, on the one hand, and the wafer orientation, on the other hand. Investigations by means of TEM [4.21] revealed that the star-like structures observed mainly at the surface of a moderately fast cooled sample doped with a high copper concentrations consist of agglomerates of small spherical copper silicide precipitates that are arranged in planes parallel to the silicon {110} lattice planes. These agglomerates are surrounded by extrinsic dislocation loops. The sizes of the agglomerates amount to 80 μm in regions of high copper concentrations and to about 0.5 μm in regions of low copper concentrations. One example for a copper precipitate colony is depicted in Fig. 4.3 presenting a low-magnification TEM micrograph. The small spherical Cu_3Si particles in planar arrangement are oriented parallel to Si{110} lattice planes. They are bounded by numerous extrinsic edge dislocations which form the star-like network and are comparable to the structure of the etch pits formed after preferentially etching copper-doped samples.

Large surface agglomerates can extend up to 40 μm into the bulk of the sample. The precipitates cross the original surface, as has already been reported for iron and nickel precipitates. This leads to defective devices due to lowered breakdown voltages of the gate oxides, as has been revealed by

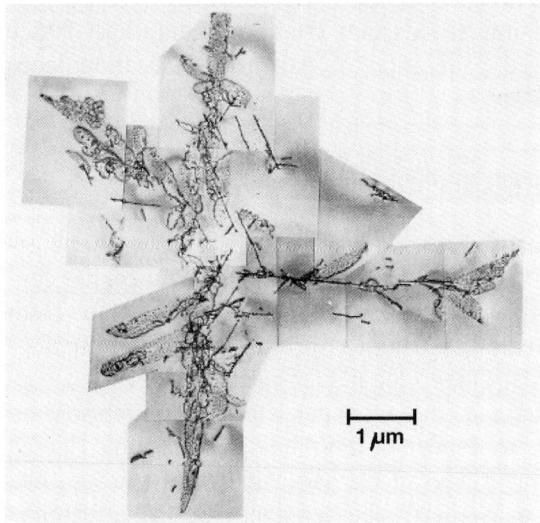

Fig. 4.3. Low-magnification TEM micrograph of a star-like precipitate colony consisting of planar arrangements of spherical Cu_3Si particles parallel to Si(110) planes and surrounded by extrinsic edge-type dislocation loops

intentional copper contamination [4.59]. If the copper precipitation takes place after growing the oxide layer the defect formation depends on the ripening process of near-surface copper precipitate colonies. In the early stage no defects were detected. After further annealing at 500°C a progressive stage shows a lens-shaped precipitate growing at the silicon-silicon oxide interface, which penetrates both regions. The defect is still covered by the oxide layer but first cracks in the oxide may be visible. In a final stage a mushroom-like defect has been observed showing a crack in the oxide surrounding the precipitate, which slightly lifted the oxide layer [4.59].

The diameters of the small and spherical silicide precipitates vary from 5 to 20 nm. They exhibit a single crystalline non-cubic symmetry with differing orientations within the silicon lattice. It is assumed that their composition corresponds to Cu_3Si which is a metal-rich silicide [4.16, 42].

In samples which were quenched from HT the observed surface precipitates were equal to those obtained after moderately fast cooling of the sample. In the bulk of the sample these agglomerates were mixed with a different type of precipitates showing platelets parallel to the {111} silicon lattice plane. Their diameter varied from 0.5 to 1 μm. Although the composition of these platelets is not yet known it is assumed that the platelets rearrange to agglomerates within prolonged cooling periods [4.21, 38].

The characteristic shape of the star-like etch pits is due to extended extrinsic dislocation loops oriented parallel to the {110} silicon lattice planes. Assuming a composition of Cu_3Si an emission of 0.5 silicon atoms is expected (Table 3.10) for the precipitation of one copper atom. This rather high emission of silicon self-interstitials into the surrounding host lattice agrees well with the observed generation of extrinsic dislocation loops [4.42].

4.3.6 Known Impurity Sources and Common Concentrations

In general, a mechanical contact of wafers with metallic copper can be excluded as a contamination source in production lines. The main copper contamination especially on as-received polished wafers may originate from copper impurities in chemicals used during wet processing such as etching, cleaning, lapping and polishing. Due to the higher electronegativity of copper (1.9r.u.) compared to silicon (1.8r.u.), copper is preferably replated on uncoated surfaces. This leads to a more-or-less homogeneously distributed copper contamination on both surfaces of the wafer in contrast to spot-like contaminations by means of mechanical metal contacts. So the contamination by replating in chemicals can easily be distinguished from mechanical contamination by means of a haze test applied to the wafer (Sects.6.3 and 7.1).

A further contamination source can be resistance-heating of the furnace tubes, as reported in connection with the discussion of sourses for nickel contamination. Copper concentrations of 10^{14} cm^{-3} and more have been measured with an increasing tendency for higher diffusion temperatures [4.41].

A quite unusual source for copper contamination has been found in the electron-beam evaporation equipment for the formation of aluminum contacts. The aluminum is kept in a copper susceptor. During electron bombardment the aluminum melts and dissolves copper from the susceptor. The copper concentration increases up to several percent in the aluminum with the increasing number of evaporation runs performed. Because of the high mobility of copper even at RT, the copper which is dissolved in the evaporated aluminum, can diffuse into the silicon substrate and may cause a variety of problems.

Copper contamination can also be caused by the gas supply system of furnaces and reactors if the pipes consist of copper. After a mechanical treatment of the gas system, or even after fluctuations of the gas pressure or current, small particles of copper are released and carried into the reactor where they may evaporate and contaminate the wafers. The smallest particles are carried by the gas stream and may reach the wafer surfaces where they cause a spot-like contamination.

4.3.7 Avoidance of Contamination

Replating of impurities on surfaces can be avoided by using clean chemicals. If the wafers are copper contaminated a subsequent cleaning will not remove this contamination from the surface. Only etch processes removing 0.1 to 0.5 μm silicon from the surface will yield clean surfaces. But even during etching, replating can take place. In the case of higher copper concentrations the etching should be performed several times always using new solutions in order to reduce the replating of copper. However, this etching disturbs the quality of the polished surface of the wafer. Therefore, it should be performed before polishing, or the wafer must be repolished.

In order to avoid copper contamination during heat treatments in a furnace tube, the tube should be cleaned periodically, as reported in the discussion of avoiding nickel contamination (Sect.4.2.7). Copper contamination during aluminum evaporation by electron irradiation can be reduced by periodically recharging the Al target. Copper contamination caused by the gas-supplying system can be reduced by employing gas filters at each point of use and replacing them periodically. These filters can eliminate the extensive store room for particles in the mostly extended and complicated supplying system.

4.4 Molybdenum

Molybdenum belongs to the group of 4d transition metals. This group has been investigated less than the 3d transition metals since almost all of the main impurities detected in wafers during device production belong to the 3d transition metals. Thus, the chemical trends of the properties of the 4d transition metals such as solubilities, diffusivities, and the fractions of interstitial or substitutional defects cannot yet be deduced just from the position of the respective metal within the sequence of the 4d transition metals.

Although the possibility for a molybdenum contamination of wafers is poor, it can be observed occasionally in epitaxial wafers and in processed wafers after plasma etching. The reason for this appearance during epitaxy and plasma etching is due to the reducing or oxygen-free atmosphere applied during these processes. Molybdenum reacts strongly with oxygen by forming an oxide which does not diffuse into the bulk of the wafer and the epitaxial deposition is one of the few reactions at HT where even residual oxygen is absolutely excluded because of its explosive mixture with hydrogen forming the atmosphere during this process.

To our present understanding the contamination source seems to be limited to contaminated valves in the gas supply system. After replacing the respective valve, the molybdenum contamination disappears and clean epitaxial layers can be grown again. Although molybdenum does not belong to the fast diffusing transition metals this impurity diffuses into the substrate material during the epitaxial process, and hence it can be detected and measured there.

Molybdenum does not form haze and therefore it is assumed that it does not precipitate by a homogeneous nucleation mechanism. However, it forms electrically active defects which can be detected by DLTS. Only one donor level was observed by several researchers. It is located in the lower half of the silicon band gap. Applying DLTS with Schottky contacts molybdenum can be measured, therefore, only in p-type wafers. To date, molybdenum precipitates have not been analysed or even observed, and it is not known exactly whether molybdenum can be gettered. There is only one remark in the literature that molybdenum cannot be gettered [4.60]. Molybdenum forms silicides, and various modifications of $MoSi_2$ have been found after heat treatments of silicon coated with metallic molybdenum at temperatures between 500 and 1100°C [4.29]. The activation energy of this reaction amounts to 3.2 eV which is high compared to those of iron (1.7 eV), nickel (1.4eV), or palladium (1.5eV) [4.29].

4.4.1 Solubility

Data for the solubility of molybdenum as a function of the temperature have not been published so far. Therefore, first experiments have recently been tried [4.3]. For this purpose a mechanical contamination on the back of a previously etched p-type wafer was followed by a diffusion in a furnace tube in an argon atmosphere at different temperatures. Finally, magnesium Schottky contacts were evaporated, and DLTS measurements made to determine the concentrations of the molybdenum. Thus far, these experiments have not provided reliable and repeatable quantitative results. It is assumed that the residual-oxygen content in the furnace tube avoids an undisturbed diffusion of molybdenum which is known to react immediately with oxygen at elevated temperatures by forming an oxide. The highest molybdenum concentration in the first series of measurements was obtained at the low temperature of 900°C, and they were observed on the wafer front surface opposite to the contamination source. This result is in contradiction to the expected behavior and has not yet been explained. The measured molybdenum concentration of $3.6 \cdot 10^{13}$ cm^{-3} is a lower limit for the solubility at 900°C [4.3]. Thus, the solubility of molybdenum may be in the region of that of iron, manganese and cobalt, which are the medium-fast diffusing 3d transition metals. This result agrees with experience and with the chemical trends of the solubilities of the 3d and 4d transition metals:

$$S > 3.6 \cdot 10^{13} \text{ cm}^{-3} \quad (900°C) . \tag{4.21}$$

4.4.2 Diffusivity

Diffusivity data for molybdenum as a function of temperature are not available today since their determination would require the measurement of diffusion profiles at different temperatures. From experience it is known that molybdenum is not a fast-diffusing and not an extremely-slowly-diffusing impurity. It is expected that its diffusivity might be of the order of magnitude of that of the group chromium, manganese and iron which are the medium-fast-diffusing 3d transition metals and form donor-acceptor pairs at RT. However, the pairing of molybdenum with boron at RT has not yet been observed. Therefore a slightly lower diffusivity could be expected since CrB pairs are well known.

4.4.3 Behavior During Heat Treatment

From experience it is known that the electrically active molybdenum is rather insensitive to the annealing processes, i.e., molybdenum does not disappear into precipitates during annealing at lower temperatures. It does not require quenching either from HT to become electrically active. This is expected for impurities exhibiting lower diffusivities which do not precipitate easily.

4.4.4 Electrical Activity

Molybdenum forms electrically active defects. It is assumed that these defects occupy interstitial sites in the silicon lattice. It has been reported that all molybdenum is electrically active even after solar-cell production [4.60], which would then exclude the formation of precipitates. Molybdenum forms only one donor level in the lower half of the silicon band gap. The activation energy averaged over the results that have bveen taken from four different research groups [4.60, 61] are listed in Table 4.8. Due to the single donor level in the lower half of the silicon band gap, molybdenum can be detected by DLTS with Schottky contacts only in p-type silicon.

4.4.5 Properties of the Precipitates

The existence of precipitates has not yet been reported.

4.4.6 Known Impurity Sources and Common Concentrations

In manufacturing only one impurity source has been detected so far. It usually leads to maximum concentrations of molybdenum in the upper range of 10^{12} cm^{-3} or the lower range of 10^{13} cm^{-3}. The source has been found in contaminated valves present in epitaxy reactors or plasma reactors to control the gas stream. One condition for the molybdenum diffusion is

Table 4.8. Properties of interstitial molybdenum [4.60, 61]

ΔE_A [eV]	a/d	σ_h [cm^2]	σ_e [cm^2]
$+0.28 \pm 0.01$	d	$6.0 \cdot 10^{-16}$	$\approx 1.6 \cdot 10^{-14}$ (estimated)

the reducing atmosphere that is used either during the epitaxial process or in the oxygen-free atmosphere of a plasma reactor, which avoids the oxidation of molybdenum. Furthermore, reactive gases such as HCl or other chlorine-containing gases may dissolve molybdenum and carry it to the wafers. Almost all other heat treatments for device production are performed in atmospheres containing, at least, a residual oxygen content which may avoid the diffusion of molybdenum because of its oxidation.

4.4.7 Avoidance of Contamination

The molybdenum contamination in epitaxy reactors can be avoided by replacing a contaminated valve by an uncontaminated one. Cleaning of the reactor is much more difficult. Cleaning with oxygen is, in general, limited to contaminated plasma reactors.

4.5 Palladium

Although palladium does not belong to the main impurities in device production it is discussed here because its characteristic haze pattern leads to the development of a measurement technique which is based on an intentionally formed palladium haze [4.62]. This measurement technique enables the investigation of gettering effects, as will be discussed in Sect. 8.2.2. In order to perform this palladium test, several details of the behavior of palladium should be known. However, palladium as an unintentional impurity has never been detected during a multitude of routine DLTS measurements. This is particularly remarkable since we prepared palladium tests and routine checks of processes and ingots in the same furnace tube for years.

After etching a wafer the unpolished reverse side is easily contaminated with metallic palladium by mechanical scratching. Applying suitable illumination the trace of the contamination is visible. It is even possible to determine the quantity of palladium which sticks to the surface by measuring the difference in weight of a wafer before and after many palladium contaminations [4.62]. During annealing the sample palladium forms silicides starting with metal-rich Pd_2Si at low temperatures ($100 \div 300\,^\circ C$), and followed by PdSi at higher temperatures ($850\,^\circ C$) [4.29]. Palladium belongs to the fast-diffusing 4d transition metals. It precipitates via a homogeneous nucleation mechanism, diffuses to the wafer surfaces during cooling the sample from HT, and forms a characteristic bright haze after preferential etching. The haze pattern of palladium does not follow a diffusion profile but disappears abruptly at a certain palladium concentration (Figs. 3.8-10).

While cooling the sample from HT, palladium precipitates almost quantitatively at the surfaces of the sample and in its bulk beneath the denuded zones. The palladium precipitates resemble those observed in copper-contaminated samples. They form colonies consisting of agglomerates of rather small silicide particles surrounded by an extrinsic dislocation network. Like copper, palladium precipitates exhibit a ripening process during slow cooling or a subsequent annealing of the sample at lower temperatures.

After quenching the sample to RT a fraction of about 1% of the respective palladium solubility remains dissolved and forms electrically active defects [4.62]. Various deep energy levels can be observed by DLTS which is discussed in Sect.4.5.9.

4.5.1 Solubility

Accurate values for the solubility of palladium as a function of the temperature have not been published to date. For an approximate estimate several concentration measurements have been performed using atomic absorption spectroscopy on palladium-diffused samples at temperatures between 950°C and 1150°C [4.62]. The highly scattered results were averaged by a method of statistical weighting, which leads to the Arrhenius function

$$S = 5 \cdot 10^{22} \, e^{(-0.73 - 1.61/kT)} \, \text{cm}^{-3} \quad (950°C \leq T \leq 1150°C) \,. \quad (4.22)$$

For the solution enthalpy a similar value of 1.56 eV was found by *Frank* [4.63]. Both values are in the region of the solution enthalpies for nickel and zinc (Table 3.1). But the solution entropy of nickel is positive as for all other interstitial 3d transition metals, whereas the entropies of substitutional zinc and of substitutional palladium are negative. As a consequence, the solubilities for zinc and palladium as a function of temperature nearly agree and exhibit a significantly smaller gradient compared to all other 3d interstitial transition metals (Figs.3.1 and 4.4).

The solubilities of palladium are plotted in Fig.4.4 as a function of the diffusion temperature together with the solubilities of several 3d transition metals and of 4d and 5d transition metals as far as they are known today. The substitutional transition metals Pd, Pt, Au and Zn form a separate group of metals with medium high solubilities. This group is situated between Ni and Cu, on the one hand, exhibiting the highest solubilities, and Mn, Fe, Co, on the other hand, exhibiting lower values. The gradient of the solubilities for substitutional metals deviates, with increasing temperature, more from one another than in the case of the interstitially dissolved transition metals.

In conclusion, palladium exhibits a large solubility which is considerably higher than those of iron and cobalt but lower than those of nickel and

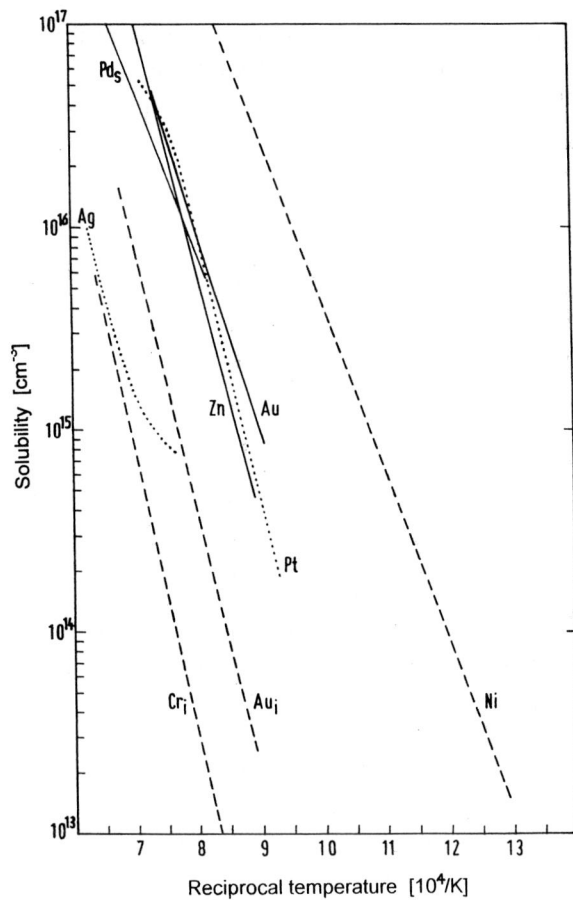

Fig. 4.4. Solubilities of several 4d and 5d interstitial and substitutional transition metals as a function of the inverse diffusion temperature

copper. The increase of the solubility with the diffusion temperature is smaller and, therefore, the supersaturation during cooling is less strongly pronounced.

4.5.2 Diffusivity

Accurate values for the diffusivity of palladium as a function of the temperature are not available to date. In order to obtain approximate values, lateral diffusion profiles of substitutional palladium have been measured by DLTS [4.62]. The samples were line-shaped contaminated with palladium and diffused at different temperatures. Then the palladium concentrations were determined as a function of the distance from the palladium source. Although

111

the deep energy level measured by DLTS is assumed to belong to a substitutional defect, the respective diffusivity is expected to be due to an interstitial mechanism. The DLTS measurements were performed in a near-surface region. Palladium diffuses rather quickly on interstitial sites and is then transferred to substitutional sites by a reaction with an intrinsic defect being a vacancy (**Frank-Turnbull mechanism** [4.64]) or a silicon self-interstitial (**kick-out mechanism** [4.65]). Both intrinsic defects need the neighboring surface in order to be generated (vacancy) or annihilated (self-interstitial). The diffusion of the intrinsic defect for a distance of several μm seems to be only a small effect compared to the interstitial palladium diffusion for a distance in the range of cm. From these experiments the following diffusivity data D_1 [4.62] were deduced. They are compared to the results D_2 obtained recently by *Frank* [4.63]:

$$D_1 = 0.08\,e^{-0.89/kT} \text{ cm}^2/\text{s} \quad (950°C \leq T \leq 1100°C), \qquad (4.23)$$

$$D_2 = 2.95 \cdot 10^{-4}\,e^{-0.22/kT} \text{ cm}^2/\text{s}. \qquad (4.24)$$

Although the parameters are quite different, the deviation at 1100°C is rather small ($< 6\%$). Larger differences are obtained at lower temperatures. At 950°C, for instance, the deviation amounts to 53%. Compared to the migration enthalpies of copper and nickel of about $0.4 \div 0.5$ eV, the value of 0.2 for D_2 appears too low and the value of 0.9 for D_1 appears too high. From haze inspections (Fig.3.8) a diffusivity for palladium is expected, which should be similar to that of cobalt exhibiting a migration enthalpy of 0.53 eV (Table 3.2). Taking the average values of D_1 and D_2 for low and high temperatures leads to

$$D = 2.56 \cdot 10^{-3}\,e^{-0.48/kT} \text{ cm}^2/\text{s} \quad (950°C \leq T \leq 1100°C) \qquad (4.25)$$

which, indeed, almost agrees with the diffusivity of cobalt. In conclusion, the diffusivities of cobalt, nickel and palladium agree within measurement error. Detailed investigations to decide whether Pd diffuses via a Frank-Turnbull [4.64] or a kick-out mechanism [4.65] are not yet available. From the precipitation behaviour the interstitial solubility is estimated to exceed the substitutional solubility by a factor of 100.

4.5.3 Behavior During Heat Treatment

Wafers can easily be contaminated by a mechanical contact to metallic palladium. On the other hand, replating of palladium from solutions containing palladium as an impurity is expected because of the high electronegativity of palladium of 2.2 r.u. compared to silicon (1.8r.u.). Contamination with palladium by evaporation seems to be negligible due to our experience with the palladium test (Sect.8.2.2). This may be explained by the high melting point of palladium and its high evaporation temperature, both exceed 1550°C.

During heat treatment metallic palladium forms silicides starting with $Pd_2 Si$ at low temperatures between 100°C and 300°C and changing to PdSi at 850°C. Due to its high diffusivity palladium contaminates large areas during common HT diffusion processes. In the cooling period palladium diffuses out to the sample surfaces and forms precipitates at the surfaces thus causing the formation of haze after a preferential etch. In addition, palladium precipitates in the bulk beneath both denuded zones due to outdiffusion. The precipitation is almost quantitative. The residual dissolved palladium amounts to $5 \cdot 10^{13}$ cm^{-3} at 950°C and $5 \cdot 10^{14}$ cm^{-3} at 1050°C which is about 1% and 2.5% of the corresponding solubility, respectively [4.62]. Although these concentrations are high compared to other impurity concentrations observed in processed wafers, palladium as an impurity in device production is technologically unimportant and has not been detected so far.

Dissolved palladium is electrically active and forms several levels in the silicon band gap, if the sample has been cooled moderately fast from HT, or quenched to RT and subsequently annealed at lower temperatures. Samples which were quenched from diffusion temperatures exhibit several deep energy levels which have not been identified so far. But it is pretty certain that palladium forms various hydrogen complexes. The acceptor level in the upper half of the silicon band gap is assumed to be due to the substitutional palladium because of its thermal stability. A donor and a double donor can be detected only if etching of the sample is avoided to prevent the formation of hydrogen complexes.

The palladium precipitates resemble those of copper. They form star-like colonies of agglomerates containing small silicide particles arranged parallel to the {111} silicon lattice planes and are surrounded by extrinsic dislocations. The plane orientation differs from that in copper precipitates which are parallel to {110} silicon lattice planes [4.36, 37].

4.5.4 Electrical Activity

Palladium belongs to the heavier 4d transition metals (Fig.1.1) which exhibit the same particular tendency to form substitutional defects as the heavier 3d transition metals. However, the diffusion process takes place on interstitial sites. The formation of the substitutional defects is then coupled to intrinsic defects. As mentioned in Sect.3.2, there are two competitive possibilities, the absorption of vacancies (**Frank-Turnbull mechanisms** [4.64]) or the emission of silicon self-interstitial (kick-out mechanism [4.65]). The diffusion profiles for the two mechanisms differ slightly. In order to deter-

Table 4.9. Properties of substitutional palladium and palladium-related defects

Defect	Δ_A [eV]	d/a	δ_M [cm^2]	T_{DLTS} [K]	Ref.
Pd$_s$	-0.21 ± 0.01	a	$\sigma_e = 1.6\cdot10^{-15}$	120	4.66-70
Pd$_s$	$+0.31\pm0.02$	d	$\sigma_h = 5.6\cdot10^{-16}$	200	4.66-68,71,72
Pd$_s$	$+0.12\pm0.01$	dd	$\sigma_h = 6.5\cdot10^{-17}$	70	4.67,68,73,74
Pd$_s$Pd$_i$	...	...	elect. inactive		4.34
Pd$_s$Fe$_i$	-0.32				4.57
PdH	>3				4.69

mine the prevailing formation mechanism, suitable diffusion profiles for the respective substitutional defect must be measured with utmost accuracy. This has been done for several transition metals such as gold, platinum, silver and zinc; but not for palladium as far as is known today.

The substitutional defect which forms during moderately-fast cooling of the sample from HT or during a subsequent annealing of a quenched sample at temperatures $> 110\,°C$ [4.57] exhibits only one acceptor level in the upper half of the silicon band gap if the sample was etched for the preparation of Schottky contacts. This and has been observed by several research groups. The mean value of its activation energy (8 groups) and the majority-carrier capture cross-section (5 groups) are listed in Table 4.9. Several researchers reported a donor state of palladium in the lower half of the silicon band gap already in the seventies, others could not confirm its existence. Finally, a double donor also situated in the lower half of the band gap was only detected in the eighties. An explanation can be found in the reaction of palladium with hydrogen due to etching the sample for the preparation, where hydrogen is introduced into the surface layer of the sample. Hydrogen can react with palladium by forming electrically active or inactive complexes. The first refer to hydrogen neutralizing the palladium acceptor and donor was given by *Pearton* [4.66]. Hydrogen complexes are known to dissociate at elevated temperatures, which explains the experimental finding of increasing the acceptor in n-type silicon and appearing of the donor in p-type silicon after annealing the sample. The other, not yet identified levels, observed in p-type silicon after quenching the sample to RT disappear. It is also known that hydrogen reacts stronger with boron in p-type than with phosphorous in n-type silicon. Since hydrogen complexes are only found in the surface region after etching they are detected mainly by DLTS measurements which are performed in the same region.

The double donor was not detected before 1984 by *Lemke* [4.67] but identified as an other donor of palladium. A confirmation of the existence of a double donor has been given in 1997 by *Sachse* et al. [4.68] using for comparison unetched palladium-doped wafers which were obtained from cleaved crystals. It would be interesting to study palladium-hydrogen complexes in more detail, which should be partly electrically active and inactive. *Lemke* [4.69] published a DLTS spectrum of an Pd-contaminated and etched n-type silicon wafer exhibiting three levels which were correlated to hydrogen-palladium complexes. However, their activation energies have not been given. Additional levels were also observed in p-type samples by other researchers.

Palladium also forms a donor-acceptor pair between an interstitial (d) and a substitutional (a) palladium atom: $Pd_1 Pd_s$ [4.34] similar to nickel and copper. In contrast to the respective copper pair, the nickel and the palladium pairs are electrically inactive and therefore passivate the substitutional levels.

After diffusing a palladium-doped sample, intentionally contaminated with iron, a new deep energy level was observed with an activation energy of E_c-0.32 eV which, however, is not identical with the level obtained after quenching pure palladium-doped samples [4.57]. It is assumed that donor-deep-acceptor pairs $Fe_i Pd_s$ are formed at higher temperatures since a RT reaction of palladium with unintentional interstitial iron has not yet been observed. At RT and common doping levels the Fermi level in n-type silicon is situated between the iron donor and the palladium acceptor, and therefore both levels are electrically neutral. As a consequence there is no Coulomb attraction to form a donor-acceptor pair. But at higher sample temperatures the occupation statistics allows that a fraction of the deep levels of iron and palladium are positively and negatively charged, respectively, and a pair formation is possible.

The activation energies of palladium and the related defects are compliled in Table 4.9, as far as they are known today.

4.5.5 Properties of the Precipitates

Palladium belongs to the haze-forming transition metals, and it precipitates via a homogeneous nucleation mechanism. The etch pits which can be revealed by preferential etching do not exhibit characteristic morphologies, as shown in Figs.3.14,15. Star-like etch pits can be revealed by very short etch durations affecting only the region of the sample near the surface [4.37]. For lower cooling rates large palladium precipitates grow at the expense of smaller ones forming denuded zones in their neighborhood, as observed in Figs.3.14,15.

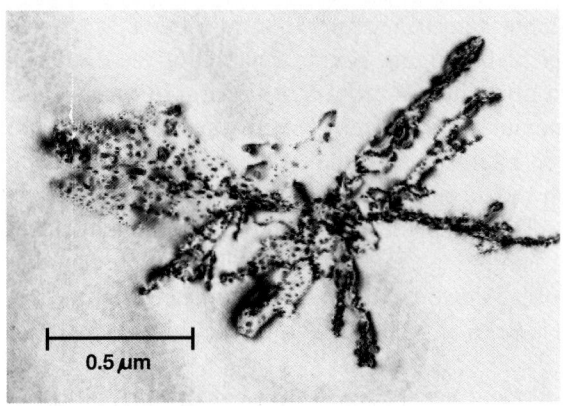

Fig.4.5. TEM micrograph of a star-like precipitate colony consisting of planar arrangements of Pd_2 Si particles parallel to Si(111) planes and bounded by extrinsic dislocation loops

Palladium forms colonies of small crystalline silicide particles arranged in star-like agglomerates parallel to the {111} silicon lattice planes (in contrast to copper) [4.37]. Their lateral extension may amount up to 20 μm, whereas the depth is less than 2 μm [4.21]. Because of these shallow precipitates the etch pits broaden considerably during prolonged etching processes and the star-like morphology is lost.

Again the precipitates traverse the original wafer surface, as has been observed for other transition metals [4.36]. The precipitates exhibit a hexagonal structure which fits well with the metastable Pd_2 Si, as deduced from electron-diffraction patterns. The c-axis of the hexagonal silicide is oriented parallel to the silicon {111} direction [4.21]. Samples cooled moderately fast exhibit similar dislocation networks, as seen from copper precipitates. One example of a TEM micrograph is depicted in Fig.4.5. Small Pd_2 Si particles in a planar arrangement parallel to silicon (111) lattice planes are bound by extrinsic dislocation loops, thus forming the star-like colonies. Samples contaminated with palladium, which were quenched to RT, show the same precipitate agglomerates but do not exhibit extrinsic dislocations. Again, the small precipitates are almost spherical in shape. The emission of 0.55 silicon self-interstitials during the precipitation of one palladium atom (Table 3.10) is of the same order of magnitude, as found for copper (0.5) and belongs to the highest values observed within the haze-forming tansition metals [4.21].

4.5.6 Known Impurity Sources and Common Concentrations

Palladium has not been observed as an unintentional impurity in wafers during device production although it is sometimes used for backside contacts, for example, in solar cells (Ti-Pd-Ag). Therefore, impurity sources and concentrations are not known.

4.6 Platinum

Platinum belongs to the heavier 5d transition metals. Because it is not a wide-spread metal it is rarely found as an unintentional impurity during device production. However, it is used as an intentional impurity to adjust a specified carrier lifetime in thyristors and high-voltage transistors. Thus, an unintentional contamination with platinum may happen by inadvertently using contaminated furnace tubes, cleaning facilities or other processes. Therefore, the behavior of platinum as an impurity in silicon should be known in order to perform intentional diffusions, on the one hand, and to avoid or to identify unintentional cross contaminations, on the other hand. Only a small amount of information is available on platinum precipitates, and detailed investigations are not known so far. Platinum can be gettered by a phosphorus diffusion.

A mechanical contact of metallic platinum to uncoated wafer surfaces is unlikely in device production. A liquid-phase contamination cannot be excluded because of the reasons mentioned above. The electronegativity of the noble platinum is 2.2 r.u. and by far exceeds that of silicon at 1.8 r.u. As a consequence, platinum will be replated on wafers and diffuses into the wafers during a subsequent temperature process. In addition, a contamination may happen during an inadvertently performed diffusion of wafers in a furnace tube which is commonly used for platinum diffusion only. The contamination takes place, although the evaporation temperature of platinum is very high, exceeding 2000°C.

Platinum diffuses quickly into a sample on interstitial sites, but at a much slower rate it occupies substitutional sites via the kick-out mechanism [4.75]. During substitutional diffusion, silicon self-interstitials are emitted into the host lattice. The substitutional diffusion velocity is limited by the diffusion of the silicon self-interstitials to the sample surfaces where they can annihilate in order to reach thermal equilibrium. Dislocations which may be present in the sample can considerably increase the platinum diffusion on substitutional sites since they enable the annihilation of the self-interstitials in the bulk of the sample. The solubility of substitutional platinum considerably exceeds that of interstitial platinum [4.75].

After cooling platinum-contaminated samples from the diffusion temperature, mainly two deep levels have been found by several researchers, an acceptor state in the upper half of the silicon band gap and a donor state in the lower half. Both levels belong to substitutional platinum. This has been proven recently after having discovered several palatinum-hydrogen complexes [4.77, 78] generated during etching the sample for the evaporation of Schottky contacts (Sect.3.3.2). The complexes are electrically active in both halves of the band gap and impair the detection of substitutional platinum by reducing their signal heights in DLTS spectra. Avoiding the formation of hydrogen complexes, for instance, by avoiding etching yields clear spectra exhibiting three deep levels of equal concentrations, the well-known acceptor and donor, and an additional double donor close to the valence-band edge [4.53]. The detection of the double donor is rendered more difficult not only by the hydrogen complexes reducing the signal heights, but also by the low temperature (50 K) of its DLTS signal. At this temperature the dopant atoms start to freeze out.

Besides the 3 substitutional platinum defects and their 4 hydrogen complexes [4.77, 78] there seem to be still other platinum-related defects which further complicate the spectra. Some of them indicated by roman numbers (I+III) [4.80] were reported by several researchers. Some of the reported activation energies fairly agree with hydrogen complexes (PtI, II), but PtII also agrees with the platinum donor-acceptor pair $Pt_s P_i$ identified by *Lemke* [4.79]. A defect with almost equal activation energy has been tentatively correlated to PtP pairs [4.81].

Platinum is often used to control the minority-carrier lifetime in devices to reduce the switching velocity of diodes and transistors. Several publications discussed these questions together with the problem of increasing generation current due to the presence of deep energy levels. In some papers the advantages and disadvantages of using platinum or gold for lifetime controlling have been compared. Because of different platinum levels obtained by different preparation techniques, the results are manifold and it is still difficult to draw reliable conclusions. However, gold always exhibits two deep levels due to different charge states of the same defect and therefore remains independent of variations in the sample preparations. Both lifetime killers exhibit more than one energy levels which render the calculation of the resulting carrier lifetime more complicated.

4.6.1 Solubility

Beyond about 1000 °C the solubility of platinum is retrograde and, therefore, it cannot be fit to a single Arrhenius equation in the whole temperature region (Fig.4.4). In the following equations (4.26, 27) S_1 [4.80] and S_2

[4.75] were deduced from two different measurements obtained by means of a neutron-activation analysis. Their validity is limited to the respective temperature regions indicated:

$$S_1 = 5 \cdot 10^{22} e^{(9.44 - 2.676/kT)} \text{ cm}^{-3} \quad (800\,^\circ\text{C} \leq T \leq 1000\,^\circ\text{C}), \quad (4.26)$$

$$S_2 = 5 \cdot 10^{22} e^{(-2.69 - 1.31/kT)} \text{ cm}^{-3} \quad (1000\,^\circ\text{C} \leq T \leq 1100\,^\circ\text{C}). \quad (4.27)$$

Although the solution enthalpies and entropies are quite different, the results at the boundary between both temperature regions (1000°C) agree well within experimental errors:

$$S_1(1000\,^\circ\text{C}) = 1.6 \cdot 10^{16} \text{ cm}^{-3}, \quad (4.28)$$

$$S_2(1000\,^\circ\text{C}) = 2.2 \cdot 10^{16} \text{ cm}^{-3}. \quad (4.29)$$

Extrapolating the results from *Lisiak* and *Milnes* [4.80] for temperatures beyond 1000°C the solubility values reported by *Hauber* [4.75] determined for temperatures >1000°C appear to be within the retrograde region of the solubility which then leads to a drastically reduced solution enthalpy. Beyond 1100°C the solution enthalpy is expected to decrease further.

4.6.2 Diffusivity

Platinum diffuses very fast on interstitial sites but the interstitial solubility is low. The effective diffusion of platinum is performed by the so-called **kick-out mechanism** where silicon self-interstitials are involved [4.75]. This diffusion mechanism is mainly determined by the annihilation of the emitted self-interstitials which must diffuse to the sample surfaces in dislocation-free crystals. As pointed out for palladium, the diffusion parallel to the sample surface can be considerably faster than perpendicular to the surface where the silicon self-interstitials generated by the kick-out process in the bulk of the sample have to pass much longer distances to reach the surface.

Quantitative diffusivity values were not reported for the interstitial or the substitutional mechanism, although diffusion profiles were measured and published [4.75]. One problem of the kick-out mechanism is the dependence of the effective diffusion coefficient on the local concentration of the diffusing impurity. This is in contrast to the **Frank-Turnbull mechanism** where vacancies are involved and where the effective diffusion coefficient is independent of the local impurity concentration [4.75]. The only result deduced from the diffusion via the kick-out mechanism is the Arrhenius plot for the self-diffusion in silicon which, however, always exhibits equal results independent of the respective metal used for diffusion.

However, in highly dislocated crystals the effective diffusivity of inter-stitial platinum $D^{eff}(Pt_i)$ can be expressed in form of an Arrhenius equation [4.82]

$$D^{eff}(Pt_i) = 2.1 e^{-1.79/kT} \quad cm^2/s \quad (950°C < T < 1200°C) \tag{4.30}$$

and

$$D(Pt_i) = D^{eff}(Pt_i) C^{eq}(Pt_s)/C^{eq}(Pt_i) . \tag{4.31}$$

The interstitial diffusivity of platinum can then be calculated if the relation between the equilibrium concentrations of substitutional platinum $C^{eq}(Pt_s)$ and that of interstititial platium $C^{eq}(Pt_i)$ is known.

4.6.3 Behavior During Heat Treatment

At elevated temperatures $(300 \div 500°C)$ metallic platinum deposited on sili-con surfaces forms two different silicides: Pt_2Si and $PtSi$. The silicides act as diffusion sources. As mentioned before, the interstitial diffusivity of pla-tinum is fairly fast but the interstitial solubility is low. Therefore, the plati-num diffuses quickly through the sample up to the reverse side. The dif-fusion of the substitutional platinum via the kick-out mechanism requires longer times, although starting from both surfaces of the sample. Inten-tional platinum diffusions to adjust a definite carrier lifetime in a sample are performed, in general, at temperatures in the region between $800°$ and $900°C$ where the solubilities of substitutional platinum are in a suitable range. The duration of the diffusion depends on the depth of the electrically active zone where the carrier lifetime should be reduced. Since the electri-cally active platinum occupies lattice sites, it is not particularly sensitive to the cooling rates applied to the samples at the end of the diffusion process.

Platium occupies mainly substitutional lattice sites. The substitutional defect exhibits four charge states by forming a double donor and a donor in the lower half of the band gap and an acceptor in the upper half. Platium is known as a lifetime killer because of the high majority-carrier capture cross sections of the donor and the acceptor states. The minority carrier capture cross section of the acceptor states is likewise very high. An interstitial de-fect of platinum at RT is not known. Although platinum belongs to the noble metals several compounds were detected and identified, others assumed and not yet identified. The substitutional defect is characterized by a high ther-mal stability and does not tend to form precipitates. As a conseqence plati-num does not form haze and does not precipitate via a homogeneous nuclea-tion mechanism.

Platinum can be gettered by phosphorus diffusion. As pointed out by *Falster* [4.83], the gettering mechanism differs from that of gold. Whereas the gold concentration correlates with the phosphorus concentration, an enhanced platinum concentration is restricted to that surface of the sample where the phosphorus diffusion is performed. The following mechanism was proposed to explain the behavior of platinum during gettering, but it may also influence the gold gettering by phosphorus diffusion. During phosphorus diffusion, silicon self-interstitials are emitted and shift the equilibrium for the kick-out mechanism towards the left side in (4.32), i.e.,

$$Me_i + Si_S \longleftrightarrow Me_S + Si_i . \tag{4.32}$$

Thus, the substitutional platinum is transferred to the mobile interstital platinum which, however, causes a reduced metal solubility in the sample. Due to its supersaturation, the platinum diffuses to the neighboring sample surface where the phosphorus diffusion has occurred. Here it precipitates, probably by forming a platinum-silicide layer.

As a substitutional defect, platinum is not very sensitive to low-temperature annealing. So annealing at 430°C does not change the amount of electrically-active platinum, whereas the annealing at higher temperatures slowly reduces the concentration of dissolved platinum. This reduction does not cause the formation of new energy levels [4.84], and hence it may be due to outdiffusion or precipitation of the platinum.

4.6.4 Electrical Activity

All the problems which arose in the past to correlate the various observed deep levels in platium-doped material could recently be solved by pointing out that platinum reacts strongly with hydrogen [4.77, 78] which has been introduced into the sample surface during etching. Etching of the sample surface is necessary to achieve better electrical properties of the evaporated Schottky contacts commonly used for DLTS measurements. It is well known that hydrogen reacts with shallow dopants by forming electrically inactive complexes. The penetration depth corresponds to the depth of the space-charge region due to the drift of ionized hydrogen within its electrical field. Since the reaction with boron in p-type silicon exceeds that with phosphorus in n-type, the partial passivation of the platinum donor in the lower half of the bad gap is more pronounced compared to that of the acceptor in the upper half. This was the main reason which induced several researchers to propose various models for the existence of different platinum-correlated defects (PtI+III) [4.80]. All these models and explanations [4.66, 85, 86] became untenable by avoiding etching or by outdiffusion of hydrogen from the sample at 300 °C for 3 h [4.77] since these spectra clearly showed only three deep levels of equal concentrations.

Table 4.10. Properties of substitutional platium and platium-related pairs

Defect	ΔE_A [eV]	d/a	σ_e [cm²]	σ_h [cm²]	T_{DLTS} [K]	Ref.
Pt_s	-0.23 ± 0.01	a	$2.9\cdot10^{-14}$	$2.6\cdot10^{-14}$	110	4.87
Pt_s	$+0.32\pm0.01$	d	10^{-16}	$8.4\cdot10^{-15}$ *	160	4.88
Pt_s	$+0.08\pm0.01$	dd	...	$3.5\cdot10^{-17}$	50	4.89
$Pt_s Pt_i$	$+0.43$					4.90
PtH-rel.	-0.18	d	$9\cdot10^{-16}$		90	4.91
PtH-rel.	-0.50	d	$3\cdot10^{-15}$		250	4.91
PtH/H₂-rel.	$+0.40$			$1\cdot10^{-16}$	210	4.91
PtH_x	$+0.31$	a		$8\cdot10^{-15}$	150	4.91

*: $\sigma_h = 1.3\cdot10^{-15}e^{-0.007/kT}$ expressing the tempearture dependence of the Pt donor

Average values of the activation energies for substitutional platinum and their majority-carrier capture cross-sections are listed in Table 4.10 together with their minority-carrier capture cross-sections, as far as they have been determined, since all these data are needed to calculate carrier lifetimes. Although platinum is currently used for carrier-lifetime tailoring the experimental results for the capture cross sections still differ considerably. For averaging, the values showing the highest deviations, were neglected but the remaining results still scatter within one order of magnitude, at least those for the majority carriers of the donor and the acceptor. Only one or two values are available for the minority-carrier capture cross-sections of the donor and the acceptor, respectively.

Furthermore, recent results on platinum-hydrogen complexes and $Pt_s Pt_i$ pairs are included, which are the only compounds which were identified fairly reliably to date. So far unidentified deep energy levels are not included. Some of them (Pt II: $E_v +0.42\,eV$ [4.80, 81]) might be identical with a hydrogen complex ($E_v +0.40\,eV$ [4.77, 78]) or with the donor-acceptor pair $Pt_s Pt_i$ ($E_v +0.43\,eV$ [4.79]), others (average: $E_c -0.32\,eV$) have been observed by several researchers [4.80, 81] but do not yet fit in the series of known defects. As mentioned in Sect. 4.6.3, interstitial defects were not observed at room temperature. The concentration of the substitutional defects agrees fairly well with the solubility at the diffusion temperature, therefore the concentration of interstitial defects must be considerably lower than that of the substitutional defects.

4.6.5 Properties of the Precipitates

To our knowledge, the structure of platinum precipitates has not yet been investigated, and therefore its properties are still unknown.

4.6.6 Known Impurity Sources and Common Concentrations

Boron-nitride wafers used as boron-diffusion sources can act as a source of platinum impurities since boron nitride is produced in platinum crucibles. Otherwise impurity sources are limited to laboratories or production lines where intentional platinum diffusions are executed. Cross contamination can be obtained by performing fault processes such as cleaning of platinum-diffused wafers in common equipment, or erroneous diffusion in a furnace tube reserved for platinum diffusion. Cross contamination can also be caused by handling or by handling systems. In these cases platinum concentrations in the range of several 10^{12} cm^{-3} up to several 10^{13} cm^{-3} can be detected through DLTS measurements. The DLTS signal obtained by a temperature scan on p-type samples may interfere with those of other possible impurities such as molybdenum at slightly higher temperatures, or gold at slightly lower temperatures. In the case of overlapping spectra in p-type samples an additional control measurement applying n-type samples is proposed.

Problems with the quantitative evaluation of spectra showing different platinum-related signals may arise since the total platinum content is the sum of all platinum-related signals if they are due to different defets. However, the donor and the acceptor levels are two charge states of the same defect, the total platinum content corresponds only to the concentration of a single defect. Problems with a quantititive evaluation of DLT spectra can also arise from inverse U-shaped depth profiles obtained after platinum dif fusion [4.92].

4.6.7 Avoidance of Contamination

Due to the limited number of processes where platinum contamination may occur, its avoidance is quite simple and can be achieved by avoiding erroneous processes and by using handling equipment for platinum-diffused wafers which are specially marked and remain strictly separated from all other equipment.

4.7 Gold

In modern technology, gold as an unwanted impurity plays only a minor role in contrast to former times where it was repeatedly found in furnace tubes and in cleaning facilities. In comparison to the common impurity concentrations of iron, copper, and nickel, gold contamination is now rarely detectable and its concentration scarcely exeeds the order of magnitude of 10^{12} cm^{-3}. However, gold is known to be a strong recombination center and therefore the upper limit for a tolerable gold impurity content is much less than 10^{12} cm^{-3}. About 20 years ago DLTS was applied for impurity control in materials, and the gold contamination could be strongly reduced. In the same period of time gold as a rear contact material was more and more replaced by other materials, and this strongly reduced cross contamination of gold. At present, gold and platinum are mainly employed for carrier-lifetime tailoring and the possible contamination sources equal those mentioned for platinum contamination, as discussed in Sect. 4.6.

Like platinum, gold belongs to the heavy 5d transition metals. It is one of the most investigated metal impurities in silicon. The contamination of wafers with gold is easily performed by a mechanical contact of metallic gold, especially to the rough backside of wafers. Furthermore, gold is replated on uncoated wafers in every solution containing trace contaminations of gold because of its high electronegativity of 2.3 r.u. compared to 1.8 r.u. for silicon. Additionally, a vapor-phase gold contamination takes place in furnace tubes due to the evaporation of gold.

Gold is easily diffused into samples even at rather low temperatures ($>$ 500°C). It diffuses fast on interstitial sites but its interstitial solubility is low in analogy to platinum. Substitutional gold is formed by the kick-out process which reduces the diffusivity of the substitutional gold. After cooling the sample from the diffusion temperature gold usually remains on its substitutional lattice site and forms two deep energy levels: an acceptor state in the upper half of the silicon band gap and a donor state in the lower half. The majority- and minority-carrier capture cross-sections of gold belong to the most determined carrier capture cross-sections of metal impurities in silicon. Gold precipitates in samples during slow cooling after diffusion or during a subsequent annealing at lower temperatures. Gold does not form haze because of the low substitutional diffusivity. It can be gettered by phosphorus diffusion and in the past often served as a monitor material to detect gettering effects.

4.7.1 Solubility

Several results for the solubility of gold as a function of the diffusion temperature have been published in the literature in the last three decades. In general, the total gold concentration or the electrically-active gold concentration (substitutional gold) were determined. The interstitial solubility was reported only in a very early publication [4.93]. The deviations between the various results amount, in general, to less than about 30 %. This is a rather small deviation compared to differences in solubilities between other transition metals such as nickel and titanium which amount to several orders of magnitude at equal temperatures. The following equations (4.33, 34) present Arrhenius plots for the solubilities of interstitial gold $S(Au_i)$ [4.93] and recent results for the total gold S_{tot} content obtained by a neutron-activation analysis [4.94]:

$$S(Au_i) = 5 \cdot 10^{22} e^{(5.1-2.56/kT)} \text{ cm}^{-3} \quad (850°C < T < 1200°C) , \quad (4.33)$$

$$S_{tot} = 5 \cdot 10^{22} e^{(3.0-1.98/kT)} \text{ cm}^{-3} \quad (800°C < T < 1100°C) . \quad (4.34)$$

The recalculated solubilities of interstitial and total gold content are plotted in Fig.4.4, and compared to solubilities of other transition metals. It is evident that the solubility of interstitial gold coincides with the group of solubilities for interstitial manganese, iron and cobalt. On the other hand, the total gold content equals those of substitutional platinum, palladium, and zinc, although the deviations in this group are more pronounced.

4.7.2 Diffusivity

For the fast interstial diffusivity of gold, an early publication from 1964 by *Wilcox* and *LaChapelle* [4.93] presents the following Arrhenius equation which must hold for dislocated crystals since at that time dislocation-free material was not available

$$D(Au_i) = 2.4 \cdot 10^{-4} e^{-0.39/kT} \text{ cm}^2/s \quad (700°C < T < 1100°C) . (4.35)$$

In a recent publication of *Bracht* and *Overhof* [4.82] another Arrhenius equation was given for the effective diffusivity of interstitial gold $D^{eff}(Au_i)$ in dislocated crystals

$$D^{eff}(Au_i) = 0.46e^{-1.70/kT} \text{ cm}^2/s \quad (959°C < T < 1290°C) . \quad (4.36)$$

As mentioned already in Sect.4.6.2, for platium the diffusivity of intersti-

Table 4.11. Comparison of recalculated diffusivities of interstitional gold $D(Au_i)$ and $D(Au_i)^{eff}$ [cm^2/s]

T [°C]	800	900	1000	1100
$D(Au_i)$, Eq.(4.35)	$3.5 \cdot 10^{-6}$	$5.1 \cdot 10^{-6}$	$6.9 \cdot 10^{-6}$	$8.9 \cdot 10^{-6}$
$C^{eq}(Au_s)/C^{eq}(Au_i)$	154	263	418	606
$D^{eff}(Au_i)$, Eq.(4.36)	$4.8 \cdot 10^{-9}$	$2.3 \cdot 10^{-8}$	$8.6 \cdot 10^{-8}$	$2.6 \cdot 10^{-7}$
$D(Au_i)$, Eqs.(4.36+31)	$7.4 \cdot 10^{-7}$	$6.0 \cdot 10^{-6}$	$3.6 \cdot 10^{-5}$	$1.6 \cdot 10^{-4}$

tial gold can be calculated from the effective diffusivity by multiplying D^{eff} with the relation between the equilibrium concentration of substitutional and interstitial gold following (4.31). Both values can be calculated from (4.33) [4.93] and (4.34) [4.94]. It should be mentioned that (4.33 and 35) originated from the same researcher, and (4.34 and 36) from the same institute. A comparison of the results is given in Table 4.11 for four temperatures within the allowed temperature regions.

It is evident from Table 4.11 that both equations result in quite different values indifferent whether (4.35) is compared to (4.36) or to (4.36) combined with (4.31). The diffusivities from (4.35) fit in the diffusivities of other transition metals, and the calculated values are between those of iron and cobalt which seems reasonable from experience. The gradient D^{eff} with the temperature is too steep and that of the combination is even unbelievable. However, both solubilities, of interstitial and of total gold content seem reasonable compared to others, as depicted in Fig. 4.4.

4.7.3 Behavior During Heat Treatment

Gold forms an eutecticum with silicon at a temperature of 370°C and a silicon content of 31 at.%. At higher temperatures this liquid phase will act as a diffusion source. Gold does not form stable silicides. At HT the silicon content in the liquid phase increases up to 100%. Because of the fast interstitial diffusion of gold, a contamination on the front side of a wafer will diffuse rapidly through the wafer and will reach the reverse side. Substitutional diffusion then starts from both sides and gives rise to the formation of U-shaped impurity depth profiles within the cross-section of the wafer. As a consequence of the diffusion via the kick-out process, the gold concentration in the middle of this U-shaped profile increases with the square root of the diffusion time, whereas the diffusion via a vacancy mechanism would result in a linear increase [4.65, 94].

During cooling the sample from the diffusion temperature, the gold impurities, in general, remain electrically active because of the thermal stability of substitutional defects due to lower substitutional diffusivities. During the application of extended cooling periods or annealing processes at lower temperatures, the electrical activity of the gold impurities is reduced and the formation of precipitates is observed. It has been shown that the precipitates consist of a metastable gold silicide of thus far unknown composition [4.95]. A ripening of the precipitates was observed with increasing annealing duration. It is accompanied by the formation of extrinsic dislocation loops in the beginning, followed by their subsequent annihilation. Gold can be gettered by phosphorus diffusion since high doping concentrations enhance the solubility of substitutional gold. Gold can also be gettered, for example, by polysilicon layers on the backside of the wafers, or by internal gettering.

4.7.4 Electrical Activity

Defects formed by quenched-in interstitial gold have not yet been investigated in detail. Several deep levels were observed in gold-doped samples after quenching [4.3]. However, their repeatability was poor and it is expected that the highly-mobile interstitial gold forms complexes of various kinds to overcome high supersaturations. From diffusion experiments in highly doped p-type silicon *Bracht* [4.82] concluded the existence of a single donor state of interstitial gold at $E_v +0.48$ eV. In p-type crystals at high temperatures interstitial gold should be positively charged.

On the other hand, substitutional gold has been investigated over several decades. Now it seems to be well established that the acceptor state in the upper half of the silicon band gap and the donor state in its lower half both belong to the same defect [4.102] which was doubted before [4.103]. Several times the existence of a third deep level for gold was assumed, but so far this could not be confirmed. However, it is well known that gold forms a variety of complexes with other transition metals, especially donor-acceptor pairs (Table 4.12) [4.15]. The mean activation energies for substitutional gold are complied in Table 4.12. They were obtained by averaging the results from different research groups [4.96, 97].

For the mean majority-carrier capture cross-sections several reasonable results have been selected from many publications in this field, eliminating extremely low or high values which may differ by almost two orders of magnitude. More recent results seem to focus on the values presented in the table. The majority-carrier capture cross-sections are temperature independent. The data for the minority-carrier capture cross-sections published so far are still poor. However, both minority-carrier capture cross-sections

Table 4.12. Properties of substitutional gold and gold-related pairs

Defect	ΔE_A [eV]	d/a	σ_e [cm²]	σ_h [cm²]	T_{DLTS} [K]	Ref.
Au_s	− 0.55±0.01	a	$1.4 \cdot 10^{-16}$	$7.6 \cdot 10^{-15}$ *	300	4.96
Au_s	+0.34±0.02	d	$2.7 \cdot 10^{-15}$ *	$2.5 \cdot 10^{-15}$	160	4.97
$Au_s Au_i$	− 0.44					4.90
AuH	− 0.19	aa	$1 \cdot 10^{-17}$		110	4.98,99
AuH	− 0.56	a	$1 \cdot 10^{-16}$		300	4.99,100
AuH	+0.21	d		$3 \cdot 10^{-15}$	115	4.99,101
AuH-rel.	+0.47[a]	a?		$5 \cdot 10^{-16}$	242	4.99,101
AuV	+0.42	d		$6.4 \cdot 10^{-16}$	200	4.15
AuV	− 0.20	a	$4.9 \cdot 10^{-16}$		115	4.15
AuCr	+0.35	d		$8.7 \cdot 10^{-16}$	170	4.15
AuMn	+0.57	d		$2.1 \cdot 10^{-15}$	250	4.15
AuMn	− 0.24	a	$4.9 \cdot 10^{-16}$		120	4.15
AuFe	+0.43	d		$2.8 \cdot 10^{-15}$	250	4.15
AuFe	− 0.35	a	$2.1 \cdot 10^{-15}$		160	4.15
AuNi	+0.48[a]					4.57
AuNi	+0.35[a]					4.57
AuCu	+0.42[a]					4.57
AuCu	+0.32[a]					4.57

[a] : Bistable

for the acceptor and donor states exhibit a temperature dependence, which has not been determined by all researchers. Averaging of the minority-carrier capture cross-sections has been performed for the temperature of the DLTS peak which is also listed.

In addition to the single substitutional gold defect discussed so far, pairs are also important. They form between the gold acceptor and other transition-metal donors during the cooling period, or during a subsequent annealing of the sample at lower temperatures. As listed in Table 4.12, pairs are formed with the 3d transition-metals vanadium, chromium, manganese, iron, nickel and copper. Mainly the gold-iron pairs can be of technological importance since both impurities can be present simultaneously in wafers after processing. These pairs form very strong recombination centers exceeding even substitutional gold [4.15]. Gold-iron pairs exhibit two deep energy levels, an acceptor in the upper half of the silicon band gap and a donor in its lower half. The values of their activation energies are listed in Table 4.12 together with their majority-carrier capture cross-sections [4.15]. The majority-carrier capture cross-sections of the acceptor and of the donor state almost agree and the minority-carrier capture cross-sections have not yet been determined.

4.7.5 Properties of the Precipitates

As a substitutional defect, gold exhibits a thermal stability which retards the formation of gold precipitates if the impurity concentrations remain low. For high supersaturation of gold the precipitation behavior and the ripening process of the precipitates were studies as a function of the annealing duration at $850\,^\circ$C [4.95]: the supersaturated gold precipitates by forming metastable orthorhombic silicide particles of still unknown composition which are semicoherent to the silicon matrix. During the starting period the supersaturation of silicon self-interstitials cause the formation of extrinsic dislocation loops which, in turn, act as nucleation centers for the precipitates. After further annealing, compact and almost spherical silicide particles develop and the dislocation loops annihilate completely. With the beginning of the precipitation, the electrical activity reduced to that of the equilibrium gold concentration at the respective annealing temperature. After phosphorus gettering of gold-doped samples and quenching the samples to RT, no gold silicide particles or other gold-related precipitates could be observed.

4.7.6 Known Impurity Sources and Common Concentrations

As mentioned before, gold-contamination sources are mostly coupled to intentional gold doping or to production lines using rear contacts containing gold layers. The gold contamination occurs by means of a mechanical contact to contaminated tweezers or handling systems or by replating in solutions containing gold impurities. A vapor-phase contamination with gold during heat treatments in contaminated furnace tubes has also been observed frequently as well as contamination in equipment used for plasma etching. Gold concentrations up to several 10^{12} cm^{-3} and sometimes even more can be observed after diffusing the surface contamination into the bulk of the wafer.

4.7.7 Avoidance of Contamination

In order to avoid gold contamination the gold-doped wafers or the wafers with rear contacts containing gold must be kept strictly separated during processing and handling. Any mechanical contact with the same tweezers or the same carriers must be prevented. Wet-chemical processes should not be performed in the same solutions and even not in the same vessels. The vacuum chamber of a plasma-etching equipment must be carefully cleaned and controlled after processing of gold-doped wafers. A special problem is the cleaning of gold-contaminated furnace tubes, since an oxygen atmosphere does not accelerate the cleaning process, in contrast to contaminations with iron, nickel, or copper. The gold contamination is only reduced during long HT processes by evaporation of the gold. In conclusion, the amount of gold contamination in production lines must be kept as low as possible, which requires a continuous control of all relevant furnaces and equipment. This can be done, for example, by DLTS measurements performed on wafers which were processed in the respective equipment (Sect.6.2.1).

5. Properties of Rare Impurities

This chapter deals with some peculiarities of transition metals, which are not found as common impurities during device production, but nevertheless may occasionally arise in new methods applied in production lines or needed for defect analysis. The respective knowledge of such properties originates from intentional impurity contamination and diffusions. The intentional doping was performed mainly to study the impurities' electrical behavior and to determine their deep energy levels by DLTS.

Progress in device fabrication has triggered the development of new techniques which may create new problems with heretofore uncommon impurities. One example is the application of silicides as a rear contact material in VLSI devices currently investigated in many laboratories. By this method, several presently rare impurities, such as titanium or tungsten, may be unintentionally introduced into wafers as cross contaminants during device production. These impurities can affect the electrical properties of the wafers and therefore reduce yield. The following listing of properties, as far as they are known today, should provide a quick overview of their possible influences on devices. This treatment should also help to detect such uncommon impurities and suggest ideas for their avoidance.

Recently, several so far unknown transition metals have been investigated by *Lemke* [5.1] who has grown crystals intentionally contaminated with still missing metals. He measured the deep energy levels and their majority-carrier capture cross-sections in order to complete the list. This enabled him to survey the chemical trends which have been studied before only for the 3d transition metals. As a consequence, almost all transition metals have been studied with the exception of 4: Yttrium, lanthan, technicium, and mercury. By comparing the segregation coefficients κ_{el} deduced from the determination of the concentration of the electrically active defects measured by DLTS with published segregation coefficients κ_{tot} (if available) *Lemke* obtained informations about the precipitation behavior of the respective impurities. This leads us to speculate on the localization of the atom in the crystal, whether it forms substitutional or interstitial defects.

Furthermore, it became possible to compare the properties of the metal in one group that exhibits equal numbers n for the s+d electrons. As may be inferred from Table 3.3, there are several similarities in one group concerning the number of charge states, their character being donors or acceptors, and the trend of shift in activation energies, which sometimes is very small (Zr-Hf, Nb-Ta, Pd-Pt, Ag-Au). It even seems possible that the transfer from the stable interstitial to the stable substitutional defect takes place between the groups n = 8 and n = 9, although, in gernal, there is a continuous shift of the solubility relations between interstitial and substitutional defects, which decrease with increasing atomic numbers. However, this assumption needs to be substantiated experimentally at both sides of this boundary.

The following list of rare impurities includes those transition metals where at least some properties are known, and it excludes others which are totally unknown concerning their behavior in silicon. The succession of the impurities follows the sequence of increasing atomic numbers starting with the 3d transition metals (Sects.5.1-7) followed by the 4d (Sects.5.8-13) and finally the 5d transition metals (Sects.5.14-20).

5.1 Scandium

Scandium is the lightest 3d transition metal and can be excluded as an unintentional impurity in device fabrication since it is scarcely found as an impurity in metals and chemicals. The properties of scandium are similar to those of titanium. Its diffusion into a sample is almost avoided by residual oxygen being present in the ambient atmosphere of a furnace tube at HT. Intentional diffusion should be performed in vacuum or in a inert-gas atmosphere, except nitrogen. Since titanium forms hydrogen complexes which deteriorate the DLT spectra, etching of the scandium-contaminated sample should be avoided if possible.

Solubility and diffusivity data of scandium are not known, but it is assumed that both are lower than the respective data of titanium. It is supossed that scandium diffuses on interstitial sites and forms interstitial defects because of the chemical trend and its similarity to titanium. Scandium is electrically active and forms two deep energy levels in the upper half of the silicon band gap, a donor and a double donor. In addition, there is another donor in the lower half of the band gap, which should be a triple donor [5.1]. Although this is the only one within the series of the transition metals, it seems reasonable with regard to the general trend from donors to acceptors, with increasing atomic numbers (Table 3.3). The activation energies of the deep energy levels of scandium have been investigated by three researchers using quite different preparation methods and resulting in different values [5.1,2,3]. The early and common techniques (contamination, diffusion in an argon atmosphere, etching, Schottky contacts, DLTS) [5.3] result in two deep levels in the upper half of the band gap. Quite different results were obtained by *Lemke*. He grew scandium-contaminated crystals and measured the etched and cleaved samples by DLTS [5.1]. In this case three levels were found, as mentioned above. Finally, for the most recent investigation by *Achtziger* et al. [5.2] radioactive titanium was implanted and diffused. Subsequent transmutation of titanium to scandium was observed by DLTS. In this case the donor and the double donor agree well with those recorded for the grown crystals. A triple donor was also observed, but its concentration did not increase with time, which is expected. Still a third level in the upper half of the band gap was found near to the double donor, which has not been identified. In addition, there are very early results by *Lebedev* recorded by photocapacitance measurements [5.4]. They exhibit 7 deep levels. Two of them agree fairly well with the donor and the double donor, another one could be identical with the third level obtained by *Achtziger*, but in the case of 7 levels some of them can agree with others within an expected measurement error of 0.03 eV and an indicated error of 0.05 for the donor level [5.4].

Scandium forms electrically active hydrogen complexes. There are at least two levels in the upper half of the band gap, as deduced from the spectra recorded on etched and on cleaved samples [5.1]. Their activation energies were not given. Hydrogen complexes could be one reason for the differences in the results. It cannot be excluded likewise that the metallic scandium used for contaminating samples in [5.3, 4] would contain some titanium impurities which are assumed to diffuse faster and in higher concentrations compared to scandium.

The measured majority-carrier capture cross-sections of the donor agree within about an order of magnitude [5.1, 2], which is not uncommon for this parameter. The capture cross-sections for the double donor cannot be compared since this donor is only characterized by the lower limit [5.1].

Precipitates of scandium have not been observed so far. Because of the low diffusivitiy combined with a low solubility precipitates are not expected except for long annealing processes at lower temperatures.

Table 5.1. Properties of interstitial scandium

Defect	ΔE_A [eV]	d/a	δ_M [cm^2]	T_{DLTS} [K]	Ref
Sc$_i$	-0.21 ± 0.01	d	$\sigma_e = 3 \cdot 10^{-14}$	90	5.1,2
Sc$_i$	-0.50 ± 0.01	dd	$\sigma_e = 2 \cdot 10^{-14}$	230	5.1,2,4
Sc$_i$	$+0.20$	ddd	$\sigma_h = 1.1 \cdot 10^{-19}$	150	5.1

5.2 Titanium

A contamination of samples with titanium cannot be excluded since titanium silicide has been investigated in many laboratories as a constituent in modern contact systems. Cross contamination may happen by processing clean wafers in the same equipment. For example, titanium is easily sputtered on wafers if present in the same vacuum chamber. A contamination of wafers by a mechanical contact to the metallic titanium has been performed intentionally. Due to its low electronegativity (1.3r.u.) titanium should not be replated on silicon (1.8r.u.) in chemical solutions, at least not in higher densities.

Problems arise with an intentional diffusion of titanium into wafers in common furnace tubes even if an argon ambient atmosphere is employed. Since titanium rapidly forms oxides and nitrides at elevated temperatures, even small amounts of residual air in the ambient atmosphere hinders its diffusion. This is the case in almost all processes at technological temperature with the exception of epitaxial reactors where a hydrogen atmosphere is used. A further exception was found in rapid thermal processing. This is usually performed in a small chamber which can be diluted by argon more effectively in short periods of time.

Titanium forms silicides at elevated temperatures: TiSi at $500°C$ and $TiSi_2$ above $600°C$ [5.5]. The silicide layer acts as a diffusion source. Titanium exhibits the lowest known interstitial solubilities S and interstitial diffusivities D of the 3d transition metals; both follow Arrhenius laws:

$$S = 5 \cdot 10^{22} e^{(4.22 - 3.05/kT)} \text{ cm}^{-3} \quad (950°C \leq T \leq 1200°C) \quad [5.6], \quad (5.1)$$

$$S = 5 \cdot 10^{22} e^{(10.17 - 3.80/kT)} \text{ cm}^{-3} \quad (950°C \leq T \leq 1250°C) \quad [5.7], \quad (5.2)$$

$$D = 1.45 \cdot 10^{-2} e^{-1.79/kT} \text{ cm}^2/\text{s} \quad (950°C \leq T \leq 1200°C) \quad [5.6]. \quad (5.3)$$

$$D = 1.2 \cdot 10^{-1} e^{-2.05/kT} \text{ cm}^2/\text{s} \quad (600°C \leq T \leq 1150°C) \quad [5.7]. \quad (5.4)$$

The solubilities and diffusivities measured by *Hocine* et al. [5.6] resulted in slightly higher values compared to those given by *Kuge* et al. [5.7]. At $1100°C$ the difference in the diffusivities amounts only to 8% which is within the expected measurement error (Table 3.2). The difference in the solubilities at $1100°C$ amounts to about 35%. So far no criteria are availabe for a decision which values should be prefered. Because of the significant differences in the enthalpy values the discrepancies increase drastically at lower temperatures. At the lower limit of the temperature region ($950°C$) the solubilities differ already by more than a factor of three. Hence, one reliable concentration measurement performed after diffusion at rather low temperatures could be helpful. However, the titanium concentration after a diffusion at 950°C is in the order of magnitude of 10^{11} cm^{-3} which is already rather low for an accurate concentration measurement.

After cooling the sample from HT, titanium is electrically active on interstitial sites [5.13] and forms three deep energy levels: a donor and an acceptor state in the upper half of the silicon band gap, and a double donor in the lower half. The mean activation energies [5.8-10] are listed in Table 5.2. Non-vanishing values of E_∞ indicate temperature-dependent majority-carrier capture cross-sections, as discussed in Sect.3.3.1. In these cases the Arrhenius plots do not result in the enthalpy values as compiled in the table but in the sum of the activation energy and E_∞. The majority-carrier capture cross-sections usually indicate the values, obtained near to the temperature where the DLTS signal is observed (T_{DLTS}). To obtain RT values the respective capture cross-sections must be calculated by using the temperature exponent E_∞, as listed in Table 5.2.

Table 5.2. Properties of interstitial titanium and titanium-related pairs

Defect	ΔE_A [eV]	d/a	E_∞ [eV]	δ_M [cm^2]	T_{DLTS} [K]	Ref
Ti$_i$	-0.08 ± 0.01	a	?	$\sigma_e = 3.5\cdot10^{-14}$	50	5.8
Ti$_i$	-0.27 ± 0.01	d	0.004	$\sigma_e = 1.3\cdot10^{-14}$	160	5.9
Ti$_i$	$+0.28\pm0.01$	dd	0.04	$\sigma_h = 1.9\cdot10^{-17}$ *	200	5.10
TiH	-0.31			$\sigma_e = 7\cdot10^{-16}$		5.11
TiH	-0.55			$\sigma_e = 1.4\cdot10^{-13}$		5.11
TiH	electrical neutral			dissociation >570 K		5.12

*: $\sigma_h = 1.5\cdot10^{-16}\exp(-0.036/kT)$, expressing the temperature dependence

Because of the high majority-carrier capture cross-sections of titanium donors and acceptors in the upper half of the band gap, a pronounced influence on the carrier lifetime can be expected if titanium is an impurity in p-type silicon used, for example, for solar-cell production [5.14]. This is still more pronounced since titanium scarcely forms precipitates in the application of common-temperature processes, as its solubility and diffusivity are rather low and therefore long annealing processes would be required to enable precipitation. From the chemical trends of the properties within the 3d transition metals it can be deduced that titanium should precipitate only via a heterogeneous nucleation mechanism. Therefore, nuclei of foreign impurities are required, in addition.

On the other hand, TiO_2 precipitates were observed in silicon crystals doped with titanium in the melt [5.15]. The precipitates reduced the concentration of the electrically active titanium in the bulk of the crystal which had been cooled applying standard low cooling rates. The titanium precipitates provided the formation of dislocations. Titanium can be gettered by phosphorus diffusion and by damaged wafer surfaces, if the respective low diffusivities are taken into account. Gettering titanium requires extended diffusion durations and low cooling rates [5.16].

Titanium forms two electrically active hydrogen complexes and a neutral one (Table 5.2). The midgap level exhibits a high majority-carrier capture cross-section and therefore could be detrimental to lifetime-sensitive devices. However, the hydrogen, in general, penetrates only the silicon surfaces up to a depth of several μm, for instance, during etching the sample (Sect. 2.2.2). In most devices this region is highly doped to improve electrical contacting, furthermore the complexes dissociate at rather low temperatures of a few hundert degrees C and become neutralized.

5.3 Vanadium

Contaminations of samples with small amounts of vanadium are frequently observed if the sample is highly contaminated with iron. A probable contamination source is found in steel containing additions of vanadium. A vanadium contamination is easily performed by a mechanical contact with metallic vanadium. A liquid-phase contamination seems unlikely since the electronegativity of vanadium (1.6 r.u.) is clearly below that of silicon (1.8 r.u.). To date, no vapor phase contamination of vanadium has been observed which, however, may be due to the rather low vanadium concentrations of the samples investigated in combination with its high evaporation temperature of 1888°C.

Vanadium forms silicides at elevated temperatures, mainly VSi_2 at 600°C [5.5]. The silicide acts as a diffusion source. So far only one recent publication, that of *Sadoh* and *Nakashima* [5.17], offers results on the solubility and diffusivity data which, however, do not fit into the chemical trends of these parameters. The respective Arrhenius equations are given by (5.5, 6).

$$S \geq 5 \cdot 10^{22} \, e^{(11 \, - \, 4.04/kT)} \, \text{cm}^{-3} \quad (950\,°C < T < 1250\,°C), \tag{5.5}$$

$$D \geq 9 \cdot 10^{-3} \, e^{-1.55/kT} \, \text{cm}^2/\text{s} \quad (600\,°C < T < 1200\,°C). \tag{5.6}$$

The solution enthalpy (5.5) exceeds that of Ti (Table 3.1) and therefore the calculated solubility is lower than that of Ti, which does not fit in the otherwise general chemical trend. Although the diffusion enthalpy is lower than that of Ti the calculated diffusivity exceeds that of Ti only by a factor of four whereas the diffusivity of the transition metal next in sequence, Cr, is about a factor of 400 higher than that of Ti. From experience and from the existence of VB pairs detected by *Ludwig* and *Woodbury* [5.18] applying EPR but not yet by DLTS [5.7], as well as VZn, and VAu pairs we would expect a diffusivity of vanadium slightly lower than that of chromium. This because those pairs have so far not been detected for titanium. In conclusion, both values (5.5,6) should be treated as lower limits of the solubility and diffusivity, which is indicated by the unequality signs. The solubility data are not included in Tables 3.1 or Fig.3.1. They must be taken with utmost care. So far the reason for the deviations are not known in detail.

After diffusion and cooling the sample to RT, vanadium on interstitial sites is electrically active and forms three deep energy levels: a double donor in the lower half of the silicon band gap, a donor and an acceptor state in the upper half. The double donor exhibits a strong temperature dependence of the majority-carrier capture cross-section and therefore yields a larger discrepancy between the activation energy determined by an Arrhenius plot, on the one hand, and by the emission and capture of charge carriers, on the other hand (Sect.3.3.1). The mean values for the enthalpies and the majority-carrier capture cross-sections [5.19-21] are listed in Table 5.3. It is expected that the solubilities of vanadium almost vanish at RT in

Table 5.3. Properties of interstitial vanadium and vanadium-related pairs

Defect	ΔE_A [eV]	d/a	E_∞ [eV]	δ_M [cm^2]	T_{DLTS} [K]	Ref
V_i	-0.18 ± 0.02	a	?	$\sigma_e = 1.4\cdot10^{-16}$	140	5.19
V_i	-0.45 ± 0.02	d	0.00	$\sigma_e = 2\cdot10^{-14}$	200	5.20
V_i	$+0.31\pm0.02$	dd	0.13	$\sigma_h = 2.0\cdot10^{-18}$ *	200	5.21
VH	-0.50			$\sigma_e = 1.6\cdot10^{-13}$		5.22
VZn	$+0.29$	d		$\sigma_h = 8.0\cdot10^{-16}$	200	5.23
VAu	-0.20	a		$\sigma_e = 4.9\cdot10^{-16}$	115	5.23
VAu	$+0.42$	d		$\sigma_h = 6.4\cdot10^{-16}$	200	5.23

*: $\sigma_h = 3.8\cdot10^{-15} \exp(-0.13/kT)$, expressing the temperature dependence

analogy to all other investigated transition metals. Therefore, the quenched-in electrically-active defects are metastable at RT. As a consequence, there is a tendency for complexing, outdiffusion or precipitation of these defects. Because of the low diffusivity of vanadium at RT the reactions are diffusion limited, and thus far activation energies have only been reported in the literature on pairs of vanadium with zinc and gold (Table 5.3). Because of the low concentrations of vanadium and gold impurities which are commonly found in samples, the respective defects are technologically unimportant. On the other hand, Zn is not found in wafers after standard diffusion processes since it evaporates almost quantitatively in the furnace tube and is therefore prevented from diffusing into the sample. In analogy to titanium, vanadium also forms an electrically-active hydrogen complex situated at midgap, exhibiting a high majority-carrier capture cross-section (Table 5.3). This, however, is likewise technologically unimportant because of the same reasons that we discussed for titanium (Sect.5.2). To date, precipitates of vanadium have not been observed or investigated, and therefore details cannot be reported.

5.4 Chromium

Chromium is an essential constituent of most of the preferential-etch solutions to be applied to reveal crystal defects in samples after processing. Therefore, a cross contamination in the production line can occasionally occur. Although replating of chromium on silicon samples is not expected because of the rather low electronegativity (1.6r.u.) compared to that of silicon (1.8r.u.), adsorption layers of chromium on the surface of wafers treated in respective solutions were detected by sensitive methods. However, after applying common diffusion processes, chromium has not been found in the bulk of wafers since its diffusion is impaired by the presence of residual oxygen in the atmosphere of the furnace tube. The formation of chromium oxides avoids its diffusion, as discussed in connection with titanium diffusion (Sect.5.2). In spite of its high melting point (1920°C), chromium exhibits a rather low evaporation temperature of 1205°C, and metallic chromium is expected to evaporate quickly during high-temperature processes. Therefore, the application of preferential etching on wafers in the production line is nearly without risk. Diffusion of strongly chromium-contaminated samples in an almost pure inert-gas atmosphere leads to an impurity concentration. The highest values, corresponding to the respective solubilities, can be obtained only by a diffusion in a reducing atmosphere, such as a forming gas, and hydrogen or in vacuum. Its diffusion in Rapid

Thermal Processing (RTA) equipment is more effective since this can be successfully diluted by an inert gas in a short time period.

The solubilities S and diffusivities D of interstitial chromium can be expressed in the form of Arrhenius equations. They are presented together with the temperature regions for which they have been determined:

$$S = 5 \cdot 10^{22} \, e^{(4.7 - 2.79/kT)} \, cm^{-3} \quad (900\,°C \le T \le 1300\,°C) \quad [5.24], \quad (5.7)$$

$$D = 10^{-2} \, e^{-0.99/kT} \, cm^2/s \quad (900\,°C \le T \le 1250\,°C) \quad [5.25]. \quad (5.8)$$

$$D = 6.8 \cdot 10^{-4} \, e^{-0.79/kT} \, cm^2/s \quad (27\,°C \le T \le 400\,°C) \quad [5.26]. \quad (5.9)$$

As shown in Fig.3.1, the solubility of chromium is situated between the lower values of titanium and the higher values of manganese, iron, and cobalt. For the diffusivities two different equations are given, an early result of *Bendik* et al. [5.25] determined at high temperatures and a more recent result of *Nakashima* et al. [5.26] that was determined for lower temperatures starting at RT. Both are depicted in Fig.3.2. Because of the different migration enthalpies both straight lines cross each other at about 1000 K which is within the temperature region where no measurements have been performed. This is in contrast to the diffusivities of manganese and iron, where the low and high temperature values are almost on one straight line and hence a mean value can be calculated (being valid in the whole temperature region). Averaging of the chromium diffusivity has not been performed since the high temperature values deviate by a factor of three from the extrapolated result. From recent detailed investigations of the diffusivity of copper (Sect.4.3.2) it can be inferred that the interstitial diffusion of transition metals especially at low temperatures can be influenced by several parameters such as the charge state of the diffusor or the oxygen content of the silicon sample [5.27]. On the other hand, it can be deduced from the data of Table 3.2 that the temperature-independent prefactors of all reliable results, with the exception of titanium, are on the order of magnitude of 10^{-3}, to be achieved by averaging (5.8 and 9). In conclusion, further investigations are required to achieve better results.

After diffusion and fast cooling the sample to RT, chromium remains dissolved on interstitial sites forming metastable electrically-active defects. Only one deep energy level was found, which is a donor state in the upper half of the silicon band gap. Therefore, chromium is only detectable in n-type samples with DLTS using Schottky contacts. The mean activation energy and the majority-carrier capture cross-section of the chromium donor [5.28] are listed in Table 5.4. Because of its metastability at RT and its medium-high diffusivity, interstitial chromium forms donor-acceptor pairs with shallow acceptors such as boron, aluminum and gallium (Table 5.4) and with deep acceptors of substitutional transition metals such as zinc and

Table 5.4. Properties of interstitial chromium and chromium-related pairs

Defect	ΔE_A [eV]	d/a	$C\delta_M$ [cm^2]	T_{DLTS} [K]	Axis Orient.	Ref
Cr$_i$	-0.22 ± 0.01	d	$\sigma_e = 7.3\cdot10^{-15}$	110		5.28
CrB	$+0.28\pm0.02$	d	$\sigma_h = 1.5\cdot10^{-15}$	150	[111]	5.29
CrAl	$+0.49\pm0.05$	d	$\sigma_h = 1.5\cdot10^{-16}$	240	[111]	5.30
CrGa	$+0.48\pm0.01$	d	$\sigma_h = 1.5\cdot10^{-15}$	240	[111]	5.30
CrZn	-0.1	a	$\sigma_e = 2.8\cdot10^{-15}$			5.31
CrAu	$+0.35$	d	$\sigma_h = 8.7\cdot10^{-16}$	170		5.31

gold (Table 5.4). Only the chromium-boron pairs may be of some technological importance. The concentrations of the other pairs are too low to be detected if the impurities are not intentionally introduced into the wafer. Since the CrB donor is situated in the lower half of the silicon band gap it is detected in p-type samples with DLTS using Schottky contacts. So chromium can be detected in n- and p-type samples as an interstitial defect and a CrB pair, respectively. For a quantitative formation of pairs at RT a sufficiently long period of time (1 to 3 days) should elapse after cooling the sample from HT, which still depends on the boron concentration in the sample ($5 \div 10\,\Omega\cdot$cm).

During co-diffusion of chromium with copper, zinc or silver, substitutional defects of chromium are formed, as has been observed in the early days of deep-level investigations by *Ludwig* and *Woodbury* [5.18] applying the electron-spin resonance technique. These defects were found to be very stable upon annealing. After application of equal preparation techniques *Lemke* [5.32] obtained two chromium-related defects which were tentatively correlated to CrZn pairs (Table 5.4), and Cr$_2$Zn pairs or substitutional chromium, respectively. A reliable correlation has not been reported. Since the respective defect concentration was low, it is technologically unimportant. Furthermore, this reaction usually does not take place because of the absence of zinc which evaporates out of the sample if not diffused in a closed ampoule.

Chromium forms precipitates after quenching the sample from HT and subsequent annealing at low temperatures, starting as low as 170°C with a 50 minute annealing duration [5.33]. Applying cooling rates from the haze program, chromium does not form haze. Since haze is coupled to a homogeneous nucleation mechanism it is deduced that chromium precipitates via a heterogeneous nucleation mechanism at least during medium-fast cooling the sample from HT. Investigations of chromium precipitates have not yet been reported, but to date chromium precipitates do not play a relevant role in device technology.

5.5 Manganese

In the Periodic Table, manganese is located in the middle of the 3d transition metals (Fig.1.1) between the light and slowly-diffusing metals exhibiting low solubilities, and the heavy and fast-diffusing metals with high solubility. Due to the chemical trend within the 3d transition metals it shows a medium-high solubility and a medium-high diffusivity, both almost agree with the respective properties of iron. However, in contrast to iron, manganese does not belong to the main impurities in device production. Therefore, the number of investigations and publications is much lower compared to those on iron.

Intentional contamination of wafers with manganese can easily be performed by means of a mechanical contact of metallic manganese to a clean surface. Strong replating of manganese on silicon in chemical solutions is not expected due to the rather low electronegativity of 1.5 r.u. compared to 1.8 r.u. for silicon. As observed for chromium, a trifling surface adsorption on uncoated wafers may also be found for manganese. Manganese exhibits a very low evaporation temperature of 980°C, and therefore vapor-phase contamination is expected even at rather low diffusion temperatures. So far an unintentional contamination of wafers in device production has not yet been observed.

Like most of the other transition metals, manganese forms silicides at elevated temperatures: MnSi at $400 \div 500°C$ and $MnSi_2$ above 800°C [5.5]. It is expected that these silicides act as diffusion sources during heat treatment. As mentioned before, the interstitial solubilities S and diffusivities D of manganese resemble those of iron and follow Arrhenius equations in the temperature regions investigated. For both parameters two slightly differing results are available. Additionally, mean values were calculated:

$$S = 5 \cdot 10^{22} \, e^{(7.32 \, - \, 2.81/kT)} \, cm^{-3} \quad (900°C \leq T \leq 1150°C) \quad [5.24], \, (5.10)$$

$$S = 5 \cdot 10^{22} \, e^{(6.9 \, - \, 2.78/kT)} \quad cm^{-3} \quad (900°C \leq T \leq 1150°C) \quad [5.34], \, (5.11)$$

$$S = 5 \cdot 10^{22} \, e^{(7.11 \, - \, 2.8/kT)} \, cm^{-3} \quad (900°C \leq T \leq 1150°C) \, [5.24, 34],$$
$$(5.12)$$

For electrically active manganese Mn_i:

$$S = 5 \cdot 10^{22}\, e^{(5.97 - 2.71/kT)}\ cm^{-3} \quad (900\,°C \le T \le 1150\,°C)\ [5.24], \quad (5.13)$$

$$D = 6.9 \cdot 10^{-4}\, e^{-0.63/kT}\ cm^2/s \quad (900\,°C < T < 1200\,°C)\ [5.34], \quad (5.14)$$

$$D = 2.4 \cdot 10^{-3}\, e^{-0.72/kT}\ cm^2/s \quad (14\,°C < T < 90\,°C) \quad [5.26], \quad (5.15)$$

$$\boldsymbol{D = 1.63 \cdot 10^{-3}\, e^{-0.71/kT}\ cm^2/s \quad (14\,°C < T < 1200\,°C)\ [5.26, 34],}$$
$$(5.16)$$

The differences between the solubilities recalculated with (5.10 and 11) amount to about 15% for a temperature of 1100°C. which is within the expected measurement error. However, the differences between the total solubility (5.10) and the maximum concentration of electrically active manganese (5.13) yields up to 39%. Hence, a considerable fraction of the dissolved manganese at high temperatures becomes electrically inactive during quenching the sample to RT, for instance, due to precipitation or passivation. The difference between the two diffusivities at 1100°C (5.14) and the extrapolated value of (5.15) amounts to about 35%. But if we compare the migration enthalpy of −0.63 eV (5.14) with the chemical trend of the other 3d transition metals listed in Table 3.2 it turns out that this value is too high to fit in the trend. On the other hand, the temperature-independent prefactor is too low compared to the surrounding values. The average of both equations, however, fit well to this trend.

Interstitially dissolved manganese can be quenched-in and forms electrically-active interstitial defects: a double donor in the lower half of the silicon band gap, a donor and an acceptor state in the upper half. In analogy to iron and chromium, the interstitial manganese reacts with other impurities and forms various donor-acceptor pairs. A peculiarity of manganese is the complexing of interstitial atoms to Mn_4 clusters.

In contrast to iron, manganese also forms substitutional defects, for instance, during co-diffusion with copper. This has already been observed during the very early investigations of *Ludwig* and *Woodbury* [5.18]. The determination of the respective deep energy levels followed much later [5.35]. Although substitutional manganese is not yet of any technological importance, the respective deep energy levels are listed in Table 5.5 for completeness together with the activation energies of the interstitial defect and the donor-acceptor pairs.

The double donor of the interstitial manganese exhibits a temperature-dependent majority-carrier capture cross-section. Therefore, an Arrhenius plot does not reveal the accurate enthalpy of the deep energy level, as found for all double donors of the 3d transition metals. As may be inferred from the table, a large number of manganese-related deep energy levels is already

Table 5.5. Properties of manganese and manganese-related defects

Defect	ΔE_A [eV]	d/a	E_∞ [eV]	δ_M [cm^2]	T_{DLTS} [K]	Ref
Mn$_i$	-0.12 ± 0.01	a	?	$\sigma_e = 3.1\cdot10^{-15}$	70	5.36
Mn$_i$	-0.43 ± 0.02	d	0.00	$\sigma_e = 3.1\cdot10^{-15}$	200	5.37
Mn$_i$	$+0.27\pm0.03$	dd	0.07	$\sigma_h = 2.0\cdot10^{-18}$ *	200	5.36
Mn$_s$	-0.43	a	$\sigma_e = 9\cdot10^{-17}$			5.35
Mn$_s$	$+0.34$	d	$\sigma_h = 2\cdot10^{-16}$			5.35
MnB	-0.55	d	$\sigma_e = 9\cdot10^{-14}$			5.38
MnAl	-0.45	d	$\sigma_e = 5\cdot10^{-15}$			5.38
MnGa	-0.42	d				5.39
MnZn	$+0.18$	d				5.32
MnAu	-0.24	a				5.32,40
MnAu	$+0.57$	d				5.32,40
Mn$_4$	-0.28	d				5.36

*: $\sigma_h(Mn_i) = 7.9\cdot10^{-17}\exp(-0.064/kT)$

known and it seems likely that there may be still other complexes which have not yet been observed.

Manganese forms precipitates during slow cooling of the sample, but it does not form haze during moderately fast cooling of the sample from HT. Like iron, manganese is able to form haze during very slow cooling of the sample or during a subsequent annealing process at lower temperatures. Little is known about the precipitation behavior of manganese. Since it does not belong to the common impurities in device production this is not of technological importance. It seems that manganese can strengthen haze that originates from other impurities which precipitate via a homogeneous nucleation mechanism, as already reported for iron (Sect.4.1). It is not yet known whether the Mn$_4$ complex is a first step for a homogeneous nucleation mechanism and what kind of preparation technique will promote its formation.

5.6 Cobalt

Cobalt is the lightest of the fast-diffusing 3d transition metals which, however, still exhibits a medium-high solubility. Unintentional cobalt contamination is very rarely found in samples after processing. A cobalt contamination of wafers can easily occur by a mechanical contact of metallic cobalt to an uncoated wafer. A liquid-phase contamination is possible since the electronegativity of cobalt equals that of silicon 1.8 r.u.. However, cobalt impurities in chemicals are unlikely. A vapor-phase contamination with cobalt seems unlikely since its evaporation temperature of 1649°C is high.

Cobalt forms silicides which exhibit a decreasing metal content with increasing temperatures, as was reported for nickel (Sect.4.2.3): $Co_2 Si$ at temperatures between 350°C and 500°C, CoSi between 375°C and 500°C, and finally $CoSi_2$ at temperatures above 550°C. It is assumed that mainly the high-temperature modification $CoSi_2$ acts as a diffusion source. The solubility S of cobalt [5.24] almost agrees with those of iron and manganese, a fact not yet well understood because of the lack of theory. For the diffusivity D two different equations were published [5.41,42] resulting in similar values but differing in the migration enthalpy and the temperature-independent prefactor. Comparing the migration enthalpy of 0.37 eV [5.41] with the other values listed in Table 3.2 it turns out that this would be the lowest enthalpy of all 3d transition metals, which contradicts general experiences. However, the average of both exhibit a migration enthalpy of 0.53 eV which is between that of iron (0.66eV) and Ni (0.47eV), and fits well to the chemical trend of the 3d transition metals. Therefore this equation is proposed until new results will be available:

$$S = 5 \cdot 10^{22} \, e^{(7.6 - 2.83/kT)} \text{ cm}^{-3} \quad (700°C \leq T \leq 1200°C) \text{ [5.24]},\quad (5.17)$$

$$D = 9 \cdot 10^{-4} \, e^{-0.37/kT} \text{ cm}^{-3}/s \quad (700°C \leq T \leq 1100°C) \text{ [5.41]}.\quad (5.18)$$

$$D = 4.2 \cdot 10^{-3} \, e^{-0.53/kT} \text{ cm}^{-3}/s \quad (900°C \leq T \leq 1100°C) \text{ [5.41,42]}.$$
$$(5.19)$$

During cooling a sample from HT the interstitially dissolved cobalt precipitates almost quantitatively because of its high mobility and its homogeneous nucleation mechanism. Cobalt forms strong haze during quenching the sample from HT. In thin samples cobalt diffuses preferentially to both sample surfaces where it precipitates and forms a multitude of small silicide particles without emitting dislocations. The structures of the precipitates show small platelets which consist of monocrystalline $CoSi_2$, epitaxially grown parallel to the silicon {111} lattice planes in a twin configuration [5.43]. One example for cobalt precipitates is shown in a cross-sectional

TEM micrograph in Fig.5.1. The rather small precipitate has formed at the surface of the wafer.

Thus far, cobalt haze has exhibited a peculiarity: whereas strong haze was observed on (111)-oriented wafers after cobalt diffusion and preferential etching, (100)-oriented wafers revealed no haze after the same preparation procedure. It was not clear whether on (100)-oriented surfaces the accumulation of precipitates did not form or the respective haze was not shown. Only recently it could be clarified that cobalt haze on (100)-oriented surfaces is not visible by a Yang etch which was usually applied in our haze investigations, but it can be revealed by a New-etch solution which was developed recently (Sect.3.4.2). The sensitivity of the Yang etch seems to be too low to reveal the very small lattice distortions of the monocrystalline surface due to the shallow cobalt silicide platelets which cross the original (100)-oriented silicon surface in the form of very small lines. This crossing of the original wafer surface was also observed and reported for other transition-metal precipitates. Furthermore, the $CoSi_2$ precipitates exhibit the best fit to the silicon host lattice besides $NiSi_2$ (Table 3.10 and Fig.3.6).

Because of the almost quantitative precipitation of cobalt even during quenching the sample after diffusion to RT, the correlation of residual electrically-active defects to defined defect structures is not self-evident. The problem becomes still more severe because, in general, several deep

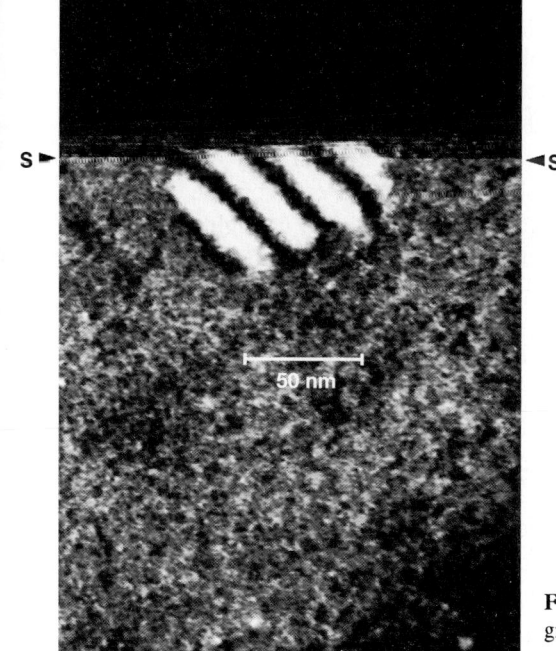

Fig.5.1. Cross-sectional TEM micrograph of a $CoSi_2$ platelet at the surface of a wafer (S)

145

Table 5.6. Properties of substitutional cobalt and cobalt-related defects

Defect	ΔE_A [eV]	d/a	δ_M [cm^2]	T_{DLTS} [K]	T_{Dis} [K]	Ref
Co$_s$	-0.41 ± 0.01	a	$\sigma_e = 2.2\cdot10^{-15}$	200		5.45
Co$_s$	$+0.41\pm0.01$	d	$\sigma_h = 5\cdot10^{-18}$	200		5.46
Co$_i$?	$+0.2$	d				5.47
CoH$_x$	-0.07	a?	$\sigma_e = 5\cdot10^{-14}$		400	5.49,50
CoH$_x$	-0.17	a?	$\sigma_e \approx 10^{-14}$		400	5.49
CoH$_x$(?)	-0.22	a?	$\sigma_e \approx 10^{-15}$		400	5.49,50
CoH$_x$	-0.26	a?	$\sigma_e \approx 10^{-15}$		400	5.49
CoH$_x$	-0.39		$\sigma_e \approx 10^{-13}$		400	5.49
CoH$_x$	-0.40		$\sigma_e \approx 10^{-14}$		400	5.49,50
CoH	$+0.22$		$\sigma_h \approx 10^{-15}$	bistable		5.50
CoH	$+0.17$		$\sigma_h \approx 10^{-14}$			5.50
CoH$_2$	$+0.09$		$\sigma_h \approx 10^{-14}$		600	5.50
Co$_s$Co$_i$	electrically inactive					5.52

levels of varying concentrations were recorded, instead of the expected two or three signals of equal concentrations, as predicted by theory [5.44]. This problem could be solved, at least partly, by the detection and identification of many cobalt-hydrogen complexes [5.49, 50]. As mentioned before (Sect. 2.2.2), hydrogen is introduced into the sample surface, for instance, by etching. This can be avoided if Schottky barriers are evaporated on freshly cleaved silicon surfaces, as proposed and carried out by *Lemke* [5.56]. Even in this case several cobalt-related deep levels can be observed.

Interstitial cobalt is not expected at RT allthough it might be predominant at high temperatures. Nevertheless a donor state has been correlated to interstitial cobalt [5.47] and compiled in Table 5.6, This, however, should be treated with caution, since, in general, all interstitial defects diffuse to the surfaces and form haze or precipitate in the bulk of the sample. A residual number of interstital cobalt atoms can also react with substitutional cobalt by forming electrically inactive donor-acceptor pairs [5.52], thus passivating the electrically active defects (Table 5.6). However, there is a tendency to form substitutional defects with increasing atomic number. A correlation of measured energy levels with theoretical results is listed in Table 3.6. There is little doubt on the correlation of an energy level in the upper half of the band gap to the acceptor state of substitutional cobalt. A couple of recent publications proposed only two levels with an additional donor state in the lower half of the band gap (Table 5.6). One recent publication suggested a second donor state in the lower half of the band gap because of its high thermal stability up to 600°C [5.26]. This, however, may not be sufficient since also hydrogen complexes show a rather high thermal stability, For example, a CoH$_2$ complex which should be stable up to 600 K [5.50]. So far it is not well understood why cobalt forms so many different hydrogen complexes. In addition, it is assumed that cobalt forms also CoB pairs after quenching highly doped samples to RT [5.48] , but their activation energies have not yet been determined.

5.7 Zinc

Although zinc can sometimes be found as an impurity on polished surfaces of as-received wafers, it has never been detected as an impurity in the bulk of processed wafers. The reason is its low evaporation temperature (343°C) below its melting point (419°C). During heat treatment of a wafer, zinc evaporates from the surface, and the vapor is diluted with the ambient atmosphere. Even implanted zinc which penetrates the surface of the wafer evaporates quantitatively during heat treatment. Intentional zinc diffusion therefore requires a closed ampoule to avoid the dilution of the zinc vapor. Therefore, zinc is not a detrimental impurity in device fabrication. However, as the heaviest 3d transition metal, it exhibits several peculiarities which may be worth reporting for a better understanding of the chemical trends within the sequence of the 3d transition metals.

Within the tendency of increasing solubilities and diffusivities with increasing atomic numbers of the 3d transition metals, zinc is the only exception showing lower solubility and diffusivity data compared to copper as the preceding 3d transition metal. The solubility of zinc has been measured recently applying a neutron activation analysis [5.54] which resulted in different values compared to early measurements [5.55]. The slope of the solubility versus the diffusion temperature now fits well to the tendency of the 3d transition metals, but the absolute values are between those of cobalt and nickel (Fig.3.1). The solubility follows the Arrhenius equation [5.54]:

$$S = 5 \cdot 10^{22} e^{(7.26 - 2.49/kT)} \text{ cm}^{-3} \quad (840°C < T < 1200°C) . \quad (5.20)$$

Recently, diffusion of zinc has been investigated in detail by varying several diffusion parameters. This resulted in new informations about the influence of the different diffusion mechanisms and the relation between the equilibrium concentrations of interstitial zinc $C^{eq}(Zn_i)$ and substitutional zinc $C^{eq}(Zn_s)$. This temperature dependence follows the Arrhenius equation [5.56]:

$$C^{eq}(Zn_i) / C^{eq}(Zn_s) = 6.3 \cdot 10^4 e^{-1.79/kT} , \qquad (5.21)$$

and amounts to 4.7% at 1200°C and 0.5% at 1000°C, revealing that the concentration of interstial zinc shrinks drastically with decreasing temperatures. At RT interstitial zinc almost vanishes which explains that segregation of zinc is negligible [5.54]. It further explains that at high diffusion temperatures (>1000°C) the kick-out mechanism (Sect.4.6.2) is predominant and zinc-correlated donor-acceptor pairs are only formed during annealing the sample at lower temperatures (600°C) [5.54].

Since zinc diffuses predominantly by the kick-out mechanism an explicit diffusivity cannot be given for dislocation-free silicon because it is concentration dependent. Only for highly dislocated crystals that exhibit sufficient annihilation points for silicon self-interstitials to reduce their local supersaturation, an Arrhenius equation for the effective diffusivity of interstitial zinc $D^{eff}(Zn_i)$ is presented [5.56]:

$$D^{eff}(Zn_i) = 0.64\,e^{-1.85/kT} \text{ cm}^2/s \quad (870°C < T < 1210°C) \text{ (dislocated Si) .} \qquad (5.22)$$

Similar results were obtained for the effective diffusivities of platinum and gold which likewise diffuse by a kick-out mechanism [5.56]:

$$D^{eff}(Pt_i) = 2.1 \cdot e^{-1.79/kT} \text{ cm}^2/s \quad (950°C < T < 1200°C) \text{ (dislocated Si) ,} \qquad (5.23)$$

$$D^{eff}(Aui) = 0.46\,e^{-1.70/kT} \text{ cm}^2/s \quad (950°C < T < 1290°C) \text{ (dislocated Si) .} \qquad (5.24)$$

It should be mentioned that in most cases these effective diffusivities of the interstitial defects cannot be applied since highly dislocated silicon is rarely found in device manufacturing today with the exception of solar cell material. This material, however, exhibits a diffusivity modified due to the presence of grain boundaries [5.58].

In addition, results on the equilibrium solubilities (C^{eq}) and diffusivities (D) of vacancies (V) and silicon self-interstitials (I) were deduced from the zinc diffusion experiments [5.57]. They, however, cannot be compared to competing results so far:

$$C^{eq}(V) \approx 1.4 \cdot 10^{23} e^{-2.0/kT} \text{ cm}^{-3} , \qquad (5.25)$$

$$D(V) \approx 3.0 \cdot 10^{-2} e^{-1.8/kT} \text{ cm}^2/s , \qquad (5.26)$$

Table 5.7. Properties of substitutional zinc and zinc-related defects

Defect	ΔE_A [eV]	d/a	δ_M [cm^2]	T_{DLTS} [K]	Ref
Zn$_s$	-0.53 ± 0.02	aa	$\sigma_h = 2.5\cdot10^{-15}$	300	5.60
Zn$_s$	$+0.32$	a		167	5.61
ZnCr	-0.10	a	$\sigma_e = 3\cdot10^{-15}$		5.23
ZnFe	-0.47	a	$\sigma_e = 1\cdot10^{-16}$		5.23
ZnV	$+0.29$	d	$\sigma_h = 1\cdot10^{-16}$		5.23
ZnMn	$+0.18$	d	$\sigma_h = 2.4\cdot10^{-15}$		5.23
ZnB ?	$+0.09$?	d			5.59
ZnCu ?					5.62

$$C^{eq}(I) = 2.9\cdot10^{24}\, e^{-3.18/kT}\ \text{cm}^{-3}\ , \tag{5.27}$$

$$D(I) = 50.1\, e^{-1.77/kT}\ \text{cm}^2/\text{s}\ . \tag{5.28}$$

At RT zinc forms two substitutional defects, a double acceptor at midgap, and an acceptor state in the lower half of the band gap,. Zinc and cadmium are the only exceptions from all transition metals so far investigated which do not form a donor state. So zinc can act as an acceptor for the formation of donor-acceptor pairs with other 3d transition metals. The mean values of the deep energy levels of zinc and zinc-related defects are listed in Table 5.7. The existence of ZnB pairs [5.59] has not yet been substantiated and seems unlikely because of the very low concentration of interstitial zinc at lower temperatures following (5.21).

5.8 Zirconium

As far as we know, the properties of zirconium in silicon have been studied only once by *Lemke* [5.63] who pulled zirconium-contaminated crystals and measured the activation energies of the deep energy levels by DLTS. Deduced from the chemical trends of the transition metals it is assumed that zirconium forms interstitial defects. Details, for example, the solubility and diffusivity, precipitation and complexing are not known. Since zirconium, in general, is not employed in technical equipments, it is technologically unimportant. Its properties, however, are useful to complete the chemical

Table 5.8. Properties of interstitial zirconium

Defect	ΔE_A [eV]	d/a	δ_M [cm²]	T_{DLTS} [K]	Ref
Zr_i	-0.13	a	$\sigma_e > 10^{-14}$	70	5.63
Zr_i	-0.42	d	$\sigma_e > 10^{-14}$	180	5.63
Zr_i	$+0.32$	dd	$\sigma_h = 1.3 \cdot 10^{-17}$	250	5.63

trends of the 4d transition metals. As shown in Table 5.8, the number and even the activation energies of the deep energy levels fit well in the n = 4 group, 4 being the sum of the outer s and d electrons.

5.9 Niobium

To our knowledge the electrical properties of niobium in silicon have been studied twice [5.63, 65]. The activation energies of the deep energy levels measured by DLTS are taken from a review of *Lemke* [5.21]. Deduced from the chemical trends of the transition metals it is assumed than niobium forms interstitial defects. Details, for example, the solubility and diffusivity, precipitation and complexing are not known. Since niobium, in general, is not employed in technical equipments, it is technologically unimportant. Its properties, however, are useful to complete the chemical trends of the 4d transition metals. As shown in Table 5.9, the number and even the activation energies of the deep energy levels fit well in the n = 5 group, 5 being the sum of the outer s and d electrons.

Table 5.9. Properties of interstitial niobium

Defect	ΔE_A [eV]	d/a	δ_M [cm²]	T_{DLTS} [K]	Ref
Nb_i	-0.28	a	$\sigma_e = 7.5 \cdot 10^{-18}$	140	5.21
Nb_i	-0.62	d	$\sigma_e > 10^{-16}$	250	5.21
Nb_i	$+0.18$	dd	$\sigma_h = 3.8 \cdot 10^{-16}$	100	5.21

5.10 Ruthenium

As far as we know, the properties of ruthenium in silicon have been studied only once by *Lemke* [5.21] who pulled ruthenium-contaminated crystals and measured the activation energies of the deep energy levels by DLTS. Deduced from the chemical trends of the transition metals it was assumed that zirconium forms interstitial defects. Details, for example, the solubility and diffusivity, precipitation and complexing are not known. Ruthenium, in general, is not employed in technical equipments and thus is technologically unimportant. Its properties, however, are useful to complete the chemical trends of the 4d transition metals. As shown in Table 5.11, the number and even the activation energies of the deep energy levels fit well in the $n = 8$ group, 8 being the sum of the outer s and d electrons.

Table 5.10. Properties of interstitial ruthenium

Defect	Activation energy [eV]	d/a	Capture cross-section [cm²]	T_{DLTS} [K]	Ref
Ru_i	-0.14	a	$\sigma_e = 1.1 \cdot 10^{-16}$	100	5.21
Ru_i	$+0.26$	d	$\sigma_h = 9.2 \cdot 10^{-16}$	150	5.21

5.11 Rhodium

Rhodium belongs to the few transition metals which form haze. From the 18 transition metals investigated so far and specially marked in Fig.5.2, only five 3d and 4d transition metals were determined which form haze during the cooling period after diffusion of the impurity at HT. These are the 3d transition-metals cobalt, nickel, copper, and the 4d transition-metals rhodium and palladium. As indicated in Fig.5.2, they are all located near the right-hand side of this section of the Periodic Table.

For the formation of haze, high diffusivities and high solubilities of the interstitial defects are required simultaneously. In general, the interstitial solubilities and diffusivities, at least of the 3d transition metals, increase with increasing atomic numbers. On the other hand, the solubilities of the substitutional defects likewise increase and they seem to also increase in the sequence from the 3d to the 5d transition metals. As a consequence, the

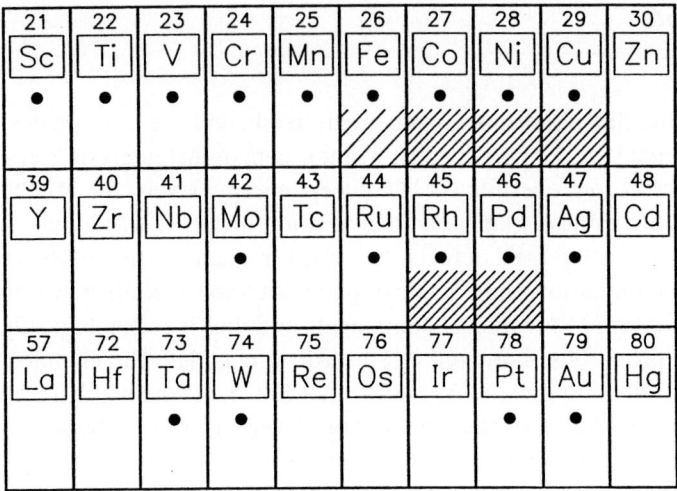

Fig. 5.2. Haze-forming metals in the section of the Periodic Table showing the transition metals. Investigated metals are marked by points, haze-forming metals by the dashed area

conditions for haze formation are only met for a rather small number of elements near the right-hand side of the sequences but not at their end, where substitutional defects prevail. Additionally, the number of haze-forming elements decreases from the 3d to the 5d transition metals for the same reason of increasing substitutional solubility. Substitutional defects exhibit a drastically reduced defect mobility.

The contamination of wafers with rhodium can be performed by a mechanical contact with metallic rhodium, and also by replating from solutions containing rhodium due to the high electronegativity of rhodium (2.2 r.u.) compared to silicon (1.8 r.u.). A vapor-phase contamination with rhodium is not expected since its evaporation temperature is extremely high (2149°C). At elevated temperatures rhodium silicides of various compositions are formed such as Rh_2Si (400°C), RhSi up to 425°C, Rh_4Si_5 (850°C), and finally Rh_3Si_4 (925°C) [5.5].

The solubility and diffusivity of rhodium in silicon are not known but both must be high since rhodium forms haze. After quenching the sample to RT rhodium forms electrically-active defects. It is assumed that the donor and the acceptor level both belong to the same defect because of equal concentrations and equal annihilation during annealing the sample [5.66]. Owing to their thermal stability and the chemical trend (Table 3.3), the deep energy levels are assumed to be correlated to a substitutional defect. The activation energies for rhodium [5.67, 68] are listed in Table 5.11. No information about the properties and the structure of rhodium precipitates are available in the literature at present.

Table 5.11. Properties of substitutional rhodium

Defect	ΔE_A [eV]	d/a	δ_M [cm^2]	T_{DLTS} [K]	Ref
Rh$_s$	-0.32 ± 0.01	a	$\sigma_e = 5.6 \cdot 10^{-15}$	150	5.67
Rh$_s$	-0.58 ± 0.01	d	$\sigma_e = 2.0 \cdot 10^{-14}$	290	5.68
Rh$_s$ Rh$_i$	electr. inactive				5.52

5.12 Silver

Although silver is employed as a rear contact material in some devices (e.g., in solar cells) and cross contamination cannot be excluded, the knowledge about its properties is still poor. A contamination of wafers with silver can be achieved by mechanical contacts with metallic silver and by replating silver on silicon from chemical solutions which contain silver as an impurity, since its electronegativity (1.9r.u.) is high compared to that of silicon (1.8r.u.).

Silver does not form stable silicides at elevated temperatures. Comparable to gold it exhibits an eutectic at a temperature of 840°C. At higher temperatures (1047°C) it evaporates, which must be taken into account for intentional silver diffusions. If the diffusion source is exhausted due to evaporation of the silver during long diffusion times, an outdiffusion to the sample surfaces and a subsequent evaporation can occur.

Silver belongs to the fairly fast diffusing transition metals [5.69], as expected from its position in the Periodic System. Its solubility was found to deviate considerably from an Arrhenius equation (Fig.4.4). Therefore, within the temperature region indicated, the Arrhenius equation (5.10) was fitted to the experimental results by adding a constant value [5.69]:

$$S = 5 \cdot 10^{22} e^{(4.67 - 2.78/kT)} + 7.1 \cdot 10^{14} \text{ cm}^{-3}$$
$$(1025°C \leq T \leq 1325°C) . \tag{5.29}$$

By comparing the solution enthalpies and entropies in Table 3.1, the values for silver almost agree with those for chromium. Due to the addition of a constant value, both solubilities approximate one another only at higher temperatures. However, the low-temperature values deviate considerably from all other solubilities (Fig.4.4). This could be caused during the sample preparation which should be verified [5.84].

Measurements of the solubilities were performed by applying a neutron-activation analysis which determines the total silver content independent of whether the defects occupy interstitial or substitional sites in the silicon host lattice. In the case of gold it has been shown that the fraction of interstitial gold to the total gold content increased with increasing temperature. However, the interstitial gold content was less than 10 % of the total gold content and therefore did not influence the solubility curve noticeably. For silver the relation of interstitial to substitutional defects is still unknown and may be quite different, as mentioned by *Rollert* et al. [5.69]. In this case, the solubility curve could be dramatically affected. This assumption could be checked by measuring the solubility of a defined electrically-active silver defect as a function of the temperature. However, at lower temperatures ($\leq 1000\,^\circ$C) the concentrations of the silver defects detected by DLTS become very small and drastically decrease with decreasing temperature [5.70]. The measured defect concentrations are more than two orders of magnitude below the respective solubility data.

In the same publication *Rollert* et al. [5.69] reported that the diffusivity of silver was too high to record diffusion profiles under their experimental conditions. Therefore, an interstitial diffusion mechanism was assumed. For a rough estimate of the diffusivity *Rollert* et al. implied similarities between interstitial silver and interstitial gold. Then, the diffusivity is approximately given by the Arrhenius equation [5.69]

$$D(\mathrm{Au_i}) \approx D(\mathrm{Ag_i}) \approx 0.6\,e^{-1.15/kT}\ \mathrm{cm^2/s}\ . \qquad (5.30)$$

A comparison of the results with those of *Wilcox* and *La Chapelle* [5.71] (Sect.4.7) yields good agreement at a temperature of about 850°C, but reveals a much higher temperature dependence that leads to deviations by a factor of four at 1100°C (Table 4.11).

Deep energy levels of higher concentrations were detected only after applying higher diffusion temperatures. Diffusion between 1000°C and 1100°C resulted in two levels of equal concentrations (between $2\cdot10^{12}$ to $2\cdot10^{13}\ \mathrm{cm^{-3}}$) which are situated in the lower half of the silicon band gap [5.70]. Different deep energy levels were observed after the application of diffusion temperatures exceeding 1200°C. They exhibit defect concentrations in the region of $10^{15} \div 10^{16}\ \mathrm{cm^{-3}}$ [5.69, 72, 73]. The results for the activation energies of the deep levels are tabulated in Table 5.12. Only identified deep energy levels are reported. It should be mentioned that silver, like many other transition metals, also reacts with hydrogen by forming complexes [5.76]. So far it is not known whether the unidentified levels can be attributed to the hydrogen complexes, to interstitial silver, or any other complexes. From the high precipitation rate and the rather low concentration of electrically active silver after cooling the sample to RT it can be

Table 5.12. Properties of substitutional silver and silver-related defects

Defect	ΔE_A [eV]	d/a	δ_M [cm^2]	T_{DLTS} [K]	Ref
Ag$_s$	-0.55 ± 0.01	a	$\sigma_e = 8\cdot10^{-17}$	300	5.72-5
Ag$_s$	$+0.37\pm0.01$	d	$\sigma_h = 8.7\cdot10^{-16}$	170	5.72-5
Ag$_s$ Ag$_i$	-0.28	a?	$\sigma_e = 3.4\cdot10^{-16}$		5.53
AgH$_x$	-0.09			50	5.76
AgH$_x$	-0.45			240	5.76
AgH$_x$	$+0.28$			200	5.76
AgH$_x$	$+0.38$			145	5.76

deduced that the interstitially dissolved silver is predominant at high temperatures. This assumption is strengthened by the detection of donor-acceptor pairs Ag$_s$ Ag$_i$ and by the fact that the concentration of the substitutional silver defects were detected only at very high diffusion temperatures. Further investigations are required to identify the low-temperature levels possibly as interstitial silver defects.

It is conspicuous that the donor and acceptor levels observed after HT diffusion of silver almost coincide with the well-known donor and acceptor levels of substitutional gold. This could strengthen the assumption that these HT energy levels may be correlated to substitutional defects of silver. The same results for HT energy levels were found by *Lemke* [5.72] on applying silver doping via the gas phase and by *Bollmann* et al. [5.73] by implanting Ag$^+$ ions into wafers. The defect concentrations after implantation could be enhanced to values exceeding 10^{15} cm^{-3} by performing a subsequent annealing process at 500°C. The defect concentrations are much higher in wafers containing dislocations compared to dislocation free material, as found in gold-doped silicon. Once more it can be deduced that both levels may be due to different charge states of substitutional silver in analogy to the respective gold levels. Even the carrier capture cross-sections resemble those of gold. During a common HT diffusion of silver, the transfer reaction from the originally interstitial silver to the electrically-active substitutional silver via a vacancy or a silicon self-interstitial mechanism may be deteriorated by the Ag-Si interface layer. This may not act as a source for vacancies or a sink for self-interstitials, as assumed by *Rollert* et al. [5.69]. This problem can be overcome by the presence of dislocations [5.69] and by ion implantation [5.73] where a multitude of intrinsic defects are generated in the sample during implantation. In both cases the concentrations of silver were considerably enhanced.

Little information is available on the precipitation behavior of silver. From the discrepancy between electrically-active defect concentrations and the respective solubilities it can be deduced that silver precipitates almost quantitatively. Only about 1% of the solubility can be detected by electrical methods. Therefore, it can be assumed that silver precipitates via a homogeneous nucleation mechanisms. On the other hand, silver does not form haze [5.74] but it is able to getter haze originating from other impurities in the sample. This indicates that silver precipitates which are present in the bulk of the sample are able to act as gettering centers for other fast-diffusing transition metals. However, the absence of haze could once more be explained by the fact that the Ag-Si surface does not act as a source or sink for intrinsic defects which are usually absorbed or emitted during the precipitation of a metal due to volume differences compared to the silicon host lattice. Although detailed TEM studies of silver precipitates are not known at present, it is deduced from the haze-gettering appearance that the precipitates should be extended similar to those of palladium and copper which again may be caused by larger volume dilatation/contraction coupled to a larger amount of vacancy/self-interstitial emission/absorption per precipitated metal atom.

5.13 Cadmium

Cadmium belongs to the rarely investigated transition metals and most of its properties remain unknown. Therefore, we only report on the deep energy levels published by three researchers, which agree fairly well. In contrast to all other transition metals investigated so far cadmium and zinc exhibit only acceptor levels. Due to the general chemical trend these levels are attributed to substitutional defects [5.77].

Table 5.13. Properties of substitutional cadmium

Defect	ΔE_A [eV]	d/a	δ_M [cm^2]	T_{DLTS} [K]	Ref
Cd$_s$	-0.45	aa			5.77
Cd$_s$	$+0.5$	a			5.77

5.14 Hafnium

To our knowledge the properties of hafnium in silicon have been studied only once by *Lemke* [5.63] who pulled hafnium-contaminated crystals and measured the activation energies of the deep energy levels by DLTS. Deduced from the chemical trends of the transition metals it is assumed that hafnium forms interstitial defects. Details, for example, the solubility and diffusivity, precipitation and complexing are not known. Since hafnium, in general, is not employed in technical equipments, it is technologically unimportant. Its properties, however, are useful to complete the chemical trends of the 5d transition metals. As may be inferred from Table 3.3, the number and even the activation energies of the deep energy levels fit well in the n = 4 group, 4 being the sum of the outer s and d electrons.

Table 5.14. Properties of interstitial hafnium

Defect	ΔE_A [eV]	d/a	δ_M [cm^2]	T_{DLTS} [K]	Ref
Hf$_i$	-0.10	a	$\sigma_e > 2 \cdot 10^{-14}$	65	5.63
Hf$_i$	-0.40	d	$\sigma_e > 2 \cdot 10^{-14}$	180	5.63
Hf$_i$	$+0.32$	dd	$\sigma_h > 5 \cdot 10^{-18}$	250	5.63

5.15 Tantalum

Tantalum exhibits an extremely low diffusivity in silicon, which contradicts the general opinion that all transition metals diffuse very quickly at HT. Accurate results on the diffusivities are not available so far, but there are rough estimates [5.78] of the upper limits for the diffusivities at the two different temperatures:

$$D < 8 \cdot 10^{-15} \text{ cm}^2/\text{s} \quad (T = 1000°C) \tag{5.31}$$

and

$$D < 1 \cdot 10^{-12} \text{ cm}^2/\text{s} \quad (T = 1300°C). \tag{5.32}$$

The diffusivities of tantalum are much lower than those of all other transition metals in silicon according to our current knowledge (Fig. 3.2). They are also much lower than those of oxygen, and carbon. If the estimation is correct, they are even lower than those of boron and phosphorus. No information is available on the diffusion mechanims.

At temperatures above 550°C tantalum forms $TaSi_2$ silicides. After growing tantalum-doped n-type silicon crystals, the electrical activity was found to be only a small fraction of tantalum incorporated in the crystals (about $5 \cdot 10^{11}$ cm^{-3}) [5.79]. The residual electrically-inactive tantalum was found to be inhomogeneously distributed in the crystal.

The activation energies of tatalum were publsihed in an early paper [5.79] where only n-type silicon was studied, and more recently by *Lemke* [5.1]. Three levels were detected as characteristic for the vanadium group (n = 5). The activation energies and the capture cross-sections listed in Table 5.15 are very similar to those of niob (Table 3.3) The activation energies determined by both researchers for donors disagree by more than 20%; therefore, only the new reults are tabulated.

Usually tantalum is not found as an unintentional impurity in device production. In former times it was used as a material for wafer carriers and for tweezers, but it was replaced by Teflon and other organic materials in modern technology.

Table 5.15. Properties of interstitial tantalum

Defect	ΔE_A [eV]	d/a	δ_M [cm^2]	T_{DLTS} [K]	Ref
Ta_i	− 0.22	a	$\sigma_e = 2.2 \cdot 10^{-17}$	140	5.1,79
Ta_i	− 0.58	d	$\sigma_e > 3.9 \cdot 10^{-15}$	250	5.1
Ta_i	+0.19	dd	$\sigma_h = 6.0 \cdot 10^{-17}$	100	5.1

5.16 Tungsten

Beside tantalum, tungsten also belongs to the 5d transition metals which exhibit very low diffusivities in silicon. They are of the order of magnitude of those of boron or phosphorus. Therefore, the statement that all transition metals are fast diffusing must be restricted to those where high diffusivities were actually measured or expected, for example, from the observation of haze formation. At a diffusion temperature of 1000°C the differences in the known diffusivities of the transition metals amount to 10 orders of magnitude. Consequently, they cannot be characterized by the properties of only several metals. The Arrhenius equation for the diffusivity of tungsten [5.80] is similar to those for boron and phosphorus, namely,

$$D = 9 \cdot 10^{-6} e^{-2.2/kT} \text{ cm}^2/\text{s} \quad (853°C \leq T \leq 1303°C) . \tag{5.33}$$

At temperatures above 650°C tungsten forms WSi_2 silicides. Tungsten and platinum silicides recently became interesting for the application as interconnecting materials in VLSI devices. Since investigations of these silicides are being performed in many laboratories, the risk of cross contamination is considerably increasing. So far, tungsten has not been detected as an unintentional impurity in wafers, which is explained by its extremely low diffusivity and its high evaporation temperature (3309°C). Replating of tungsten on uncovered samples in solutions is unlikely since the electronegativity of 1.7 r.u. is lower than that of silicon (1.8 r.u.). In addition, a mechanical contamination is also not expected since tungsten exhibits a high hardness compared to silicon, and would mainly destroy the silicon surface.

After a long diffusion at HT and slow cooling to RT tungsten forms electrically active defects. But they can also be studied in tungsten-contaminated grown crystals to avoid unintentional contamination during extremely long diffusion processes. Several researchers have investigated the deep energy levels that are formed by tungsten, starting already in the sixties. Almost all of them detected several levels $(2 \div 5)$ which, however, occupy nearly the whole silicon band gap. This could be explained by a high chemical reactivity of probably interstitial tungsten in spite of its low diffusivity. In case of hydrogen complexes the diffusivity of tungsten does not play any role since the hydrogen atoms are mobile. On the other hand, many unwanted impurities can be diffused into the samples; this might be the case in the very early investigations. More recent measurements resulted in $1 \div 3$ levels. In agreement with *Lemke* [5.1] we propose only a single donor in the lower half of the band gap, which fits to the chemical trend that the chromium group $n = 6$) exhibits only one donor state. This is well established for chromium and for molybdenum. It is also assumed that this donor is due to an interstitial defect. The data have been compiled in Table 5.16.

Table 5.16. Properties of interstitial tungsten

Defect	ΔE_A [eV]	d/a	δ_M [cm²]	T_{DLTS} [K]	Ref
W_i	+0.40	d	$\sigma_h = 5.0 \cdot 10^{-16}$	200	5.81

5.17 Rhenium

Several transition metals are difficult to diffuse in silicon, for example, Sc, Y, Mo, Ru, Ta, Tc, Re, Os; with some of them there are even difficulties to grow crystals such as Ru, Tc, Re and Os [5.82]. One reason may be the very small diffusivity, but the main reason is an extremely small segregation coefficient k_{el} in the order of magnitude of 10^{-8} or even below [5.1]. This means that the electrically active impurity concentrations in a grown crystal cannot be detected or separated from unintentional impurities. For rhenium the segregation coefficient k_{el} amounts to about $5 \cdot 10^{-9}$ [5.1] which is the smallest value of all transition metals studied by *Lemke*. The lowest segregation coefficients k_{el} were observed (if measured at all) for the 5d transition metals from La to Os which are assumed to form interstitial defects. In spite of these problems *Lemke* determined a donor and an acceptor state. A double donor state which is expected in the lower half of the band gap (p-type silicon) from the similarities with manganese in the same group $n = 7$ could not be found because of too low signals. The 4d transition metal T_c of this group is unstable [5.1]. The data, as far as they could be determined, are listed in Table 5.17. Further properties of rhenium are not known.

Table 5.17. Properties of interstitial rhenium

Defect	ΔE_A [eV]	d/a	δ_M [cm²]	T_{DLTS} [K]	Ref
Re_i	− 0.07	a	$\sigma_e = 8.7 \cdot 10^{-16}$	45	5.1
Re_i	− 0.35	d	$\sigma_e = 5.1 \cdot 10^{-16}$	200	5.1
Re_i	+?	dd			5.1

5.18 Osmium

Like rhenium, osmium belongs to the 5d transition metals which exhibit very small segregation coefficients k_{el}, namely $1.5 \cdot 10^{-8}$ [5.1]. Therefore, crystals grown from osmium-contaminated liquids exhibit DLTS signals of low concentrations. Like rhenium, the corresponding 4d transition metal of the iron group ($n = 8$), osmium exhibits one defect with two deep energy

Table 5.18. Properties of interstitial osmium

Defect	ΔE_A [eV]	d/a	δ_M [cm^2]	T_{DLTS} [K]	Ref
Os$_i$	− 0.22	a	$\sigma_e = 4.6 \cdot 10^{-17}$	130	5.1
Os$_i$	+0.30	d	$\sigma_h = 8 \cdot 10^{-16}$	200	5.1

levels: a donor in the lower half of the band gap and an acceptor state in the upper half. The data are compiled in Table 5.18. Additional properties of osmium are not known so far.

5.19 Iridium

Iridium is the 5d transition metal of the cobalt group (n = 9), the first group which forms stable substitutional defects. Although there is a smooth transition of the segregation coefficient k_{el} from very small values of the interstitial 5d transition metals (La to Os) of about 10^{-8} to the rather high values of gold ($5 \cdot 10^{-5}$), the difference between the segregation coefficient of osmium and that of iridium amounts to two orders of magnitude. A further indication for stable substitutional defects is the detection of donor-acceptor pairs between atoms of the same metal Ir$_s$ Ir$_i$ [5.53].

Table 5.19. Properties of substitutial iridium

Defect	ΔE_A [eV]	d/a	δ_M [cm^2]	T_{DLTS} [K]	Ref
Ir$_s$	− 0.24	a	$\sigma_e = 9.1 \cdot 10^{-15}$	150	5.1
Ir$_s$	− 0.62	d	$\sigma_e = 7.2 \cdot 10^{-14}$	290	5.1,83
Ir$_s$ Ir$_i$	+0.14				5.53

Like rhodium the corresponding 4d transition metal, iridium exhibits two deep energy levels, a donor state near midgap and an acceptor in the upper half of the band gap. Agreement of the more recent results [5.1] with an earlier investigation [5.83] was observed only in the activation energy of the donor state. Both researchers recorded an additional but different deep level which has not yet been identified and therefore is not listed in Table 5.19. Other properties of iridium are not known.

5.20 Mercury

For capacitance-voltage measurements to control charge carrier concentrations of wafers, for example, on epitaxial layers, the wafer surface can be contaminated with mercury which is used to form the required Schottky barrier. During a subsequent cleaning process, mercury can be replated on clean wafers because its electronegativity of 1.9 r.u. exceeds that of silicon at 1.8 r.u. It is expected that mercury forms deep energy levels like the other transition metals. The reason why those deep levels were not yet measured, is the low evaporation temperature of mercury (48°C) which prevents its diffusion into the bulk of the wafer during heat treatments. Mercury evaporates quantitatively from the surface like the corresponding neighboring 3d and 4d transition metals zinc and cadmium that exhibits evaporation temperatures of 343°C and 264°C, respectively. Therefore, these metals do not deteriorate the device performance, if not diffused in closed ampoules. Since mercury is not of technological interest and problems arise with the preparation of mercury-doped samples, its properties such as solubility, diffusivity, electrical activity, and precipitation are still unknown.

6. Detection Methods

It is not within the scope of this monograph to treat the methods for investigating the behavior and the properties of impurities in silicon in detail. Reviews on analytical measurement methods do already exist [6.1]. Frequently these techniques are very involved and require a sophisticated handling by trained specialists. However, it seems useful to briefly discuss methods which can be applied to routine controls in industry and in scientific laboratories. With the aid of these methods, unwanted impurities in the bulk of wafers and on their surfaces can be detected, as well as impurities in faulty devices. Most of these techniques also enable the determination of the respective impurity concentrations.

For this purpose, different methods must be applied to obtain the required information since the impurities can be dissolved in the sample, they can be precipitated, and they can be accumulated on the surfaces of the wafers or at any interface. At the present time, no universal facility for the detection and quantitative determination of the various kinds of defects is available.

Therefore, this chapter is divided into three parts presenting techniques for the detection of the total impurity content, of the dissolved and electrically-active fraction and, finally, for the detection and characterization of precipitates in the bulk of a sample and on its surfaces. Methods or instruments for scientific investigations to determine, for instance, the structure of defects or to elucidate complex impurity reactions will not be included since these techniques will, in general, be restricted to specialized research laboratories which are often not located in industrial areas.

6.1 Detection of the Total Impurity Content

In wafers, the detection of the total impurity concentrations requires equipment which is, in general, located in chemical analytical laboratories, for example, mass spectrometers, atomic absorption or emission spectrometers, as well as insruments for X-ray fluorescence and nuclear-activation analysis. Most of the techniques are expensive and their application must therefore be universal and not only restricted to the analysis of impurities in semiconductors. A problem common to most of these instruments is their limited sensitivity. The desirable sensitivity is very high and equal for all impurities since even very low impurity concentrations, in the order of magnitude of less than 10^{12} cm^{-3}, can considerably affect the electrical behavior of devices. However, this requirement is not accomplished. Most of the instruments reported on in the following exhibit sensitivities which are not sufficiently high to solve all of the technological problems. Furthermore, their sensitivities sometimes vary by orders of magnitude with the impurity to be detected. Therefore, each analytical problem should be discussed with trained specialists before expensive investigations are started, in order to clarify whether the effort under consideration appears reasonable. In the following sections several standard analytical methods are discussed briefly to give a short survey of their suitability to aid in solving of impurity problems in silicon samples.

6.1.1 Neutron Activation Analysis

The Neutron Activation Analysis (NAA) [6.2] is, in general, not a suitable technique for the routine control of the impurity contents in wafers before and after processing, although it may be useful for special investigations. This concept needs a neutron-radiation source and an isotope laboratory. Therefore, the analysis has to be performed in a specialized laboratory and cannot be done on short time scales. It usually takes weeks or even months until the results are available from external laboratories. The method is expensive, and its sensitivity varies considerably for each type of impurity. The price of a single analysis depends on the number of elements to be analyzed. For the necessary sample preparation, a trained analyst with experience in the preparation technique of silicon specimens is required to achieve low measurement errors. In conclusion, this method should be restricted to special problems separated from common routine control measurements.

6.1.2 Mass Spectrometers, Secondary Ion Mass Spectrometers

Mass spectrometers and, in particular, Secondary Ion Mass Spectrometers (SIMSs) [6.3] are occasionally found in semiconductor laboratories. Their application to the mentioned routine controls is problematic for several reasons:
- The determination of the total concentration of a defined impurity is difficult since the impurity can appear in different charge states and different combinations of ions which must be summed up to obtain the

total impurity concentration. Different isotopes of the same impurity result in different signals if the resolution of the mass spectrometer is sufficiently high.

- A number of signals can be interpreted in different ways because of mass interferences, for example, the mass of N_2 equals that of Si.
- For the determination of impurity concentrations, calibration standards are needed for every impurity in the concentration range to be determined.
- In general, the sensitivity of SIMS is not sufficient to detect common impurity concentrations of the order of magnitude of 10^{11} to 10^{13} cm^{-3}.
- Finally, SIMS is time consuming and expensive, and therefore not suitable for the purpose of routine measurements.

6.1.3 Atomic Absorption Spectroscopy

Atomic Absorption Spectroscopy (AAS) is frequently applied to detect impurities in wafers and to determine their concentrations. The method allows the detection of one defined element at a time. The sensitivity changes between impurities. AAS requires sample preparation. For its detection, the impurity is dissolved together with the matrix element, and the impurity concentration is determined in this solution. The corresponding impurity concentration in the bulk of the wafer must be recalculated taking into account the amount of silicon or SiO_2 dissolved. Due to chemical treatments, the blank of the impurity concentration to be measured will be more-or-less enhanced and must be repeatedly controlled. For the measurement of low impurity contents the solution can be concentrated but the blank increases simultaneously. The mean detection limit is found in the range of ppm or sub-ppm.

Some improvements have recently been proposed to enhance the sensitivity of AAS. To determine the surface contamination on top of wafers below the ppb range, *Corradi* et al. [6.4] developed a suitable solution technique which removes Cr, Mn, Fe, Co, Ni, Cu, and Zn from the silicon surface by applying nitric acid at elevated temperatures (60°C) for five mintues. In this way the common technique of dissolving the silicon matrix (by adding HF to the etchant) is avoided.

6.1.4 Vapor-Phase Decomposition

Vapor-Phase Decomposition (VPD) is another technique to enhance the impurity concentration in the solution [6.5]. The surface of a wafer is exposed to HF vapor to dissolve the native or thermal surface oxide and simultaneously remove the contaminated transition metals. The HF, condensed in droplets on the sample surface, is collected and prepared for AAS analysis. Once more, a dissolution of the silicon matrix is not required. Good results have been achieved for the detection and measurement of iron and nickel. However, the determination of copper contamination is considerably affected by its replating on the silicon surface. Therefore, only reduced copper contents were measured.

6.1.5 Inductively-Coupled Plasma Spectroscopy

Inductively-coupled plasma spectroscopy is an alternative method to the atomic absorption spectroscopy, it is based on the emission spectroscopy of solutions. In this case, several impurities can be detected subsequently by using the same solution. The sensitivity of the method varies again with the respective impurity and differs, in general, from that of the absorption spectroscopy.

6.1.6 Total-Reflection X-Ray Fluorescence Analysis

For many applications, the sensitivity of X-ray fluorescence spectroscopy will be too low. Therefore, a modification was recently developed that results in a higher sensitivity for the detection of surface contamination layers on top of polished wafers: the Total-Reflection X-ray Fluorescence analysis (TXRF) [6.6]. The method is based on the spectroscopy of X rays which are totally reflected at the sample surface due to very shallow angles of the incident radiation with respect to the wafer surface. Then, the penetration depth of the X ray is small and consequently the signal originating from the surface contamination layer will be enhanced. This concept can be improved by concentrating the impurities within a small surface region of the wafer through the application of the VPD preparation technique (Sect. 6.1.4) [6.7]. The HF droplets which are formed by the condensed HF vapor are collected in the center of the wafer, and are allowed to dry in on the surface [6.8]. In this way the sensitivity of the method is considerably improved and enables the detection of impurity densities on top of wafer surfaces in the range of 10^{10} to 10^{11} cm^{-2}.

At present, the TXRF technique seems to be the best method to detect total surface contaminations on silicon surfaces, although it is more-or-less limited to investigations of the polished front side. Alcaline etched reverse sides of wafers avoid the reflection of X rays under extremely shallow angles because of the enhanced roughness of the surface. Similar problems arise if structured wafers are to be investigated after performing respective technological processes. However, on top of plane polished wafers, the sensitivity of the method is adapted to the requirements of modern technology and yields fast results suitable for process control. The equipment, although expensive, is commercially available. Furthermore, there are specialized analytical laboratories which perform measurements for customers and also yield a fast turn-around of results in spite of the required sample shipment.

The data obtained by this method, however, cannot be directly transferred to concentrations in the bulk of a wafer after diffusion of the respective impurity. As discussed in Chap.5, there are several impurities which do not diffuse into the bulk of a sample at all, even if they were previously present at the sample surface. Some of these impurities (Zn, Cd, Hg) evaporate quantitatively during heat treatment. Other impurities (Cr, Mn, Fe, Cu, Ag) partially evaporate coresponding to the process parameters applied. Finally, several impurities (Sc, Ti, Cr, Fe, Mo) form stable oxides or nitrides with the residual oxygen or nitrogen in the ambient atmosphere of the furnace tube, which avoids their diffusion into the bulk of the sample. An overview of the transition metals which do not diffuse quantitatively into silicon samples may be inferred from Table 6.1 together with the effective mechanism, the estimated fraction which does not diffuse, and the evaporation temperature.

As a consequence, the impurity concentration in the bulk of the sample after heat treatment can be much lower than the calculated concentrtion which results from the respective original surface concentration divided by

Table 6.1. Transition metals which do not diffuse quantitatively into silicon specimens due to evaporation and oxidation of the metal

Metal	Evaporation	Oxidation	Fraction at $1000 \div 1100°C$	Evaporation temp. [°C]
Sc		X	considerably	
Ti	...	X	considerably	1546
Cr	X	X	considerably	1205
Mn	X	X	partially	980
Fe	(X)	(X)	little	1447
Cu	X		partially	1273
Zn	X		totally	343
Mo	–	X	considerably	2533
Ag	X		partially	1047
Cd	X		totally	264
Hg	X		totally	48

the sample thickness. The amount of the diffused impurities depends on the process conditions which renders its estimation still more difficult. In conclusion, the impurities detected by this method on top of polished surfaces and their concentrations must be critically considered as to whether they may be detrimental to the devices to be produced by means of the various processes. Furthermore, the reverse side of the as-received or processed wafer may exhibit other impurities or other concentrations of the same impurities, as compared to the polished front side. In this case the calculation of the respective impurity concentrations in the bulk of the wafer becomes almost impossible.

An estimation of bulk contamination in wafers becomes still worse if it is calculated from measured impurity contents in chemicals which have been replated on wafers in a solution, and subsequently diffused into the wafer during heat treatments. In this case, two non-quantitative processes overlap, the replating of the impurity in the solution and the subsequent diffusion into the bulk of the wafer. The best way to overcome this problem is to process a wafer in the respective solution, to diffuse the impurities into the bulk of the wafer and, finally, to detect them in the bulk either by means of their electrical activity or by analyzing the precipitates. The analysis of electrically-active defects and precipitates will be discussed in the sections to follow.

6.2 Detection of Dissolved Impurities

As far as known today, all transition metals which are dissolved in silicon are electrically active. This is not self-evident since, for example, oxygen dissolved in silicon on interstitial sites and carbon dissolved on substitutional sites are both electrically inactive. Furthermore, all dissolved transition metals have been studied thus far via deep energy levels, in contrast to the elements of groups III and V which are known to form shallow energy levels. Therefore, dissolved transition metals can be detected and characterized by measuring their electrical properties in the concentration range in which they are detrimental to the device performance.

There are three main bulk properties of semiconductors which are influenced by a sufficiently high amount of electrically-active impurities:

- The charge carrier concentration,
- the capacity of the space-charge region, and
- the minority carrier lifetime.

These parameters are sensitive to all electrically-active deep impurities. In addition, there are other methods such as Electron Spin Resonance (ESR) or related techniques which are only sensitive to impurities exhibiting a non-saturated electron spin. Further examples are the photoluminescence which exhibits high sensitivites only for a limited number of defects, and infrared absorption spectroscopy which is sensitive to binding energies that exhibit a dipole momentum. The restricted application and the sometimes high technical expense of these methods render them less suitable for routine measurements to detect metal impurity contaminations in semiconductor materials before and after performing technological processes.

The Hall effect was a wide-spread method in early investigations of impurities. It is based on the determination of carrier concentrations as a function of the sample temperature. However, this method has rarely been used recently. High technical cost combined with extended measurement times needed to characterize a single sample make the Hall-effect measurement unsuitable for the tasks of controlling modern production lines.

6.2.1 Deep Level Transient Spectroscopy

More suitable are methods to measure the transient capacitance of various modifications, which were developed in the past decade and are summarized by the abbreviation DLTS (Deep Level Transient Spectroscopy), or similar modifications. The concept is based on the earlier Thermally Stimulated Capacitance (TSCap) or Thermally Stimulated Current (TSC) measurements where the deep energy levels in a space-charge region were emptied by applying a reverse voltage. After cooling the sample to cryogenic temperatures these levels are filled by applying a short electrical or optical pulse or by switching off the reverse voltage for a short time. During the slow increase of the sample temperature, the deep levels empty again by emitting charge carriers. The emission rate of the charge carriers as a function of the sample temperature depends on the activation energy of the respective deep energy level. The emission of the majority carriers causes a small change in the capacity of the space-charge region, which can be monitored and measured with high sensitivity. In the temperature region in which a defined deep energy level is emptied, the space-charge capacitance increases or decreases in a step-like manner. The height of the step is a function of the respective impurity concentration relative to the respective carrier concentration. Samples containing various deep energy levels of different kinds (electron and hole traps) and of different concentrations, yield TSCap spectra that appear to be complicated stairways revealing steps of different heights going up and down at different temperatures. The evaluation of the spectra is difficult since there is no defined zero level, and various steps may overlap.

It was the merit of *Lang* [6.9] who modified the technique of recording the capacitance change due to majority-carrier emission. The step-like increase or decrease of the capacitance of the space-charge region with increasing temperature recorded in TSCap spectra was transferred to an emission or absorption spectrum that exhibit maxima or minima of the signal combined with a defined zero-line. This was achieved by applying periodic pulses to a rectifying contact while cooling the sample to cryogenic temperatures or when heating it. In the temperature region in which the re-emission of the majority carriers takes place, the capacitance of the space charge region changes with the time between two applied pulses. This so-called **capacitance transient** is monitored by an oscilloscope and measured by a boxcar integrator at two selected time intervals. The difference of both capacitances as a function of the sample temperature results in the desired signal form showing maxima or minima at defined temperatures. These signals are due to the first derivatives of the original TSCap steps. This correlation is also obtainable by means of a lock-in amplifier [6.10].

A series of capacitance transients recorded at increasing sample temperatures are plotted in Fig.6.1. They exhibit a fast decay versus time at high sample temperatures and an almost vanishing decay at low tempera-

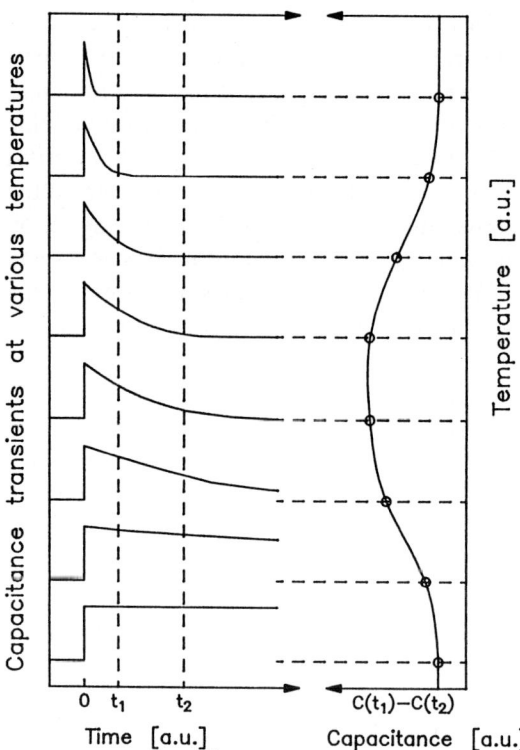

Fig.6.1. Conversion of capacitance transients at various sample temperatures to a DLTS signal by means of a boxcar integrator following *Lang* [6.9]. The signal is the difference between the measurement at t_1 and t_2

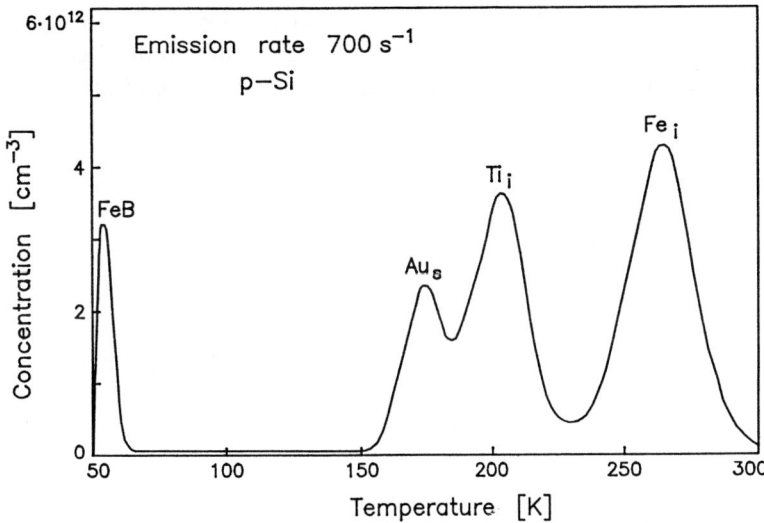

Fig. 6.2. DLTS temperature scan on a p-type sample with a Schottky contact containing iron, titanium and gold. Iron appears as an interstitial defect (Fe_i) and as iron-boron pairs (FeB). The total iron content is the sum of both concentrations

tures. This figure also demonstrates the transfer process from the capacitance transients in a DLTS spectrum which is plotted with the temperature axis in the vertical direction. The DLTS signal is the difference between the transients at times t_1 and t_2.

Figure 6.2 illustrates an application of this method showing the DLT spectrum obtained on a p-type sample with an evaporated Mg Schottky contact. The sample contains titanium, iron and gold impurities. The maximum heights of the different signals correspond to the respective impurity concentrations. The concentrations of the interstitial iron Fe_i with its maximum at about 260 K and of the iron-boron pairs FeB situated at about 60 K must be summed up to obtain the total iron content in the sample, since the signals are caused by two different defects. The double donor of the interstitial titanium Ti_i and the donor of the substitutional gold Au_s overlap to some extent. By selecting different time intervals t_1 and t_2, the DLTS signals will be shifted to slightly different temperatures. Therefore, the emission rate used for the measurement should be indicated at the spectrum for a better comparison of the results with others obtained for different emission rates.

In order to identify DLTS signals and correlate them to defined defects, a calibration must be performed prior to the measurement. This was done by many researchers in the early 1980s using samples which were intentionally doped with known impurities [6.11-16]. Usually a deep energy level is characterized by its activation energy (enthalpy or entropy), and the majority- and minority-carrier capture cross-sections together with their temperature dependences, as discussed in Sect.3.3.1. For routine measurements it is more straightforward to determine the temperature of the maximum DLTS signal (for definite standard emission rates) since this includes both the activation energy and the majority-carrier capture cross section. For a fast identification of common impurities in samples before and after performing technological processes, a list of the temperatures of the maximum DLTS signals in n- and p-type samples is very useful. An example is tabulated in Table 6.2 where the temperatures for maximum DLTS signals are correlated to the respective defects for frequently observed deep energy levels, and separated for n- and p-type samples. The listed temperatures are valid for the selected time intervals of $t_1 = 1$ ms and $t_2 = 2$ ms in equipment using boxcar integrators, corresponding to an emission time constant of 1.42 ms and an emission rate of 700 s^{-1}.

The temperatures listed in Table 6.2 exhibit a small measurement error due to a non-ideal thermal contact of the sample to the cooling finger where the temperature is measured by means of a calibrated silicon diode. The non-ideal thermal contact is caused by an electrical isolation between the sample and the cold finger. However, if the DLTS spectra were always recorded in the same direction during the cooling period of the sample, this measurement error is repeatable since the cooling rate is constant due to the power of the equipment. Therefore, the identification of the defects is cor-

Table 6.2. Temperatures of maximum DLTS signals (for emission rates of 700 s^{-1})

n-type Si			**p-type Si**		
T_{DLTS} [K]	Defect	d/a	T_{DLTS} [K]	Defect	d/a
52	Ti_i	a	58	$Cu_s Cu_i$	d
70	Mn_i	a	62	FeB	d
120	Cr_i	d	70	FeAl $\langle 100 \rangle$	d
124	V_i	a	88	FeGa $\langle 100 \rangle$	d
130	Pd_s	a	114	FeAl $\langle 111 \rangle$	d
132	Pt_s	a	136	FeGa $\langle 111 \rangle$	d
168	Ti_i	d	160	CrB	d
186	FeAu	a	177	Au_s	d
222	Mn_i	d	182	Pt_s	d
236	V_i	d	187	Mo	d
298	Au_s	a	208	Ti_i	dd
			215	FeAu	d
			226	Mn_i	dd
			267	Fe_i	d
			273	V_i	dd

rect although the absolute temperature for the maximum DLTS signal is shifted to lower temperatures by a few degrees. However, using the same equipment for measuring the spectra during the heating period of the sample would yield different maximum temperatures which still depend on the heating rate.

The detection of impurities in a semiconductor sample and the determination of their concentrations is only one application of the DLTS technique. In addition, the activation energy of the deep energy level and the majority- and minority-carrier capture cross-sections can be determined as a function of the temperature. The influence of hydrostatic pressure and high electric fields on the carrier emission can be studied. Finally, the lateral distribution of impurities, including their depth profiles, in a semiconductor sample can be determined.

The vast literature on modifications of the DLTS technique concerning the charging of the deep energy levels, various measurement and evaluation methods, and corrections for measurement errors would require a special review. Today several types of equipment are commercially available which are based on the original boxcar method or on the application of the lock-in principle which enables a high detection sensitivity [6.10]. This method enables the application of isothermal frequency scans to get a fast overview over a wide range of deep energy levels at a fixed sample temperature. This technique is also suitable to record Arrhenius plots with the aid of computer evaluation [6.17].

A further modification is the Deep-Level Fourier Spectroscopy (DLFS) which allows determination of all parameters from one single temperature scan with high accuracy [6.18-20]. A modified technique is the Constant Capacitance method (C^2 DLTS) where the reverse voltage is recorded as a function of the temperature by controlling the capacitance of the space-charge region. In comparison to the common method this technique allows the determination of a higher impurity concentration in relation to the carrier concentration of the sample [6.21, 22].

Further improvements were proposed to enhance the spectral resolution by applying double pulses (DDLTS) [6.23], by using the admittance, in particular, the optical admittance spectroscopy for studying deep energy levels [6.24, 25], and finally by applying photo-excitation or neutralization to determine minority-carrier capture cross-sections [6.26, 27]. Various papers discuss problems arising from evaluation techniques for the determination of the impurity concentration and depth profiles [6.28-30]. Finally, a comparison of results obtained by applying DLTS and other competitive methods was published by *Eichinger* [6.31].

In order to detect impurities by means of DLTS, these impurities must be dissolved in the bulk of the sample. Therefore, as-received wafers must be heat-treated and quenched to RT before recording the DLT spectrum.

This heat treatment transforms a surface contamination into a bulk contamination by in-diffusion. In wafers without a pn junction, a Schottky barrier must be evaporated onto the sample surface before DLTS measurements can be performed. Suitable barriers can be formed, for example, by titanium or magnesium layers on p-type samples, and by palladium or gold layers on n-type samples. Before evaporation the surface must be etched to remove the native oxide or any other unwanted layers. A possible sample contamination by the etch process has been discussed in Sect.2.2.2. In order to achieve low leakage currents in the Schottky barriers, the vacuum during evaporation must be sufficiently good ($<10^{-5}$ mbar).

From the various modifications of the DLTS technique which were proposed in the literature, the original temperature scan between 320 and 50 K yields the best survey of the incorporated impurity content in the sample to be studied. Using Schottky contacts, only one half of the silicon band gap can be investigated: the lower half in p-type and the upper half in n-type silicon. For investigations of the total impurity content, n- and p-type samples must be studied (e.g., for process control or calibration of unknown impurities) or a pn junction is needed which enables injection of majority- and minority-charge carriers. The injection of minority-charge carriers by pulsing the space-charge region in the forward direction allows us to study deep energy impurities which are present in the opposite half of the silicon band gap (upper half in p-type silicon, and vice versa). The respective DLTS signals in the spectrum appear (if at all) in opposite directions and can therefore be distinguished easily, in general.

Problems can arise in detecting the DLTS signal of minority-carrier traps if the respective carrier capture cross-section for minority carriers is very low with respect to that for majority carriers. In this case the charging of the trap remains incomplete and the DLTS signal is small or even vanishes. Factors of 100 between the capture cross-sections for majority and minority carriers are known for several transition metals.

As mentioned above, several transition metals form different charge states of the same defect, which are often situated in both halves of the silicon band gap. The most important exception is iron with its single donor state in the lower half of the band gap. Another exception is molybdenum which also forms only one donor in the lower half of the band gap. Chromium forms only one donor state in the upper half of the band gap. However, in boron-doped p-type samples, chromium forms chromium-boron pairs that exhibit a donor state which is situated in the lower half of the band gap. Therefore, chromium can also be detected in p-type silicon if pairing has taken place after a sufficiently long storage time at RT elapsed after quenching the sample. In conclusion, p-type silicon samples are more suitable for a routine control in samples without pn junctions where Schottky barriers are needed.

To achieve good Schottky contacts which exhibit low leakage currents, it is recommended to evaporate only one type of material in the vacuum chamber in a given period of time. If changing of the evaporation material is required, it is proposed to evaporate the new material for the first time without a sample in the sample holder. In this way the neighborhood of the evaporation source is completely covered with the new material. Schottky contacts are extremely sensitive to impurities which may evaporate simultaneously from the surroundings of the heating source.

It takes about one hour to perform one temperature scan from RT to 50 K applying a closed-cycle helium cryo-refrigerator. This period includes the determination of the impurity concentrations, heating the cryostat to RT and changing the sample. It is a rather long measurement time for the purpose of routine measurements. The commercially available equipment includes automatic recording instruments. Service is usually not required during the cooling period in which the spectrum is recorded. Therefore, several items of the equipment can be run simultaneously to enhance the number of measurements performed in a period of time.

Although more complicated, it should be possible to mount several samples on one cold finger and to record the spectra simultaneously by switching repeatedly from one contact to the other at almost the same sample temperature. Corresponding equipment is in development for commercial instruments. Another possibility would be a sample cooling by using liquid nitrogen (77K) which would cool the sample very quickly. However, at a temperature of 77 K it is not possible to record the technologically important iron-boron pairs which appear at a temperature near 60 K. To overcome this problem, a measurement frequency of about 400 MHz would be required to record the iron-boron pairs in a frequency scan performed at 77 K. This has successfully been demonstrated with non-commercial laboratory equipment [6.32].

In general, modern commercial instruments are computerized and generate a fit of the recorded spectrum. The fit enables the automatic reading of the temperature of the maximum DLTS signal and an estimated activation energy. This activation energy can deviate considerably from the results obtained by performing an Arrhenius plot at different fixed sample temperatures, especially if the sample temperature exhibits a hysteresis, as mentioned before. The measurement error due to the temperature hysteresis is drastically reduced for isothermal recording of frequency scans which are usually used for Arrhenius plots.

The Arrhenius plots are presented on the monitor if at least two frequency scans (and fits) were performed at two sufficiently different temperatures. The respective activation energy is calculated and indicated together with an estimated majority-carrier capture cross-section. This technique yields good results for the activation energy if the DLTS signal does not interfere with other impurities forming overlapping DLTS signals. However, the indicated majority-carrier capture cross-section can deviate considerably from more precise determinations by a variation of the filling pulse length. To apply this technique, rather short pulses which still exhibit accurate wave forms, are needed, especially to determine fairly high capture cross-sections. If sufficiently short electrical pulses are not available, the samples should be changed to others exhibiting lower carrier concentrations, which is possible, e.g., for process control. In this way the capture time constant τ_{cap} is enhanced since the capture rate $c_{p/n}$ and the capture cross section $\sigma_{p/n}$ are proportional to the inverse carrier concentration:

$$c_{p/n} = \frac{1}{\tau_{cap} N} \quad \left[\frac{cm^3}{s} \right], \tag{6.1}$$

$$\sigma_{n/p} = \frac{1}{\tau_{cap} N v_{th}} \quad [cm^2] \tag{6.2}$$

with N symbolizing the majority-charge carrier concentration, and v_{th} the thermal velocity of the majority-charge carriers. Several researchers prefer to determine the capture rates instead of the capture cross-sections since the calculation of v_{th} requires knowledge of the respective effective mass which still includes some uncertainties. However, in general, the measurement errors for the determination of the capture rates are large compared to the uncertainties obtained by calculating v_{th}.

The experimental determination of the *minority*-carrier capture rates and cross-sections are still more complicated, and are therefore rarely performed. For their determination, minority carriers must be injected into the semiconductor sample. This can be achieved by irradiation with light pulses at suitable wavelengths instead of electrical pulses. At present, only poor information is available on the minority-carrier capture cross-sections of most of the transition metals.

The determination of lateral impurity profiles by means of DLTS is simple but time consuming. Since the diameters of standard Schottky contacts are small (in the range of 1mm) compared to the diameters of the wafers, the lateral resolution is high if a series of sequential contacts are measured, which can be evaporated simultaneously through a mask. For the determination of depth profiles within the near-surface region, the applied reverse voltage can be increased up to values below the breakdown voltage or up to instrumental limits. For measurements in extended depths the Schottky contact can be removed and the sample can be thinned by etching or polishing. Subsequently, a new Schottky barrier is evaporated and a new DLTS measurement can be performed. However, this procedure is time consuming. If the depth profile is not combined with a lateral profile, sequential Schottky contacts can be evaporated simultaneously on top of an angle-polished sample. This transforms the depth profile into a lateral profile.

6.2.2 Carrier Lifetime

A further concept in common use to detect dissolved impurities in the bulk of a sample is to measure the minority-charge carrier lifetime since this is more or less strongly affected by the presence of deep-level impurities. In contrast to DLTS measurements where the chemical nature and the concentration of different impurities can be determined separately, the carrier lifetime is a cumulative parameter which does not allow determination of the chemical nature of the defect nor its concentration. Furthermore, it is, in general, not possible to distinguish whether a carrier-lifetime degradation results from a single impurity or from different impurities, since the inverse total lifetime is the sum of the inverse partial lifetimes:

$$1/\tau_{tot} = 1/\tau_1 + 1/\tau_2 + ... + 1/\tau_n . \tag{6.3}$$

Short carrier lifetimes can result from high concentrations of defects exhibiting small capture cross-sections, but they can also be due to low concentrations of defects with high capture cross-sections since the lifetime is proportional to the inverse product of both, according to (2.4). In conclusion, the carrier lifetime is an electrical parameter which indicates whether a sample contains defects in its bulk, acting as recombination centers and decreasing the minority-carrier lifetime. Since the capture cross-sections for the majority and the minority carriers may be quite different for the same defect, the respective minority-carrier lifetimes in n- and p-type samples can differ considerably for the same impurity and the same concentration in both samples.

Carrier lifetimes do not allow one to distinguish between the presence of dissolved impurities or precipitated impurities in the bulk of the sample since both affect the carrier lifetime in the same direction. In spite of all these disadvantages, carrier lifetime and diffusion-length measurements are increasingly applied to materials and process control in device production. The only reason for their application is the short measurement time, ranging between seconds and minutes. In comparison, DLTS measurements require about one hour to record one spectrum with additional time needed for sample preparation such as etching, evaporation of Schottky barriers, and preparation of the rear contact. Thus, a large number of lifetime measurements can be performed during the same period of time required to record one DLTS spectrum. Therefore, lifetime measurements are highly suitable for process control in a production line. If the carrier lifetime exceeds a previously specified lower limit, the impurity content in the sample may be expected to be still acceptable. Only in those cases where this lifetime limit is not reached may it be of interest to determine the chemical nature of the respective impurities and their concentrations, providing information about their origin. For these tests the higher expenses required for DLTS measurements are justified since they aid in determining the sources of failure.

Another problem which arises with carrier-lifetime and diffusion-length measurements in thin samples is their dependence on the surface recombination velocity of the respective sample. In thin wafers of high purity, almost all excess-charge carriers which were generated in the bulk of the sample by irradiation of light recombine at both wafer surfaces. Therefore, an upper limit of the effective bulk-carrier lifetime or diffusion length is caused by the thickness of the sample for each surface recombination velocity. This surface recombination velocity can vary by orders of magnitude due to different surface treatments. Lapped surfaces exhibit surface recombination velocities in the range of 10^4 cm/s whereas thermally oxidized surfaces yield values of the order of magnitude of less than 100 cm/s. However, during oxidation of the silicon sample, the bulk-carrier lifetime can be considerably affected by various defect reactions. Therefore, an oxidation process to reduce the surface recombination velocity of a sample is problematic. The easiest way to achieve a fairly low surface recombination velocity is to dip the p-type samples into a HF solution immediately before performing the lifetime measurement. However, the results obtained on thin samples will still depend on the respective sample thickness, and the lowering of the surface recombination velocity disappears with the time elapsed after performing the HF dip.

The increase in diameter of silicon crystals in the past decade led to larger thicknesses of standard wafers to avoid losses of wafers due to breakage during handling and manufacturing. With the increasing wafer thicknesses, the upper limit for lifetime measurements has increased as well. Consequently, carrier-lifetime measurements on as-received and processed wafers likewise grew in importance. For 3 inch wafers with standard thicknesses of 380 μm, the upper lifetime limit due to surface recombination amounts to about 6 μs. In 100 mm and 125 mm wafers, standard thicknesses of 625 μm are available and the upper limit for reliably corrected carrier lifetimes increases to about $40 \div 50$ μs. This already improves considerably the possibility for characterizing the cleanness of processes and even the purity of as-received CZ-grown wafers. This may be one of the most important reasons for the mounting interest in lifetime and diffusion-length measurements observed recently.

Various methods can be applied to determine the minority-charge carrier lifetime. One of the most reliable methods is the PhotoConductive Decay (PCD) which is widely used to measure the recombination carrier lifetime. This technique was improved by the application of InfraRed Emitting Diodes (IRED) and semiconductor lasers to generate excess carriers in the bulk of the sample and by the application of modern electronic sampling techniques to enhance the signal-to-noise ratio [6.33, 34].

Usually samples with dimensions of $25 \times 5 \times (0.6 \div 5)$ mm^3 were employed for the measurement where the sample thickness ($0.6 \div 5$ mm) determines the upper limit of the carrier lifetime, which can still be reliably recorded and corrected for high surface recombination velocities. The sample preparation required for the measurement is restricted to the formation of ohmic contacts at both ends of the specimen. The contacts can be achieved by rubbing emery with gallium at slightly elevated temperatures to keep the gallium in the liquid state. The measurement performed following that described in [6.34] is now evaluated with the aid of a computer which stores the photoconductivity decay curve, fits an exponential decay function, and prints the calculated carrier lifetime. The printout includes a bulk lifetime which is corrected for sample size and surface recombination velocity. The correction can be based on an infinite surface recombination velocity, which is approximated in the case of lapped or sawed surfaces. A single measurement requires about 2 minutes. The method can be applied to characterize silicon ingots by cutting samples with the mentioned size from delivered wafers if they are sufficiently thick. It can also be applied to process control where samples cut from float-zone crystals with high carrier lifetimes were heat treated in the respective furnace tubes before measuring the carrier lifetime [6.33]. The results for n-type samples vary between about 2 and several hundred μs and thus indicate whether the furnace tube was more or less contaminated with impurities which reduce the carrier lifetime after diffusing into the bulk of the sample.

Commercial equipment for determining the minority-carrier recombination lifetime without contacts by means of the PDC method using microwave reflection to monitor the resistivity decay has recently become available. They enable the determination of the carrier lifetimes at many points of a wafer and present a lifetime mapping of high lateral resolution (≥ 1 mm) on the computer screen, which can also be printed. These instruments are highly suitable for process control since the mapping of the results indicates regions on the wafer which have been contaminated during processing. The results of many measurements on one wafer surface can be averaged and enable the application of Statistical Process Control (SPC). This technique is more and more applied to guarantee the fabrication of high-quality devices. Once more, the surface recombination velocity of uncoated wafers limits the height of measurable carrier lifetimes even in wafers with thicknesses of 625 μm. In recent years the carrier lifetimes increased continuously because of improved materials and processes. Furthermore, the automatic reading instruments generate excess carriers in the sample by means of semiconductor lasers of high intensities. Consequently, medium high-level minority-carrier lifetimes are measured, which exceed, in general, the lifetimes achieved by low-level injection especially in the case of iron impurity contamination in p-type silicon. One possibility to overcome the problem of a high surface recombination velocity is the oxidation of the wafer to be measured. However, this oxidation must be performed in very clean furnace tubes in order to avoid indiffusion of additional impurities which would reduce the carrier lifetime. On the other hand, carrier lifetimes which are reduced by iron impurities in the bulk of the sample can be increased by an oxidation process due to gettering of the iron impurities within the oxide layer. Thus, the preparation of silicon oxide to reduce the surface recombination velocity can increase and decrease the carrier lifetime of the sample depending on the purity of the furnace tube applied and the nature of the impurity contamination in the sample to be measured. From experience obtained by recording lifetime mappings of thousands of wafers, it was deduced that the carrier lifetime in as-received and in processed p-type wafers is limited mainly by residual iron contamination. Interstitial iron is a very effective lifetime killer, especially at low injection levels. The carrier lifetime increases drastically with increasing injection level. The formation of FeB pairs in p-type silicon contaminated with iron increases the low-level carrier lifetimes and decreases high-level lifetimes, thus forming a crossover of the two injection dependences measured on the same sample.

Besides the recombination carrier lifetime, the generation carrier lifetime has frequently been measured to control the impurity contents in materials or to determine the amount of contamination during technological processes. This method was repeatedly applied to check the gettering efficiency of phosphorus diffusion, backside damage or any other gettering process (Sect.8.2). The generation carrier lifetime is measured on Metal Oxide Semiconductor (MOS) structures which, however, require an oxide layer on the wafer surface. The formation of the oxide layer needs a heat treatment which, in turn, can considerably change the carrier lifetime of the sample. Thus, this technology is not without problems for a quantitative characterization of the impurity content in samples. Although by comparing results on different samples prepared in the same way one may draw conclusions about the lifetimes. This, for instance, has been done by comparing generation carrier lifetimes with and without the application of a gettering process. Since the lifetime is measured at the position of the rather small MOS capacitor, the results will scatter more-or-less over the area of the wafer. For a more accurate comparison of results a larger number of measurements must be performed and averaged to achieve statistically relevant values.

6.2.3 Diffusion Length

According to (2.1) the squared diffusion length L^2 of the minority-charge carriers is correlated to the respective carrier lifetime τ [instead of t in (2.1)] with the diffusion coefficients D of the minority carriers (electrons or holes) in the sample as a proportionality factor. Therefore, measurements of the minority-carrier diffusion length can replace carrier-lifetime measurements.

A common method to determine the diffusion length is the Surface PhotoVoltage (SPV) technique. This method can be applied to wafers before and after the processes and, in general, does not require sample preparation. The measurement can be performed following the guidelines given by the American Society of Testing and Materials (ASTM) [6.35]. In contrast to PCD, this method requires polished or at least bright-etched surfaces. The sample surface must especially be prepared if the SPV signal is not high enough [6.35]. In general, a correction of the results considering a higher surface recombination velocity is not possible because this parameter is not known quantitatively for etched or polished surfaces. Consequently, the measured diffusion length will only fairly agree with carrier lifetime determined on the same sample and transferred to a diffusion length by applying (2.1). Furthermore, the injection level applied in the two methods may differ considerably (Sect. 6.2.2).

The method is based on the generation of excess charge carriers in the near-surface region of the sample by irradiation of light at various wavelengths. Due to band bending at the sample surface the minority carriers will be collected, and a photovoltage can be measured between the irradiated surface and the non-irradiated backside of the sample. By varying the wavelength of the irradiated light, the absorption depth is changed and the minority carriers have to diffuse more-or-less from the point of generation to the sample surface. By covering long distances the charge carriers can recombine and consequently reduce the measured photovoltage. The diffusion length can be determined by recording the light intensity for a constant photovoltage as a function of the penetration depth, which is proportional to the reciprocal absorption coefficient of silicon.

Besides commercial equipment measuring the minority-carrier lifetime, there is other equipment which determines the minority-carrier diffusion length by means of the SPV method. This equipment likewise enables a mapping of the results and is applicable for process control. Since all these instruments have been developed recently, a comparison of their results, and checks for their suitability for process control, have just begun.

A special scanning-photocurrent technique has been developed by *Föll* et al. [6.36] to monitor the contamination pattern of dissolved impurities of a whole wafer surface by recording a lifetime mapping. The measurement is based on the photocurrent which is generated by a scanning laser beam serving as a measure for the minority-carrier diffusion length. During recording, the wafer is positioned in an electrolytic cell. The revealed contamination pattern helps to identify possible contamination sources, as discussed in more detail in connection with the haze test in Sect.6.3. In comparison to common SPV measurements this method yields, in general, more-or-less enhanced diffusion lengths. This is explained by the higher excitation rate of excess carriers due to the employed laser [6.37] since high-level carrier lifetimes usually exceed more-or-less low-level carrier lifetimes [6.33]. In addition, the HF solution in the electrolytic cell reduces the surface recombination velocity.

As mentioned before, there are two main parameters which principally affect the measured carrier lifetime or diffusion length of a sample in addition to its contamination level: the height of the injection level for the excess-carrier generation and the surface recombination velocity.

The **injection-level dependence** of the carrier lifetime/diffusion length is a function of the lifetime-affecting impurity and its majority- and minority-carrier capture cross-sections. The slope of this function is a property of the respective impurity but the amount of the injection level is a measurement parameter which can often be varied within a given region of the light intensity. Since the injection level is the amount of the generated excess minority carriers $\Delta P_+(\Delta N)$ in relation to the (majority-) carrier concentration N of the sample without light, it depends also on the silicon sample used for the investigation and is varied by selecting another sample with different carrier concentration using the same light source intensity. The slope of the injection dependence can be calculated following the Shockley-Hall-Read model if the majority- and the minority-carrier capture cross-sections are known. The calculation becomes more complicated if the defect exhibits more than one deep energy level in the band gap. Both capture cross-sections of all energy levels must be known. For the comparison of carrier lifetimes of a sample measured by two or more types of equipment it is absolutely necessary to adjust the same injection level for all measurements or to select a sample with a negligible injection dependence of the carrier lifetime, for example, FeB-contaminated p-type silicon [6.38]. Otherwise the resulting carrier lifetimes can differ considerably (e.g., in Fe_i-contaminated samples). For very small injection levels ($\Delta P/N \leq 10^{-4}$) and for very high injection levels, which, however, can rarely be achieved, the carrier lifetimes approach a constant value. But not all commercial equipments enable measurements at very low injection levels because of a strongly decreasing signal height. Only in the case of a very low injection level the simple equations (6.1 and 2) are valid and the minority-carrier recombination lifetime in n-type silicon amounts to

$$\tau = \frac{1}{\sigma_h \nu_h N} \qquad (\Delta P/N \to 0) \qquad (6.4)$$

with the capture cross-section for holes, σ_h, the temperature-dependent hole velocity ν_h, and the majority-carrier concentration N. A corresponding equation holds for the minority-carrier recombination lifetime in p-type silicon.

The **surface recombination velocity** (SRV) can vary between about 10^4 cm/s for sawed or lapped surfaces and values below 1 cm/s for passivated surfaces. Oxidized surfaces exhibit values between 100 cm/s and below 1 cm/s depending on the cleanness and the orientation of the silicon surface. Since the SRV limits the maximum carrier lifetime which can be measured in clean silicon samples of limited thickness, as mentioned in Sect. 6.2.2, it should be reduced as far as possible to achieve accurate carrier lifetimes or diffusion lengths in thin samples. The following preparation techniques have been developed recently :

1) The sample is immersed in a diluted hydrofluoric acid during the measurement. The SRV is only partially reduced, there is still an residual influence if clean wafers of standard thickness are measured in the as-received state. This procedure is applied in the commercial equipment ELYMAT. It complicates the measurement of wafers with different diameters, since different cells are required. After removing the sample from the solution a high SRV is restored in short times.

2) Similar problems for the wafer handling arise with the immersion of the wafer in a solution of iodine in ethanol during the measurement. The reduction of the SRV seems to be more effective (<10 cm/s). This preparation technique was proposed for measurements using the LIFE TIME SCANNER WT-85 by *Horanyi* et al. [6.39]. For the measurement each wafer must be enclosed in a transparent plastic cover containing the solution and the wafer.

3) A common preparation technique is the thermal oxidation of silicon wafers which is one standard process in device fabrication and therefore must often not be performed in addition. In general, this technique achieves good results for (100)-oriented wafers and less satisfactory results on (111)-oriented wafers. Often, SRV exhibits remarkable inhomogeneities on the wafer surfaces generated during oxidation especially on (111)-oriented wafers. These can be improved by an additional techniques developed recently.

4) The main effect in oxidized wafer surfaces seems to be the repulsive force of the charged surface layer which is stable for extended times of weeks or even more. This charge can be enhanced by an additional corona charging of the oxide layers on both sides being effective especially on (111)-oriented wafers which become also more homogeneous as shown by *Schöfthaler* et al. [6.40]. Values for the SRV < 1 cm/s were reported. However, as mentioned already in Sect.6.2.2, an oxidation may considerably change the carrier lifetime in both directions. An unwanted additional contamination during the oxidation process will reduce the original carrier lifetime, on the one hand. On the other hand, a gettering process of iron in the grown oxide layer where the solubility of iron is enhanced compared to silicon, will increase the carrier lifetime. This gettering effect can be nicely demonstrated if several oxidation processes are performed successively interrupted only by lifetime control measurements. If iron is the main lifetime killer in the wafers the result can be considerably improved by performing several short oxidation processes. Finally, a constant lifetime will be achieved and it is also possible that this value is reduced again after further oxidations when additional contamination takes place during heat treatment.

5) In order to achieve the positive effect of passivating the wafer surface by simultaneously avoiding the negative effect of gettering and handling the sample in a liquid, we recently started to study various RT coatings of bare silicon surfaces. Several transparent and commercial varnishes of different composition and purpose were applied and some of them were found to increase the lifetime because of decreasing the SRV. As a consequence of the chemical analysis a new varnish was mixed which was composed of colophony with up to 10 % of iodine dissolved in 2-propanol [6.41, 42]. The application of this varnish yields much better results than the simple oxidation, and even better values than obtained with corona-charged oxides. In a direct comparison of both techniques performed on the same wafers in succession it turned out that in some cases the charging durations (estimated by experience) were not sufficiently long, resulting in much better values with the varnish technique (results were presented in [6.42]). So far the only remaining problem is the degradation of the surface passivation after enhanced storage times which is not too detrimental to the measurement of a single lifetime mapping but avoids the automatic measurement of series of wafers in succession, for example, over night. The handling of the wafer is rather simple. The varnish can be painted on the etched and rinsed wafer or the wafer is dipped in the varnish. Even spraying of the varnish is possible. After a rather short drying period

the wafer can be put on a sheet of filter paper to avoid that the sample holder of the equipment gets dirty and the measurement is started.

Most of the commercial equipment recently developed for the mapping of carrier lifetimes use fairly strong laser excitation and therefore do not determine low-level carrier lifetimes. All the equipment is expensive but exhibits the advantage of revealing the resultant pattern for the whole wafer within fairly short measurement times. All instruments are computerized and enable several statistical evaluation techniques, which includes the application of SPC to enable improved process control. The results, including the mapping, can be stored, which helps to control the development of the contamination with time in various equipment.

In order to obtain more information about the chemical nature of incorporated transition-metal impurities in a sample, *Bergholz* et al. [6.37] reported on a characteristic influence of the iron-boron pairing reaction to diffusion-length measurements. Interstitial iron exhibits a much larger capture cross section than iron-boron pairs. Since the pairs can be dissociated, for instance by a thermal treatment at 200°C for several minutes [6.43], the diffusion length in a p-type sample can be measured before and after this heat treatment. If the diffusion length decreases by about a factor of two due to heat treatment, it is most likely that the lifetime is mainly determined by an enhanced iron contamination of the sample. The respective iron concentration can be calculated if the capture cross section of iron is known or, in a more direct way, if the diffusion length was previously calibrated against the iron concentration determined by applying DLTS. The latter method is more reliable since the change in the carrier lifetime and hence in the diffusion length caused by pairing is highly dependent on the excitation level during measurement. It can be positive or negative, or even independent on pairing, as mentioned before.

6.3 Detection of Precipitaties

Some of the main impurities which are introduced during manufacturing can hardly be kept dissolved in silicon even by quenching the wafers from HT (Cu, Ni) (Sects.2.1 and 3.4). They precipitate almost quantitatively and, therefore, they must be detected as precipitates. For this purpose, methods could be applied, which were discussed in connection with the detection of the total impurity content (Sect.6.1). However, it turned out that these methods are expensive and time consuming, which contradicts the requirements for material and process control in a production line [6.38].

6.3.1 Haze Test

An inexpensive and quick method to reveal precipitates is the so-called **haze test**. The technique consists of a rather short heat treatment of the wafer at about 1050°C for several minutes followed by a medium-fast cooling period and a subsequent preferential etch of the wafer. During the cooling period, fast diffusing interstitial transition metals which are able to form precipitates via a homogeneous nucleation mechanism diffuse to the sample surfaces where they precipitate. The haze-forming transition metals are Co, Ni, Cu, Rh, and Pd. The precipitates accumulate at the wafer surfaces and can be revealed by the preferential etch, and inspected visually in a spot light. The incident light is scattered in non-reflecting directions at the accumulated shallow etch pits (also called **saucer- or S-pits**) formed at each surface precipitate. Therefore, contaminated wafer areas appear bright, whereas uncontaminated areas remain dark. For a semi-quantitative evaluation of haze, the strength of the scattered light can be measured. It is a function of the density of etch pits on the surface of the wafer and thus correlated to the impurity concentration in the wafer [6.43]. In order to obtain repeatable and comparable results, the light scattering on a clean polished wafer without haze is adjusted to 100 arbitrary units. The scattering of light on hazy surfaces can then amount up to several hundred arbitrary units. By this method the sensitivity of the measurement approximates that of visual inspection. So haze characterized by a light-scattering value of 100 a.u. cannot be visually detected, whereas haze characterized by 101 units can already be seen in a spot light. Although not yet accurately determined, this sensitivity corresponds to a detectivity of less than 10^{12} cm^{-3} for nickel, and less than 10^{13} cm^{-3} for copper [6.37, 43]. For the present requirements of modern technology, this sensitivity is sufficiently high to control the production of haze-free wafers (less than 101 to 102 a.u.) and promises no yield losses due to nickel or copper precipitates in the present state of microminiaturization.

Similar to carrier lifetime measurements for the detection of disolved impurities, the haze test does not allow determination of the chemical nature of the precipitates or their accurate concentration. Nevertheless, this test is applied frequently to control materials before and after technological processes since it is inexpensive and fast, and enables sufficiently reliable results. From the limited number of haze-forming transition metals (Co, Ni, Cu, Rh, Pd) only copper and nickel are known as wide-spread impurities in device production. Iron as a further main impurity does not form haze during the heating cycle used for the haze test without the presence of heterogeneous nucleation centers formed by other impurities such as copper or nickel. However, after performing a subsequent annealing process, for instance, at about 500°C for 30 minutes, about 80% of the iron precipitates and strong haze can be observed [6.44]. For high impurity concentrations the etch pits of copper and of nickel haze can be distinguished by their microscopic structure (Figs. 3.12, 13, 18, 19). For low impurity concentrations they can be distinguished by applying the New etch (Sect. 3.4.2) with different compositions to reveal haze.

In a production line the chemical nature of the contamination, which leads to the formation of detrimental precipitates in the electrically active zone of a device, is not really important. The main problem is to discover the origin of the impurity contamination and to eliminate it. For this purpose the distribution of haze on the whole surface of the contaminated wafer can be helpful in identifying the proper process where the contamination takes place, due to their characteristic finger-prints. Replating of impurities from contaminated chemicals, for instance, causes a homogeneously distributed haze on the whole wafer surface. A contamination by handling systems causes characteristic finger-prints of haze structures which, in general, can easily be identified. Furthermore, contaminated edge rounding systems will cause ring-shaped haze, which is limited to the edge of the wafer including the flat portion. In all of these cases, and in many more, it is not really important whether the haze is due to nickel or copper contamination, or to both. However, it is important to avoid this contamination source by cleaning the respective system or replacing the contaminated chemicals by pure chemicals. Then the haze test can be applied once more to verify the success. Beside measuring the light scattering at the shallow etch pits, the etch-pit density can be counted by means of a microscopic inspection. More conveniently, the Light Pit Density (LPD) can be determined with the aid of laser scanning methods which are commercially available in various modifications.

It was shown nicely by *Hourai* et al. [6.44] that the etch-pit density after an intentional and definite contamination of samples with Fe, Ni and Cu, and a subsequent diffusion, strongly depends on the respective surface metal concentration. The etch-pit densities and, after a second oxidation process, the densities of oxidation-induced stacking faults which obviously formed at the nuclei of the metal precipitates, exhibit characteristic threshold surface impurity-metal concentrations, where they abruptly increase by several orders of magnitude. These critical surface impurity-metal concentrations agree for the formation of shallow etch pits and for the formation of surface stacking faults. They amount to about 10^{11} cm^{-2} for Ni, 10^{12} cm^{-2} for Cu, and 10^{13} cm^{-2} for Fe. From this experimental finding it can be deduced that upper limits for stacking fault densities require different upper limits for Ni, Cu and Fe contents in the bulk or at the surface of a sample. However, if we determine the density of haze, which is directly correlated to the density of shallow etch pits due to Cu and Ni , we need only one definite limit to keep the final density of stacking faults at a sufficiently low level for avoiding detrimental effects on the device performance. This specification for haze density is correlated to the preceding heat treatment which therefore must be specified in addition.

As mentioned before, the haze density versus the metal concentration exhibits a proportional and a retrograde slope. The etch pit density first increases with the metal concentration and then decreases at higher concentrations due to ripening effects of the larger precipitates at the expense of the smaller ones in their neighborhood. This was also observed by *Hourai* et al. [6.44] for Ni. But in spite of the decreasing etch pit densities at higher metal concentrations the densities of stacking faults still increased with increasing metal concentrations.

There are a few other defects that form "haze" in a more general sense. They interfere with haze in the restricted definition used in this monograph as an accumulation of etch pits at the wafer surfaces due to precipitation of fast diffusing transition metals (Sect.3.4.2):

(i) The most common interference is "haze" due to oxygen precipitates. In contrast to the "real" haze which is observed only at the surface of a wafer after heat treatment, the oxygen precipitates, if present, can be observed in the whole volume of the wafer without performing an additional heating cycle. Because of the inhomogeneously incorporated oxygen in pulled crystals, the etch pits of the oxygen precipitates appear striated on the surface of the wafer. Oxygen precipitates are usually not observed in ingots, but they can be formed in wafers during extended heat treatments. Therefore, oxygen precipitates visualized on the surface of a wafer by preferential etching can easily be distinguished from transition metal haze.

(ii) A second interference can be observed in highly doped crystals with resistivities in the region of $m\Omega \cdot cm$. Approaching the limit of solubility of the doping elements, precipitation can occur during the pulling of the crystal from the melt. These precipitates can also be visualized by preferentially etching the wafers. Once more, the precipitates are observed in the bulk of the wafer and on its surfaces, even without performing the heating cycle necessary for revealing transition-metal haze. In this case the etch pits of the precipitates are very small and well separated from one another. They are statistically distributed on the surface of the wafer and can hardly be detected by measuring the scattered light. Therefore, this kind of "haze" can also be distinguished from real transition-metal haze, especially by preferential etching of as-received wafers without any heat treatment.

(iii) After plasma etching, wafers can exhibit hazy surface structures if silicon has been removed during this process which should remove only the oxide layer. In general, the process is known for process control and, therefore, it should be possible to determine whether silicon has been removed or not. To obtain a better distinction of transition-metal haze from any other hazy appearance, the wafer can be scratched intentionally, for instance by means of a diamond, before heat treatment is performed. This is often done to mark the respective wafer with a number for its identification. After heat treatment and preferential etching the transition-metal haze will be gettered in the area surrounding the scratches, however, this "haze" structure will not be gettered.

(iv) A final phenomenon which leads to an interference with transition metal haze is still not understood well. On wafers which were heat treated in furnace tubes exhibiting cracks in the tubes or in the gas-supply system, a very strong haze of 600 to 700 a.u. can be observed after preferential etching. This kind of "haze" appears on the whole wafer surface or on large areas often showing characteristic stream-lineshaped patterns. It is not gettered, for example, by scratches due to wafer markings, and it is not reduced by cleaning the furnace tube. In every case observed so far, this kind of "haze" was caused by cracks in the furnace tube or in the gas system, even if the cracks could not be found before replacing the tube. This "haze" disappeared immediately after changing the tube or repairing the gas-supply system. It is assumed that eventually atmospheric CO_2 reacts with the silicon surface at HT by forming silicon carbide which is then revealed by preferential etching. In general, this "haze" is much stronger ($600 \div 700$ a.u.) than common transition-metal haze (<180 a.u.) and it is additionally distinguished by a strange pattern. After preferentially etching the wafer, the front side of the wafer cannot be distinguished from its back side. This "haze" is a reliable criterion for the detection of broken furnace tubes and leaky gas-supply systems.

Since there is no universal method which allows detection of all kinds of impurities with high sensitivity, to identify their chemical nature, and to determine their concentration and distribution in the wafer in short measurement times, various methods must be used to obtain the required information. A comparison of results obtained by applying different techniques to solve the same problem is summarized in the review paper by *Bergholz* et al. [6.37]. Further details on the requirements of modern technology and the application of the various measurement concepts are discussed in Chap. 7.

7. Requirements of Modern Technology

With the increasing size of highly integrated circuits due to their ever increasing complexity – bearing in mind the diminishing dimensions of structures, and diffusion depths, in VLSI and ULSI – the detrimental effect of lattice distortions within the very small and shallow electrically active zones has increased considerably. Most of the lattice distortions are caused by precipitated transition-metal impurities. They are introduced into the wafers by diffusion of unintentional surface contaminants during the large number of process steps required to fabricate a modern device. In order to achieve sufficiently high yields, the number of faulty structures must be reduced drastically. Therefore, the tolerable impurity concentrations shrunk considerably. As a consequence, new specifications limiting the concentrations of impurities in device fabrication were set up. They concern mainly the concentrations of iron, nickel, and copper. In order to control these new specifications the suppliers of polished wafers and their customers had to agree on suitable measurement techniques.

The requirements on the purity of the wafers before and after processing are quite different for the various devices and depend on the respective electrical parameters and demands. Examples can be found in the literature. Even very low concentrations (10^{11} cm^{-3}) of mid-gap impurities have been found to result in a total yield loss in power devices [7.1], gold in concentrations of less than 10^{12} cm^{-3} caused faulty photodiodes [7.2], and titanium severely degrades the performance of solar cells if the concentration exceeds 10^{11} cm^{-3} [7.3]. All these examples suggest maximum concentrations for dissolved impurities, which should not be exceeded to avoid yield losses. However, it is much more difficult to determine maximum concentrations of those impurities which precipitate during cooling of the wafers after a HT process, such as iron, nickel and copper. These are the main impurities in device production. The process of impurity precipitation depends considerably on the cooling conditions and on the concentration of nucleation centers in the bulk and at the surfaces of the wafer. As a consequence, a large amount of small precipitates or a smaller number of extended precipitates can be formed in wafers with equal impurity concentrations according to the last thermal treatment. In many production lines, more or less unintentional gettering phenomena such as internal gettering or phosphorus gettering are incorporated. They drastically reduce the precipitation of fast-diffusing transition metals in the critical electrically-active zone of a device. If processes are changed by using modern equipment and techniques, the unintentional gettering can be affected and the tolerable impurity concentrations may decrease. A typical example is the replacement of phosphorus diffusion by phosphorus implantation, followed by a rapid thermal processing. This usually reduces the gettering effect and, as a consequence, the tolerable impurity content.

In order to determine the upper limit of the tolerable impurity content, an intentional contamination of a few test wafers is not successful. During HT processes several impurities such as copper or iron evaporate from the contaminated wafers and contaminate the furnace tubes and environments. As a consequence, the mean impurity level is increased. This increase is much less for a few contaminated wafers than expected after processing a large number of contaminated wafers. Therefore, the specifications for tolerable impurity contents must be derived from experience, mostly from faulty devices, unintentionally produced and subsequently analysed.

From the total losses of yield in earlier days, caused by highly iron-contaminated wafer surfaces which passed into the production line without control, it is known that iron concentrations in the region of 10^{13} to 10^{14} cm^{-3} are not tolerable even for bipolar integrated circuits. They are not very sensitive to impurity contaminations. As a consequence, the upper limit for the iron contamination, at that time, was tentatively fixed to 10^{13} cm^{-3} after surface contaminants had diffused into the bulk of the wafer. Since iron precipitates almost quantitatively during common technological diffusion processes, the upper concentration limits for copper and nickel, which likewise precipitate after diffusion, have been set at the same order of magnitude. Since that time, periodical checking of the impurity contents in wafers before and after processing was performed and resulted in a better knowledge of common impurity levels.

For the purity of De-Ionized (DI) water, detailed specifications and guidelines have been developed for the devices to be fabricated and the intended microminiaturisation [7.4]. The respective impurity concentrations in ppb concerning the transition metals which are usually found in ultrapure water, are listed in Table 7.1.

Specifications or guidelines for the purity of wafers have been established for the present and future device fabrication (Chap.9). As a rule, the accurate impurity concentrations for the various transition metals are not known in detail. The acceptable and attainable impurity concentrations for the present fabrication of megabit memory devices are estimated to be of the order of magnitude of $<10^{12}$ cm^{-3} iron in the bulk of the wafer.

In the past, when transition-metal impurity contents were controlled, a significant reduction of the number of different contamination sources was achieved. This will be discussed in more detail in Sect.7.1. Another method

Table 7.1. Specified metal concentrations (specs.) in DI water [7.4]

Metal	1985 specs. for 256K DRAM [ppb]	1988 guidelines for 1M DRAM [ppb]	guidelines for 4M DRA VLSI [ppb]
Cr	0.1	0.05	0.002
Mn	0.5	0.05	0.002
Fe	0.1	0.1	0.002
Ni	...	0.05	0.002
Cu	0.1	0.05	0.002

to reduce the impurity concentration in the electrically-active zone of a device is to enhance the gettering efficiency. Problems which arise with gettering transition metals, the application of various gettering mechanisms which were developed in the past, and finally the determination of the gettering efficiency are discussed in Chap. 8.

7.1 Reduction of Contamination

To reduce contamination or to eliminate the contamination sources we need to search for the origin of the respective impurities. For a manufacturer of electronic devices who purchases polished wafers from specialized suppliers, there are two main groups of contamination sources: the as-received polished wafer and the processes which will subsequently be applied to device production. Due to past experience, control of both the wafer ingots and the processes is recommended as much as possible. Since the sample preparation differs in both groups, they are discussed separately in the following two sections.

7.1.1 Control of Ingots

The transition-metal impurity concentration in the *bulk* of silicon crystals is, in general, below the detection limit of DLTS (about 10^{-4} of the respective charge-carrier concentration) and hence it will be of the order of magnitude of 10^{11} cm^{-3}. To determine this bulk contamination in a wafer, it must be etched to remove all surface contaminants. Subsequently, the wafer must be heat treated to dissolve all precipitated transition metals, and it must be quenched to RT to keep the impurities dissolved if possible. During quenching the sample from HT, only the fast-diffusing transition metals which are the haze-forming impurities, such as Co, Ni, Cu and Pd, precipitate again. All other transition metals remain dissolved and can be detected by DLTS within the detection limit mentioned above. Additional preferential etching of the wafer reveals haze which, however, should also be below the detection limit (Sect.6.3). The only transition metal which is occasionally observed in the bulk of CZ-grown crystals is iron, exhibiting concentrations, in general, below 10^{11} cm^{-3}. But it cannot be absolutely excluded that a small iron content may sometimes originate from a vapor-phase contamination in the furnace tube during the heat treatment which is needed for the sample preparation. In conclusion, present-day crystal-pulling technology for growing silicon single crystals guarantees material of high purity. A regular routine control of the bulk contamination in as-received wafers is therefore not required at the moment and would be too expensive.

A contamination of the *surface* of the wafers begins with the mechanical treatment of the crystal rod by rounding, flat grinding, sectioning of wafers, edge rounding, and finally lapping, etching and polishing of the wafer surfaces. Although intermediate etching processes should quantitatively remove these surface contaminants, a replating or cross contamination cannot be excluded. Various contamination sources, mainly of iron, copper and nickel, were detected in the past decade, such as replating of iron during the ultrasonic cleaning process, replating from pumping systems for chemicals, replating of copper from impure chemicals, mechanical contamination during etch rounding and during handling the wafers.

Several companies detected and eliminated these contamination sources in the past by applying a combination of haze test and DLTS measurements. In contrast to the sample preparation needed for measuring the bulk impurity concentration, the wafers must not be etched before diffusion since this would remove the surface contaminants. For sample preparation a diffusion process at 1050°C for 20 minutes in an inert-gas atmosphere is proposed, followed by a subsequent cooling period to RT within 4 minutes. Within the diffusion duration, all transition metals which diffuse as fast or faster than iron are able to diffuse from the reverse side of the wafer to its front side since the respective diffusion lengths of the impurity atoms exceed a common wafer thickness. During the cooling period, the haze-forming transition metals diffuse to the surfaces of the wafers and form precipitates which can be revealed by a subsequent preferential etching. After this etching the wafer is then visually inspected for haze in a spot light. If haze is observed, it is evaluated, for example, by measuring the scattered light, as reported in Sect.6.3.

Before or after this haze inspection, a segment is cut from the edge region of the wafer for a second heat treatment at 1100°C for 10 minutes

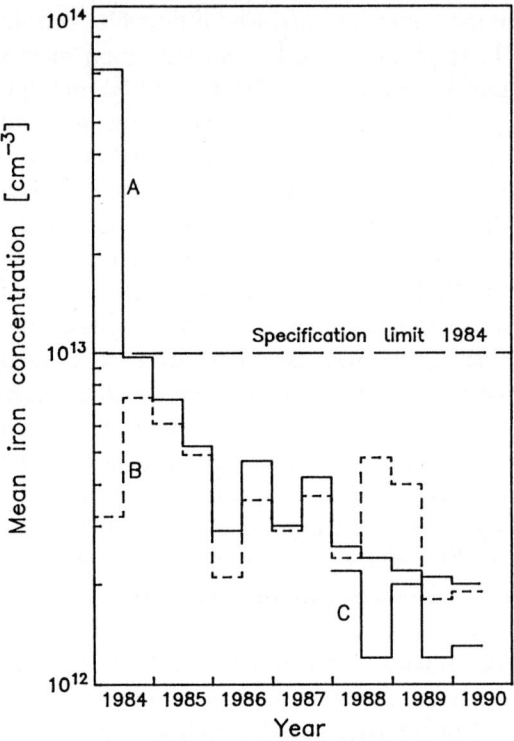

Fig. 7.1. Mean iron concentrations determined by DLTS after diffusion of as-received polished wafers originating from three different suppliers (A, B, C) as a function of the year of delivery

to dissolve all of the precipitated transition metals. During quenching this sample to RT all transition metals except the haze-forming metals are kept in solution for their detection by DLTS. The samples are etched to remove several μm from the sample thickness, and Schottky contacts are evaporated. From the recorded temperature scan the concentration of iron is determined in p-type wafers. In general, impurities other than iron remain below the detection limit of DLTS.

The iron contamination on top of the as-received wafers decreased considerably in the past decade after the introduction of control measurements, and a specification for upper limits for the iron concentrations. The functional dependence of the decrease of the iron content in the bulk after diffusion along with the year of delivery is depicted in Fig. 7.1. Mean results obtained by averaging the measurements within half a year are plotted versus the date of delivery for three different suppliers. Since 1985 the mean concentrations of iron are below the specified upper limit of 10^{13} cm^{-3} valid at that time. But from time-to-time single results exceeded this limit, and therefore a continuous ingot control is still required. In general, a decreasing tendency of the mean iron contamination is observed up to the more recent results which exhibit the lowest iron concentrations since the beginning of the routine measurements. Today, the mean values for the iron concentration in a material from all three suppliers are below 10^{12} cm^{-3}. From time to time it will be necessary to reduce the specified upper limit of the iron-impurity content in order to reach the requirements of future technology, as discussed in more detail in Chap.9 (Fig.9.1).

A method to establish suitable specification limits is the application of Statistical Process Control (SPC) [7.5]. From the statistics of a series of measurements, Upper Control Limits (UCL) for the average of the measured parameter UCL_x, which, for example, may be the iron concentration in as-received wafers after diffusion (x) and from the respective standard deviations (s), UCL_s can be calculated. The control chart together with the calculated process capability index c_{pk} reveals a survey of the development of the parameter (iron content) with time, and enables one to decide whether a process (e.g., the production of polished wafers) is well-controlled or not.

Figure 7.2 depicts the iron concentrations in as-received wafers from one supplier after diffusion of surface contaminants into the bulk of the wafers monitored for the period from 1990 until the middle of 1991. In the starting period at the beginning of 1990 the iron concentrations were still high with single results exceeding the specification limit of 10^{13} cm^{-3}. Reference number *1* indicates that these results have been claimed. In the interim period (February 1990) the contamination source could be detected and avoided. The following results (not connected to the previous ones) are well below the specification limit and lead to a calculated upper control limit for the average iron content of $2.1 \cdot 10^{12}$ cm^{-3} at the end of the first control chart (September 1990). In the second control chart, the upper control limits are indicated and the average values of the iron concentrations are well below the respective upper limit. However, the standard deviation in one point exceeds the UCL_s limit in April 1991. This point is marked by the reference number *2*. The reason was that the iron content measured in one point of the wafer exceeded the UCL_x limit with $2.6 \cdot 10^{12}$ cm^{-3} but not the presently valid specification limit of 10^{13} cm^{-3}. Since the iron content for the second measurement point was low, the average did not exceed the UCL_x limit. Although the second control chart could not be completed, it seems likely that both the new UCL_x and the UCL_s limits will be smaller than those calculated for the first chart. The new results computed from the 20 measurement points, indicated in chart 2, reduce to $1.7 \cdot 10^{12}$ cm^{-3} ($2.1 \cdot 10^{12}$ cm^{-3} in chart 1) for UCL_x and the new UCL_s shrank to 0.77 (0.85 in chart 1).

In conclusion, SPC helps to define new specification limits which can be accomplished by the supplier. In our example this would be an upper iron

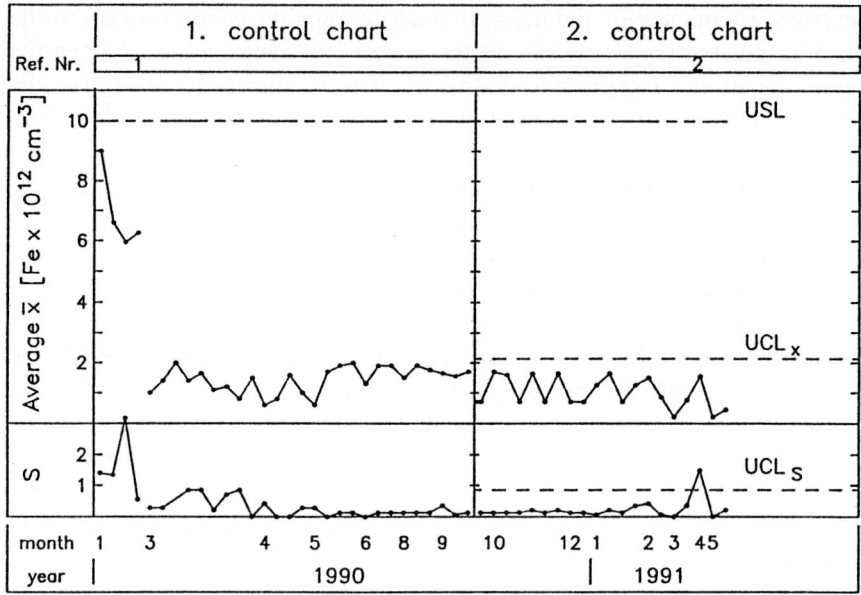

Fig. 7.2. SPC control charts for iron concentrations measured by DLTS in as-received polished wafers from supplier D after diffusion of surface contaminants into the bulk of the wafers. Results from the period 1990/1. USL: upper specification limit, UCL_x and UCL_S: upper control limit for the iron content x and the standard deviation S, respectively, as calculated from the 30 results recorded in the first control chart (*left side*). The Cpk value (not indicated) amounts to 8.75

concentration of $2 \cdot 10^{12}$ cm^{-3}. A lowering of the upper specification for impurity concentrations results in an immediate reaction of the customer in case of exceeding this limit and helps to avoid a longer period of unnoticed increasing of the impurity contamination. This has repeatedly been observed in the past at different companies (Fig. 7.1 supplier B, 1988/ 89). By applying SPC for a contamination control it might be possible to reach future demands for strongly reduced metal contamination, which is discussed in more detail in Chap. 9 (Fig. 9.1).

If haze is observed on as-received wafers it often originates from spot-like contamination sources near the edge of the wafer as indicated, for example, in Fig 3.20. These contaminations are mostly due to handling equipment, vacuum tweezers, or carriers of various design which were contaminated with time and should be cleaned periodically. In many cases the respective contamination sources can be directly identified by their typical finger-prints. These characteristic haze patterns are better revealed if the duration of the heat treatment applied to form the haze is shortened and the wafer is then quenched to RT. After preferential etching the finger-prints are clearly visible due to the reduced lateral outdiffusion of the impurities.

In this case the light scattering cannot be determined quantitatively by means of our equipment since the hazy areas are considerably reduced in size and only partially cover the window of our optical arrangement (about 1 cm in diameter). The problem can be overcome if a second sample is used for the inspection of the fingerprints after a shorter heat treatment. The second sample would also show the repeatability of this kind of contamination.

After performing the haze inspection and sectioning the samples for the DLTS measurement, the same wafers can be used to determine the oxygen concentration in CZ-grown crystals by infrared absorption spectroscopy since this parameter is mostly specified, too. The method reported in the new version of DIN 50438 part I (1994) which corrects the scattering of the infrared radiation at the unpolished back side of the wafer also corrects the additional scattering, which appears due to haze on the polished front side.

A problem arises with the sampling procedure for the ingot control. Since the reported method destroys the wafers, their number must be limited but they should still be representative for the whole shipment. A convenient compromise is to check two or three wafers of a shipment. If the iron contamination exceeds the specified value, a second Schottky contact is measured, in addition to controling the homogeneity of the iron contamination. If one of the wafers or both do not meet the requirements of the respective haze specifications, two additional wafers are investigated to check the repeatability of enhanced contamination. From the four or five results it is decided whether the shipment is accepted or refused. In the case of refusal, the supplier will, in general, once more control whether the refusal is accepted or not, which again increases the number of results.

In the past, control measurements of the minority-carrier recombination lifetime on silicon ingots were performed only if this parameter was specified for the fabrication of special devices such as solar cells or high-power transistors and thyristors. In former times thicker test wafers were separately prepared for a reliable measurement of the lifetime. With the recent development of automatic-reading equipments which enable to record mappings of the lifetime/diffusion length in short times an increasing application of these instruments for ingot- and process control is observed. So it is possible to detect the traces of various contamination sources in the lifetime mapping. They could be identified due to their characteristic finger-prints which formerly was done only by haze tests. In many cases a wafer preparation is not necessary, which allows to inspect a higher number of wafers. Since the lifetime scale can be adjusted at convenience a very high sensitivity can be achieved even without reducing the SRV.

From experience it is known that in most cases iron is the lifetime-limiting impurity in as-received wafer ingots. If the wafer is suitably prepared it is possible to determine even its iron content from a lifetime measurement due to modern software, just by pressing a bottom. In this way the DLTS measurement can be replaced by lifetime mappings which require less time and yield much more information. This is because it records the whole wafer area and not only one selected point corresponding to an area of about 1 mm^2.

The haze test, however, could not be replaced by a lifetime mapping so far. The carrier lifetime is not very sensitive to precipitated impurities and depends only on their density and not on the impurity concentration. Haze mappings do not agree with those mappings of the carrier lifetime. This was tested on many wafers. So far only one exception was observed: if the whole wafer surface was implanted with argon preceding to the haze test both mappings agree completely. It is assumed that the precipitated impurity in this case reacts with implantation defects by forming new defects which considerably reduce the carrier lifetime. However, this preparation technique would be too expensive to replace the simple haze test.

In general, the reported tests will be sufficient to decide whether the surface contaminations of the as-received wafers meet the requirements for the intended production of electronic devices. The tests should be performed in an environment of utmost cleanness to avoid any cross contamination during the required sample preparation. Usually the same facilities, especially the furnace tube or the rapid thermal processing equipment, will be used to perform the sample preparation for process control. This increases the possibilities for cross contamination since impurities may now be present, which are not found in as-received wafers. Often the sample preparation for controlling as-received wafers and processes are carried out simultaneously to reduce the preparation time. In this case some experience is required concerning possible impurities in processed material and their common concentrations to avoid faulty results due to cross contamination. All equipment, including furnace tubes, vessels for etching, tweezers for handling and all carriers, must be carefully cleaned periodically. The results obtained on as-received wafers may often be a criterion for the cleanness of the equipment employed for sample preparation since these wafers are often free of haze and do not exhibit dissolved impurities, with the exception of small quantities of iron.

7.1.2 Process Control

In principle, the same methods which were employed for the control of ingots can be applied to process control. In contrast to the preparation for the ingot control, now as-received polished wafers with suitable specifications (p-type silicon, $5 \div 20\,\Omega\cdot\text{cm}$) are briefly etched to remove native oxide layers and surface contaminations which would interfere with process-induced contaminations. Subsequently, the wafer is exposed to the process to be controlled, for example, a temperature process in a definite furnace tube. The following sample preparation equals that described for use in ingot control.

The haze inspection provides information about the contamination with fast diffusing impurities, including their distribution on the wafer surface. From DLTS measurements the contamination with moderately fast and slowly diffusing impurities is deduced. The DLT spectra enable a determination of the chemical nature of the impurities and their concentrations.

Additional carrier-lifetime measurements performed may complete the information or they may replace the time-consuming DLTS measurements if the resulting lifetimes are sufficiently high. Instead of lifetime measurements diffusion-length measurements can be carried out on the processed wafers, for example, by applying surface photovoltage measurements. Both carrier-lifetime and diffusion-length measurements can be performed by employing modern automatic reading equipment, which saves measurement time.

Carrier-lifetime and diffusion-length measurements enable the detection of iron in p-type samples if measurements are performed before and after a heat treatment which dissociates the iron-boron pairs [7.6]. The respective procedure has been discussed in Sect.6.2.2. Furthermore, carrier lifetime or diffusion-length mapping [7.7] can be recorded to control the distribution of the impurity contamination within the whole extension of the wafer (Sects.6.2.2, 3).

After the detection of a contamination source its elimination will, in general, be not too difficult. At least it should be possible to reduce the amount of impurity contamination. Depending on the respective process, the chemical nature of the impurity and its distribution on the wafer will usually be sufficient to find out the real contamination source. Common contamination sources are chemicals containing impurities which can be replated on silicon from their solution, contaminated handling systems, equipment, furnace tubes, and, finally, unintentional sputtering or evaporation from the surroundings.

If as-received wafers are already contaminated, the manufacturer of the wafers has to perform corresponding process controls to detect the respective contamination source or sources. The application of additional detection methods such as TXRF (Sect.6.1) may be helpful to limit the number of processes which can be responsible for this contamination. Thus, it may be interesting, for instance, to determine whether an impurity originates from the contaminated front sides (by means of TXRF), from the back side, or from both sides (by applying the haze test).

Almost all processes can become sources for unwanted wafer contamination and therefore periodical control measurements are recommended. Well-known contaminations originate from the cleaning procedure, from furnace tubes, rapid thermal processors, epitaxial reactors and plasma-etching equipment, implanters, from any handling system and, finally, even from interim control measurements. In many cases the cleaning of the respective equipment will reduce this contamination. Sometimes modifications in new systems are necessary to avoid contamination. This was observed, for instance, in the early implanters and plasma-etching equipment where sputtering effects from the surrounding material lead to high iron contaminations. In this case the cleaning of the equipment does not help to reduce the contamination. The source of the iron contamination must be found in the receptor, which gives rise to the sputtering of iron onto the sample. In the same way, molybdenum contamination in epitaxial reactors cannot be reduced by cleaning the system, since the contamination originates from valves containing molybdenum, which must be replaced by other clean valves. Contamination originating from chemicals containing impurities require the application of other cleaner chemicals.

In conclusion, for each contamination, suitable consequences must be realized for its reduction. Occasionally, there may be several effective contamination sources in one process, which must be subsequently eliminated one by one.

8. Gettering of Impurities

Since the early 1960s gettering has been a common method to reduce unintentional metal impurities in the electrically-active zones of semiconductor devices. Starting with the extrinsic gettering by means of phosphorus diffusion [8.1] various alternative methods were developed, mostly by utilizing a trial-and-error technique. The effectiveness of a gettering process was usually estimated from improved reverse current-voltage characteristics, from increased minority-carrier lifetimes, or from increased yields in device production. Since that time various gettering processes were developed and applied in device production such as boron or gallium diffusion [8.2], lattice distortion on the reverse side of the wafer by lapping [8.3], grinding [8.4], scratching [8.5], sand-blasting, sound stressing [8.6], laser irradiation [8.7, 8], and ion implantation of various elements [8.9].

In the early period, intentional gettering techniques were needed because the unintentional gettering by grown-in dislocations in silicon crystals was lost after the introduction of dislocation-free grown crystals. The gettering efficiency of various lattice defects were compared with each other [8.9] and the gettering effect of sample annealing [8.10] with the addition of chlorine [8.11], HCL [8.12, 13], or TCE (TriChlorEthylene) [8.14-16] to the ambient atmosphere was discovered. In addition, the gettering effects of surface layers such as nitrides [8.17, 18] and polysilicon [8.19] were investigated. Finally, internal or intrinsic gettering [8.20-22] was developed applying various modifications of the high-low-high temperature pre-annealing processes. This internal gettering is based on the formation of gettering centers in the bulk of a wafer outside of the electrically-active zone. For this purpose three temperature processes are usually applied to wafers originating from CZ-grown crystals which contain higher amounts of oxygen. The following three steps of heat treatments are commonly used to generate the gettering centers:

(i) A HT process to create an oxygen-denuded zone beneath the polished front side of a CZ-grown wafer where the active zone of the device is located.

(ii) A low-temperature annealing of the sample to create a sufficiently high number of nuclei for oxygen precipitation by a homogeneous nucleation mechanism.

(iii) Finally, a second HT process to grow oxygen precipitates at the nuclei created by (ii). During oxygen precipitation the silicon lattice is stressed in the vicinity of the growing precipitates due to the larger volume of the oxide compared to the silicon host lattice. As a consequence, excess silicon self-interstitials are emitted. In the case of higher supersaturations, the self-interstitials give rise to the formation of bulk extrinsic stacking faults. Gettering of impurities is then effected by stressed oxygen precipitates or by stacking faults. Both can act as nucleation centers for metal impurities if they become supersaturated during cooling the wafer from HTs.

In addition to the multiplicity of gettering methods, combinations of different techniques were applied simultaneously or subsequently during device fabrication [8.23, 24]. Furthermore, the effectiveness of different gettering techniques were compared [8.25-27], and their influence on defined metal impurities was investigated [8.28]. Many researchers checked the effectiveness of gettering processes by measuring the reduction of unintentional impurities introduced into the wafers during processing [8.29]. In other investigations defined metal impurities have been introduced intentionally [8.30] to study their reduction by gettering processes. The physical mechanism of gettering by phosphorus diffusion was recently investigated in more detail, yielding many new results [8.31-33]. The structures of precipitated impurity metals were investigated and identified by TEM in several recent publications [8.34]. The formation and behavior of oxygen precipitates were studied [8.35] since they are important for a better understanding of the internal gettering mechanism.

8.1 Gettering Mechanisms

The myriad of published technological details on gettering in silicon cannot be reviewed in this monograph. However, some basic statements such as the definition of gettering and some general deductions from this definition may help to provide a better understanding of the physical mechanism of gettering and its connection to the properties of transition-metal impurities [8.36]. In this monograph, gettering is defined as follows: *Gettering is the dissolution of unwanted metal impurities followed by their diffusion and precipitation in an area of the sample where they do not deteriorate the device performance. This area can be a heavily doped zone or any region outside of the electrically-active zone where the impurity precipitates do not affect the electrical behavior of the device.*

From this definition the following conditions for the gettering process can be deduced [8.36]:

(i) Gettering is always coupled to a temperature process. The temperature of the gettering process must be sufficiently high to dissolve all previously precipitated impurities. The respective solubilities of the impurity metals which are a function of the diffusion temperature must exceed their concentration in the sample.

(ii) The cooling period at the end of the temperature process is very important for the effectiveness of most gettering processes, which is still not well known. Most of the previously listed gettering techniques are effective only during this cooling period when the impurity metals become supersaturated as a first supposition for their precipitation. In this period the impurity atoms must be able to diffuse to the gettering zone where the nuclei for precipitation have previously been created. The required diffusion time depends on the distance between the electrically-active region in the device and the location of the gettering centers. As a consequence, the cooling period applying gettering by back-side damage must be longer than the cooling period required for internal gettering, since the respective distances differ by about one order of magnitude. Common denuded zones extend to $20 \div 50$ μm beneath the front surface whereas the wafer thickness is of the order of magnitude of $500 \div 600$ μm. In addition, the necessary diffusion time for the impurity metals depends on the squared diffusion length of the respective impurity metal, see (2.1). The diffusivities of transition metals may differ by several orders of magnitude from one metal to another at the same diffusion temperature. This has been discussed in detail in Sect. 3.2. According to (2.1) the diffusion time must be increased, for example, by a factor of four to double the diffusion length.

(iii) After applying a gettering process the intended impurity precipitates can be located in the highly-doped zone in the case of phosphorus diffusion gettering, in the bulk of the sample beneath the denuded zone for internal gettering, or at the back side of the sample for gettering by a back-side damage. If the electronic function of the device is extended to the bulk of the sample as, for instance, in solar cells, internal gettering cannot be applied. The gettering centers and the surrounding impurity precipitates in the bulk of the sample would reduce the minority-carrier diffusion length [8.37] and as a consequence decrease the solar-cell efficiency [8.38].

(iv) In most cases, gettering is based on the heterogeneous precipitation of metal impurities at nucleation centers which are present or were created in the sample. For all transition metals dissolved on interstitial sites and investigated so far, the heterogeneous precipitation is favored compared to the homogeneous precipitation. Therefore, all these impurity metals can be gettered if they are able to diffuse to the gettering centers and to form precipitates within the period of cooling the sample from the diffusion temperature.

In conclusion, the properties of the impurities should be known for a better understanding of gettering and for optimizing the respective process conditions. As mentioned before, most of the impurity metals forming deep energy levels and precipitates belong to the group of transition metals. The properties of these transition metals have been reported in the preceding chapters. Mainly the ever-present impurity metals in device production, namely iron, nickel, and copper, must be gettered to enhance the yield. They belong to the 3d transition metals which were discussed in more detail. Whereas nickel and copper are fast-diffusing impurities, iron belongs to the moderately-fast-diffusing metals. Therefore, iron can be taken as a trace impurity for gettering studies. If iron is gettered, for instance, during a defined cooling period, nickel and copper will certainly be gettered, too.

Two principles of gettering mechanisms in silicon which differ in the active period of time when gettering takes place are known at present:

(i)   In extrinsic crystals exhibiting high doping concentrations, much higher solubilities of many metal impurities were observed compared to those in intrinsic crystals [8.39]. According to the distribution law of thermodynamics, the equilibrium impurity concentrations in two different phases follow the respective different solubilities. As a consequence, the equilibrium impurity concentrations in samples that exhibit diffused pn junctions can be quite different in the substrate with its lower dopant concentration and in the heavily doped zone. The heavily-doped zone, for example, during phosphorus diffusion into a p-type sample, will accumulate impurities due to its high doping concentration combined with the observed emission of excess silicon self-interstitials during phosphorus diffusion [8.31]. This redistribution of the impurity concentration in the two different zones causes a gettering effect which is termed as **extrinsic gettering**. This kind can appear during any diffusion forming sufficiently high doping concentrations in a limited diffusion zone. The gettering effect due to the impurity redistribution takes place *during the high-temperature process*. In the cooling period at the end of the diffusion process the supersaturated impurities in the highly-doped zone can additionally precipitate, preferably at the silicon-silicate glass interface [8.32]. However, this is not the cause for the gettering effect but its consequence, since, at lower sample temperatures the supersaturation of the impurities is higher in the highly doped zone. Gettering centers are usually not created intentionally by applying this technique.

(ii) All other gettering techniques developed so far and listed above are based on the heterogeneous precipitation of impurities at intentionally created gettering centers. The precipitation takes place during the cooling period *at the end of the high-temperature process* and is caused by the supersaturation of the impurities at lower temperatures. In general, crystal defects such as dislocations, stacking faults, or other precipitates act as gettering centers for the impurity precipitation. The various gettering techniques differ in the method applied to create the lattice defects.

Unfortunately, there is no uniform and consistent nomenclature relating to the two quite different mechanisms. The contrary of "extrinsic gettering" would be **intrinsic gettering** but this term is particularly used for a special technique where nuclei such as oxygen precipitates, and bulk stacking faults are formed during the precipitation of oxygen. On the other hand, "extrinsic gettering" is not only used for dopant diffusion gettering but also for gettering techniques which operate by the heterogeneous precipitation of impurities.

To overcome this problem, **dopant diffusion gettering** will be used here for the mechanism based on the distribution law which can sometimes be found in the literature in the special form of, for instance, phosphorus diffusion gettering. All other gettering techniques will be termed as **precipitation gettering**, which indicates that this technique is based on the intentional formation of gettering centers where the impurities are expected to precipitate by a heterogeneous nucleation mechanism outside the electrically-active zone of the device and during the cooling period after the HT process. Recently *Schröter* et al. [8.31] proposed the term **relaxation gettering** for the group of processes summarized here by the technique of precipitation gettering. These terms are synonymous.

In the case of **precipitation gettering** or **relaxation gettering** the mechanism is based on the heterogeneous precipitation of the impurities at nuclei formed by lattice defects and generated intentionally in selected zones of the sample. Various techniques for forming lattice defects have been published. A wide-spread method is the distortion of the crystal lattice at one surface of the wafer, which is mostly the unpolished back side. Different mechanical treatments, irradiation of laser light, or implantation of various ions have been applied. During a subsequent heat treatment, for instance an oxidation process, lattice defects are formed such as dislocations or extrinsic stacking faults which then act as gettering centers. This formation of the proper gettering centers induced by the lattice distortion at the sample surface needs a HT process of sufficiently extended duration.

If gettering centers are present they act as sinks for vacancies or silicon self-interstitials. The intrinsic defects are generated during the precipitation of the impurities in the form of metal silicides, since this is usually coupled with a change in volume (Table 3.10). The impurity precipitates can be revealed by TEM near oxide precipitates or near lattice defects such as dislocations or stacking faults. In general, their morphologies resemble those discussed in Sect. 3.4 and Chaps. 4 and 5.

In the case of **dopant-diffusion gettering** or **segregation-induced gettering** [8.31], the mechanism is different and the precipitates may be different in their microscopic structure and composition as well. Due to the higher solubilities of several impurities in extrinsic silicon, the impurity concentrations can be considerably enhanced in the heavily-doped zones. So higher supersaturations can be expected during the cooling period at the end of the diffusion process. Intentionally formed gettering centers are not present in this technique. In addition to the unwanted impurites, there are high concentrations of dopant atoms in the heavily doped zone. So, for example, SiP precipitates can be formed at HT. They are situated predominantly at the silicon-phosphosilicate glass interface and penetrate both phases [8.32]. Because of their enhanced volume compared to that of the original silicon host lattice, high concentrations of silicon self-interstitials are emitted during the formation of silicon phosphide. The emission of silicon self-interstitials was also observed during the precipitation of oxygen, which is the basis of internal gettering. As may be inferred from Table 3.10, silicon self-interstitials are absorbed during the formation of iron, cobalt and nickel silicides. Therefore a physical correlation of phosphorus diffusion gettering and internal gettering was assumed by *Bourret* and *Schröter* [8.33]. This assumption is based on the supersaturation of silicon self-interstitials in both techniques. On the other hand, it is known that copper is also gettered by phosphorus diffusion gettering although it should emit silicon self-interstitials during the formation of the silicide, as evident in Table 3.10 [8.40].

In conclusion, the precipitation mechanism is not yet fully understood in detail. The progress achieved in recent years in understanding the mechanism was summarized in [8.31]. It seems necessary that lattice distortions in the neighborhood exist, which act as sinks for supersaturated intrinsic defects (vacancies or silicon self-interstitials) formed during the precipitation of the impurities.

Because of poor knowledge of the microscopic mechanism of gettering, its efficiency cannot yet be calculated or even estimated. Therefore, reliable measurements are needed to determine the quantity of impurities which can be removed from the electrically-active zone of a device in order to enhance the yield of devices. The quantity of gettered impurities must be reasonably correlated to the impurity content which is present in the sample due to the contamination during processing. The problem of measuring gettering efficiencies is discussed in the following sections.

8.2 Control of the Gettering Efficiency

As mentioned before, the gettering effect depends on the properties of the respective impurity and on the process parameters such as diffusion temperature and cooling rate. In order to quantify the reduction of impurity concentrations in a sample or in a device, the term **gettering efficiency** is introduced. For the discussion of the problem within this monograph the following definition of the gettering efficiency is used [8.36]: *The gettering efficiency is the difference of the impurity concentration in the electrically-active zone of a sample measured before and after the application of the gettering process. It is correlated to a definite impurity, diffusion temperature, and cooling rate.*

In this definition the gettering efficiency is based on a difference. Therefore, it exhibits the dimensions of concentration. One reason for this definition was the uncertainty whether the gettering efficiency still depends upon the impurity concentration which is present in the sample. The other reason for this kind of definition is the advantage of a direct comparability of the gettering efficiency with the impurity concentration to be gettered, which may be known by order of magnitude from experience.

In order to determine the gettering efficiency a definite impurity of known concentration must be introduced into the sample, and its concentration must be re-measured after applying the gettering process. In spite of the large number of publications concerning gettering in silicon, there are only a few where definite impurities were introduced intentionally into the sample, and concentrations were determined before and after gettering. Therefore, only a few impurities are known by chemical nature, which can be removed by different gettering methods. Furthermore, it is rarely known which impurity concentrations can be removed by the various gettering techniques. On the other hand, it is likewise mostly not known which impurity concentration can be tolerated to achieve acceptable yields of definite devices. Hence, it is still difficult to fix reasonable upper specification limits for impurity contents in the bulk of wafers or on top of their surfaces.

8.2.1 Conventional Methods

Conventional methods to check the effectiveness of various gettering methods are manifold but they rarely allow the determination of the amount of the highest possible efficiencies. So it is difficult to compare different techniques and to optimize respective processes. In many experiments minority-charge carrier lifetimes due to unintentional contamination during processing were determined with and without the application of gettering processes. This requires a larger number of samples to be measured in order to obtain statistically relevant results for the gettering effect since the local carrier lifetimes usually scatter considerably in one wafer. If different deep-level impurities are present in the sample simultaneously, the reciprocal partial carrier lifetimes of the impurities sum up, as was discussed in Sect. 6.2.2. The minority-carrier capture cross-sections of the various metal impurities can differ by several orders of magnitude. Thus, a detailed knowledge of the impurity content in the sample cannot be obtained in this manner. The chemical nature of the respective impurities remains unknown as well as their concentrations. It cannot even be deduced from the results whether the impurities are dissolved or precipitated since both kinds of defects reduce the carrier lifetime. Furthermore, the reduction in carrier lifetime depends on the accidental impurity content in the sample which is caused by an unintentional contamination during processing and may therefore fluctuate considerably. Similar problems arise if other device parameters are used to determine the effectiveness of gettering, such as the breakdown voltage, yields, or leakage currents.

One main problem in determining gettering efficiencies is the lack of suitable methods to measure the total impurity content in a sample zone within a reasonable time. Conventional methods such as neutron-activation analysis or Rutherford backscattering (Sect. 6.1), although used occasionally for special investigations [8.9, 23, 24], cannot be utilized as an in-process control since they are usually not available in industrial laboratories. Measurements performed in foreign laboratories are expensive and require too much time.

With the knowledge of the behavior of the various transition-metal impurities in silicon we can now restrict the necessary measurements of the total impurity concentrations either to those of dissolved impurities or to those of precipitated impurities. For this purpose we have to select suitable trace impurities which precipitate almost quantitatively during the cooling period after their diffusion or which remain quantitatively dissolved during this cooling period. Both kinds of methods have been developed [8.30, 41] and applied to check the efficiencies of various gettering techniques [8.30, 36, 42, 43]. The respective measurement techniques are reported in Sects. 8.2.2, 3.

8.2.2 Palladium Test

The palladium test [8.41] is based on the detection of the amount of precipitated palladium in the form of haze on the top of polished wafers. The wafer is intentionally contaminated with palladium as a trace impurity. The method of detecting the non-gettered palladium content in the sample is the haze test. This method is rapidly, inexpensively and easily performed, and the equipment is of low cost.

Palladium has been selected for this test for the following reasons:

- The rough back side of commercial wafers can easily be contaminated with metallic palladium. This contamination can be performed with high repeatability.
- Palladium causes the formation of the brightest haze pattern compared to all other haze-forming metals, and is clearly visible even without the aid of a spot light.
- The palladium haze exhibits sharp boundaries independent of the cooling rate applied at the end of the diffusion process. This is in contrast to, for example, nickel haze which drops gradually corresponding to the expected lateral diffusion profile (Fig. 3.10). These sharp haze boundaries of palladium facilitate the quantitative evaluation of the haze, as will be shown later.

Experimental Procedure: Before performing the palladium contamination, the wafer to be studied must be acid-etched to remove the native oxide and any unwanted impurity contamination from the surfaces of the wafer. The subsequent intentional contamination is performed on the rougher back side of the wafer by means of a piece of metallic palladium wire. For a better handling, the piece of wire can be mounted in a pencil-like holder. Spot-like contaminations at different positions distributed on the whole surface of the wafer have been proven to be best for averaging and for the control of the lateral homogeneity of the gettering efficiency. In order to achieve a sufficient amount of palladium acting as an infinite diffusion source, the *spot-like* contamination was done by scratching three neighboring lines of about 1 mm in length resulting in an almost quadratic structure of about 1 mm^2. The weight of the adherent palladium was determined to an average value of 190 ng. This weight has been determined from the difference in weight of an etched wafer before and after contamination with hundreds of equal palladium scratches. It acts as an infinite source for the following diffusion process. The diffusion into the bulk of the wafer is typically performed at 1100°C for 20 minutes in a quartz furnace tube with an argon or nitrogen ambient atmosphere. The cooling period typically lasts between 1 and 5 minutes. Then the wafer is preferentially etched applying one of the solutions reported in Sect. 3.4.2, for instance, a Yang etch. Under these preparation conditions, the circular haze patterns revealed on wafers without a gettering effect exhibit a diameter of about 20 mm. The diameter of the haze pattern still depends, to a minor extent, on the cooling rate applied. Wafers with any kind of backside gettering exhibit haze patterns with a

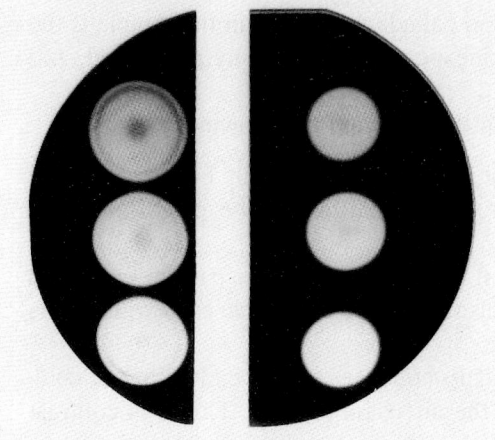

Fig. 8.1. Example for revealing gettering effects due to mechanical backside damage applying the palladium test. Left-hand segment: FZ-silicon reference sample without backside damage, right-hand segment: reduced diameters of haze patterns due to gettering. Pd-diffusion at 1050°C for 20 min, cooled by a pull velocity of 900 mm/min from the furnace tube

more-or-less reduced diameter due to a reduced Pd concentration and consequently a lower outdiffusion profile.

Experimental Results: One example of backside gettering is illustrated in Fig. 8.1 with larger haze patterns on the non-gettering FZ-silicon reference sample, and smaller haze pattern due to gettering on the sample exhibiting a mechanical backside damage. The conditions for the simultaneous diffusion of both samples are reported in the figure caption.

For a quantitative evaluation of the gettering efficiency, a reference sample without the gettering effect must be heat treated simultaneously in order to determine the reduction in the diameter of the haze pattern after equal diffusion conditions have been applied. From a known (calculated or measured) lateral diffusion profile for palladium in a sample without gettering, the local palladium concentration at the haze boundary can be determined. It can be expected that the palladium concentration at the haze boundary in the sample with gettering is equal to that in the reference sample. These concentrations are indicated in Fig. 8.2 for both wafers and for two different diffusion temperatures. The difference in the palladium concentrations measured at the radius of the haze patterns on the reference sample, on the one hand, and on the gettered sample, on the other hand, results in the gettering efficiency if this is assumed to be independent of the palladium concentration. Two calculated lateral diffusion profiles for the temperatures of 1050°C and 1100°C with a diffusion duration of 20 minutes are shown in Fig. 8.2 together with the palladium concentration for the haze boundary, as deduced from the radius of the haze pattern on the FZ-silicon reference sample (at about 10^{14} cm^{-3}). The indicated radii of the haze patterns of the wafer BSD1 with back-side damage measured after applying two different diffusion temperatures resulted in almost equal gettering ef-

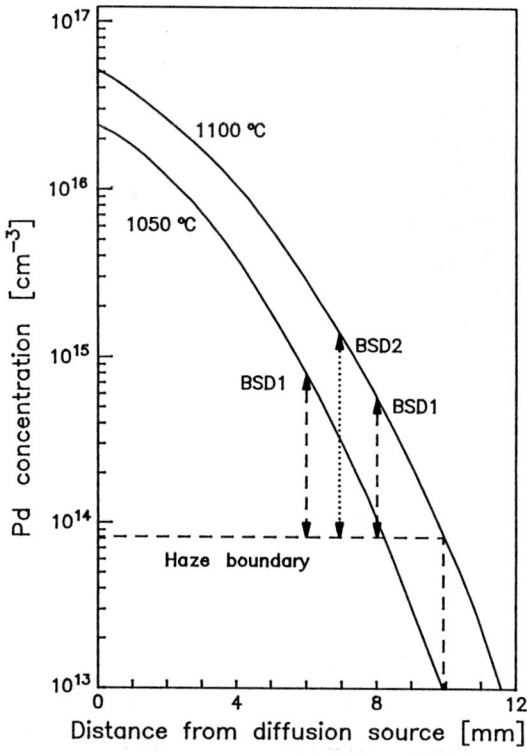

Fig. 8.2. Calculated lateral palladium profiles for two different diffusion temperatures (1050 and 1100 °C) and a duration of 20 min. Measured radii of haze patterns are indicated for the reference sample and two wafers with different backside damages (BSD1, BSD2)

ficiencies of about $5 \cdot 10^{14}$ Pd/cm^3 [8.41]. The gettering efficiency of the wafer BSD2, exhibiting another strength of backside damage, exceeds 10^{15} Pd/cm^3.

Problems concerning the accuracy of the evaluated gettering efficiency arise with the calculated diffusion profiles. For their calculation the diffusion duration must be accurately known and therefore the sample should be quenched to RT. However, during quenching, the palladium atoms may not quantitatively reach the reverse side of the wafer and gettering remains incomplete. Therefore, smaller cooling rates must be applied which, in turn, render the calculation of the laterial diffusion profile more inaccurate. Another way of overcoming this problem is to measure lateral diffusion profiles for standard processes by applying DLTS. This is possible since about 1% of the total palladium content remains electrically active and the high solubility of palladium at HT enables the measurement of this electrically-active fraction, as discussed in Sect.4.5. The maximum palladium concentration at the diffusion source must then be fitted to the solubility of palladium at the respective diffusion temperatures.

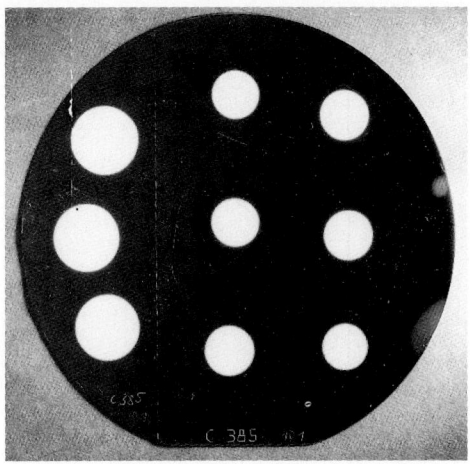

Fig. 8.3. Palladium test applied to a wafer with polysilicon on the reverse side. The reference sample (left side) was cut from the same wafer but the polysilicon was etched off. Diffusion at 1050°C for 10 min, pull velocity 150 mm/min. Remarkable reduction of the diameters of the haze patterns due to backside gettering

A second example for the palladium test applied to a wafer that exhibits gettering due to a deposited polysilicon layer at the reverse side of a wafer [8.44] is displayed in Fig. 8.3. Before heat treatment the smaller reference sample at the left-hand side was cut from the wafer, and the polysilicon on the reverse side was etched off. The samples were diffused simultaneously at 1050°C for 10 minutes and withdrawn from the furnace tube applying a pull velocity of 150 mm/min. After preferential etching the samples were photographed in a spot light. The reduction in diameter of the haze patterns in the sample with back-side gettering compared to that in the reference sample is strongly marked and results in a gettering efficiency of about $1 \cdot 10^{15}$ Pd/cm^3. This simple experiment clearly reveals that the polysilicon layer itself, and not the thermal treatment during the depositing of this layer, is needed to achieve the gettering effect. The same result has recently been published [8.45]. Yield measurements were carried out after equal preparation techniques had been applied, which, however, required much more effort.

Another example for the application of the palladium test is illustrated in Fig. 8.4. Two half wafers of a CZ-grown crystal were arsenic diffused simultaneously while the right quarters remained oxide-masked during this diffusion. On the upper half wafer a three-step internal gettering pre-annealing was applied before arsenic diffusion, the lower half wafer remained without internal gettering pre-annealing. On each half wafer a single spot-like palladium contamination was performed near to the boundary of the

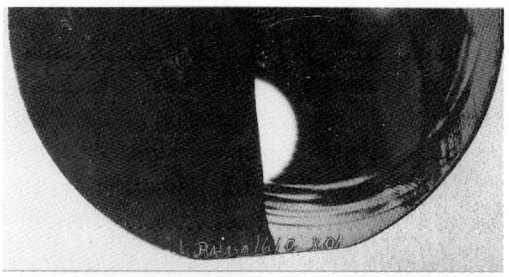

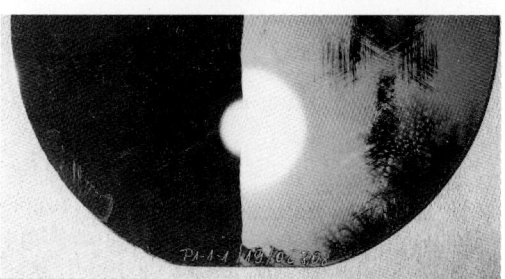

Fig. 8.4. Palladium test applied to two half wafers with internal gettering (*upper half*) and without (*lower half*). Both left-hand sides of the wafers were arsenic diffused while the right-hand sides remained oxide-masked during diffusion. The palladium contamination was performed near to the boundary of both regions. See text for details

oxide masking. After diffusion and preferential etching, the following four different gettering efficiencies can be distinguished by visual inspection:

(i) An almost vanishing gettering efficiency on the quarter wafer without internal gettering (lower half wafer) and without arsenic diffusion gettering (right side: oxide masked). In addition to the Pd-haze pattern, homogeneous haze from unintentional impurities is revealed in extended areas of the surface. In some areas this unintentional haze is gettered by slip lines causing narrow, black, and parallel lines.

(ii) The low gettering efficiency for the wafer with internal gettering but without arsenic diffusion gettering (upper right quarter). The additional haze due to unintentional impurities is reduced but still visible and the palladium haze pattern is slightly reduced in diameter.

(iii) The strong gettering efficiency for the wafer without internal gettering but with arsenic diffusion gettering (lower left quarter). The palladium haze pattern is considerably reduced in diameter, and no haze from unintentional impurity contamination is revealed.

(iv) The highest gettering efficiency in the wafer with the combination of internal gettering and arsenic diffusion gettering (upper left quarter). Haze due to unintentional impurities is not revealed and the palladium haze pattern disappears completely.

The arsenic diffusion was carried out at 1200 °C for 5 hours, and the palladium test was performed at 1100 °C for 20 minutes with a subsequent quenching of the sample to RT. Since both gettering layers, due to arsenic diffusion and due to internal gettering, are located at or near to the polished front side of the wafer, the quenching of the sample does not do too much harm to the gettering. The fast-diffusing palladium atoms have to cover only the thickness of the denuded zone to be gettered by internal gettering. This can take place even during quenching the sample from HT. From this simple experiment the sequence (8.1) for the gettering efficiencies in the present example can be deduced; it is valid for the special process conditions mentioned above:

$$(As + IG) > As > IG > 0 \qquad\qquad (8.1)$$

where As symbolizes the arsenic-diffusion gettering efficiency, IG the internal gettering efficiency, and 0 no gettering effect.

The cooling rate at the end of the palladium diffusion influences the strength of the palladium haze pattern to some extent even for internal gettering where the distance between the gettering zone and the wafer front side is only the thickness of the denuded zone of some tenths of μm. This is depicted in Fig.8.5 on the two segments of the same wafer after internal gettering pre-annealing where the lower segment was quenched to RT after palladium diffusion and the upper segment was moderately-fast cooled to RT within a period of 5 minutes. In each sample the palladium contamination was performed in the middle region. In the moderately-fast-cooled segment the palladium haze was completely gettered. In the quenched sample a hazy area interrupted by oxygen-induced striations was revealed showing the diffusion zone of the palladium contamination. The diffusion lengths of

Fig.8.5. Palladium test performed on two segments of a wafer after internal gettering pre-annealing [8.20-22]. Upper segment moderately fast cooled, lower segment quenched to RT. Reduced gettering in the quenched sample. Oxygen striations are revealed within the Pd diffused region. Details are in the text

the palladium atoms in the short period of quenching the sample enabled gettering of palladium atoms in the surface region only if they were in the direct neighborhood of the strongly-gettering nuclei. These nuclei are accumulated in the form of striations exhibiting higher oxygen contents. Thus, the oxygen striations which are situated beneath the denuded zone are reproduced at the wafer surface by their decoration with non-gettered palladium precipitates. The non-gettered bright lines correspond to the spaces between the oxygen striations, and represent the areas of lower oxygen content.

The technique of revealing gettering efficiencies by means of an intentional contamination of wafers with palladium can be extended to other haze-forming metals such as iron, cobalt, nickel, copper and rhodium; as was proposed in [8.41] and recently performed by *Falster* and *Bergholz* [8.46]. They investigated unintentional internal gettering after applying a sequence of heat treatments simulating the production of CMOS (Complimentary Metal Oxide Silicon) devices. The examination was performed by means of intentional mechanical contamination of the wafers with iron, cobalt, nickel, copper and palladium scratches followed by their diffusion, preferential etching and a subsequent inspection of the respective haze patterns. Their results will be discussed in the next section in connection with an alternative method to determine gettering efficiencies more quantitatively.

8.2.3 Iron Test

An alternative method to determine the gettering efficiency more quantitatively than by means of the palladium test is called the **iron test** [8.30]. In contrast to the palladium test this scheme is based on the detection of the interstitially dissolved and electrically-active iron measured by means of DLTS. Whereas palladium is almost never found as an unintentional impurity in processed wafers, iron belongs to the main impurities in device fabrication, as already mentioned. Furthermore, iron is the heaviest 3d transition metal which can be kept dissolved in silicon almost quantitatively after diffusion and quenching the sample to RT. The reason is its considerably lower diffusivity compared to cobalt, nickel and copper. Therefore, the selection of iron as a tracer for the gettering efficiency has the advantage that gettering of iron takes the longest time of all main impurities in silicon-device production. Consequently, if iron is gettered in a definite cooling period, the other main impurities such as copper and nickel will certainly be gettered.

However, this afore-mentioned feature is also a disadvantage in the selection of iron for determining the gettering efficiency. Iron exhibits only one donor state in the lower half of the silicon band gap. It can therefore be detected by means of DLTS via a Schottky barrier only in p-type wafers. So the iron test is most suitable for investigating gettering in bipolar processes since this is, in general, performed on p-type wafers of convenient resistivities. However, gettering processes which are applied to n-type wafers can only be investigated via p-type monitor wafers subject to the same processes, or n-type silicon wafers exhibiting pn junctions. The application of alternative detection methods such as, for example, the electron-spin resonance technique [8.47], would be possible but would require additional technical equipment. Furthermore, this technique does not enable depth profiling to determine the iron content in the denuded zone for characterizing the gettering efficiency after applying internal gettering.

Experimental Procedure. The iron test requires the following sample preparation [8.30, 43]:

After etching the wafer to remove the native oxide and any unwanted contamination layer from its surfaces, the entire back side of the wafer is sputtered with iron to a thickness of about 30 nm. The thickness of the iron layer is not critical since a tenth of a monolayer of iron on a wafer surface would already be sufficient to reach the solubility concentration of iron in the wafer at diffusion temperatures and durations proposed in this method. In order to maintain repeatable diffusion conditions it is recommended that one forms a homogeneous iron-silicide layer on the reverse side of the wafer during diffusion, which requires the deposition of at least several monolayers of iron. The iron concentration in the bulk of the wafer is adjusted by the solubility of iron in silicon as a function of the diffusion temperature. A diffusion temperature of $980\,°C$ results in a mean, electrically active iron concentration of about $2\cdot10^{14}$ cm^{-3} (solubility: $2.7\cdot10^{14}$ cm^{-3}). This is almost the upper limit of an impurity concentration which can accurately be determined by DLTS in wafers exhibiting carrier concentrations of $2\cdot10^{15}$ cm^{-3} ($5\div10\,\Omega\cdot$cm). A diffusion duration of 30 minutes is sufficient to yield a homogeneous distribution of iron within the bulk of the wafer at the diffusion temperature. The cooling rate at the end of the iron diffusion can be varied by applying different pull velocities with which the sample is withdrawn from the furnace tube. The pull velocity adjustable in automatic loading systems is taken as a measure of the cooling rate in the experiments reported below.

After cooling the sample to RT the wafer is etched in a mixture of HNO_3/HF, for instance. 10:1, to remove the iron-silicide layer from the reverse side and to achieve good rectifying contacts on top of the polished front side of the sample. The Schottky contacts are formed by evaporating magnesium or titanium through a mask containing circular holes of suitable diameters ($1\div2$mm). During evaporation the vacuum should be sufficiently good (considerably better than 10^{-5} bar) to enable low leakage currents of the resulting Schottky contacts. Both magnesium and titanium exhibit remarkable getter effects during their evaporation by forming oxides for magnesium, and oxides and nitrides for titanium, thus improving the vacuum. Therefore, a preliminary evaporation with a shielded wafer surface significantly improves the vacuum. For preparing the ohmic backside contact after evaporation of the rectifying contacts, the backside of the samples are contaminated with gallium. This is performed by rubbing emery in the presence of liquid gallium on the still uncoated, rough reverse side of the sample at the slightly elevated sample temperatures $(30° \div 40°C)$ needed to keep the gallium liquid.

The residual dissolved iron concentration in the region near the surface is determined by DLTS summing up the interstitial iron content and the FeB pair concentration. In order to enhance the repeatability of the measurement it is suggested to store the sample for one or two days at RT before recording the spectrum. After this storage time all iron is paired with boron, and only a single determination of the iron concentration has to be performed. In order to obtain accurate iron concentrations the effective carrier concentration of the sample at the temperature of the FeB signal (at about 56K) must be determined experimentally together with the built-in voltage of the Schottky contact, which is also temperature dependent. Both values result from a Capacitance-Voltage (CV) measurement at the respective low temperature. For this measurement the area of the capacitor must be accurately known since the capacitance depends on the squared area of the depletion layer. However, the accurate diameter of the circular Schottky contact is difficult to be determined. This is because the evaporated metal can diffuse beneath the mask during evaporation and thus more or less enlarge the diameter, depending on the thickness of the evaporated metal. Furthermore, the edge of the Schottky contact may not be accurate and abrupt. To overcome these shortcomings the RT resistivity of the sample is measured by means of a four-point probe before evaporation takes place. After evaporation, a CV analysis is performed at RT and the measured carrier concentration is fitted to the carrier concentration calculated from the respective resistivity by adjusting the effective area of the Schottky contact. Usually, the area determined by this technique is slightly larger than that calculated from the diameter of the Schottky contact. The CV measurement at a low temperature is then carried out taking into account the area which was determined at RT. Although this procedure takes more time, the results are more reliable and justify the increased expense.

Experimental Results. Some characteristic results obtained from iron tests applied to different samples are displayed in Fig. 8.6. The measured concentration of electrically-active iron after diffusion at 980°C is plotted

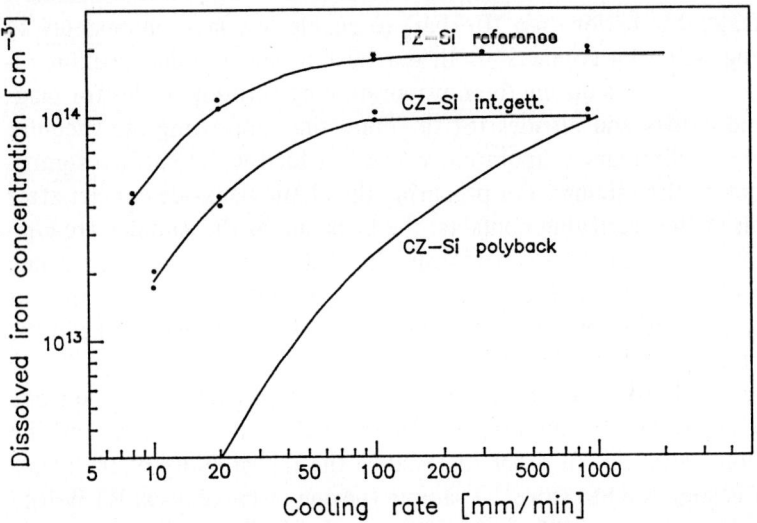

Fig. 8.6. Iron test applied to samples without gettering effect (reference, *upper curve*), CZ-grown sample with internal gettering (*medium curve*), and sample with polysilicon on the reverse side (polyback, *lower curve*). Dissolved iron concentrations plotted versus the cooling rate measured by the pull velocity from the furnace tube

versus the cooling rate. A measure for the cooling rate is the pull velocity with which the samples were withdrawn from the furnace tube. Three samples with different gettering behavior were investigated, and the results are shown and compared in the same figure.

The upper curve termed "FZ-Si reference" represents the results of measurements obtained on several wafers of as-received FZ- and CZ-grown crystals without applying intentional gettering processes. The maximum iron concentration between 2 and $2.1 \cdot 10^{14}$ cm^{-3} at high cooling rates is slightly below the solubility of iron at 980°C taken from the literature and measured by means of a neutron activation analysis [8.37] ($2.9 \cdot 10^{14}$ cm^{-3}), or by calculating the solubility from the data listed in Table 3.1 (2.7×10^{14} cm^{-3}). This iron content almost agrees with the results obtained by *Weber* who determined the electrically-active iron concentration by means of electron-spin resonance technique [8.47] ($2.15 \cdot 10^{14}$ cm^{-3}). As can be inferred from Fig. 8.6, the maximum iron concentration is nearly independent of the cooling rate down to pull velocities of about 100 mm/min, corresponding to a duration of the cooling process of about 7 minutes. For lower cooling rates the concentration of electrically-active iron decreases considerably with decreasing cooling rate. This is in agreement with the experimental finding that iron precipitates during slow cooling of the sample or annealing it between 300° and 700°C for several minutes [8.48].

The reason for the iron precipitation at lower temperatures is not yet known in detail; it will be discussed below. There are two possible reasons for iron precipitation during low cooling rates or during a subsequent annealing at lower temperatures:

(i) Although iron probably precipitates via a heterogeneous nucleation process during cooling of the iron-doped sample from HT and the application of rather high cooling rates, the supersaturation of iron at lower temperatures increases drastically, and may then enable a homogeneous nucleation mechanism in analogy to oxygen in silicon. The formation of haze at lower temperatures would justify this explanation.

(ii) If iron is believed to precipitate only via a heterogeneous nucleation mechanism, it may be assumed that low-density intrinsic defect clusters consisting of vacancies or self-interstitials, act as nucleation centers where iron can precipitate during slow cooling of the sample. However, during high cooling rates only a few iron atoms reach these nucleation centers which are far apart because of the low density. This assumption is supported by the experimental finding that the maximum iron concentrations in CZ- and FZ-silicon during the early days of the iron test (1987) were found to be about a factor of two lower, and still exhibited a slightly increasing tendence with increasing cooling rates [8.30]. Even today, non-processed FZ- and CZ-grown crystals can be found, which exhibit at least partly reduced iron contents at high cooling rates [8.43, 49]. This experimental finding will be discussed again in connection with further results in this section. In addition, iron formed haze in the early days of haze inspection [8.50] but it does not form haze today using wafers grown by modern pulling methods [8.42, 51]. This could be explained by the still higher concentrations of intrinsic defect clusters in early material. The existence of intrinsic defect clusters of vacancies and silicon self-interstitials is out of question. The type of cluster being formed of vacancies or of self-interstitials depends on the pulling velocity during crystal growth.

The middle curve shown in Fig. 8.6 represents typical results obtained with CZ-grown wafers after the three-step internal gettering pre-annealing process had been applied. The concentration of the dissolved iron and the gettering phenomenon at higher cooling rates are again independent of the cooling rate. In this region the gettering efficiency expressed as the difference to the respective results in the FZ-Si reference sample (upper curve) amounts to about 10^{14} cm^{-3}. At lower cooling rates, both the iron concentration and the gettering efficiency decrease proportionally to the concentration of dissolved iron in the reference sample. Although not directly visible on the semilogarithmic scale, the gettering efficiency of internal gettering expressed as the difference to the upper reference curve decreases considerably with a decreasing cooling rate.

Finally, the lower curve in Fig. 8.6 represents the results obtained with a CZ-grown wafer having a polysilicon layer deposited on its reverse side for backside gettering. It is obvious that the influence of the cooling rate on the gettering efficiency, in the case of back-side gettering, is marked more strongly than for internal gettering. During the cooling period the iron atoms must cross the whole wafer thickness to diffuse to the gettering centers where they can precipitate. In the range of investigated cooling rates there is no region where the gettering efficiency remains constant. By extrapolating the measured curve to higher cooling rates, the point where the getter effect vanishes completely can be estimated. This may be accomplished at pull velocities of the order of 5000 mm/min corresponding to a cooling period of about 10 seconds. Corresponding cooling rates can be achieved by manually pulling the sample from the furnace tube and quenching it to RT. In conclusion, wafers coated with polysilicon layers on the reverse side exhibit high iron gettering efficiencies at moderately low cooling rates (compare Fig.8.4 for the palladium gettering efficiencies) but this gettering vanishes completely if the cooling rates are too high. This phenomomon is still more important if the gettering of slowly diffusing transition metals such as Ti, V, or Cr are investigated.

The iron concentrations shown in Fig.8.6 are mean values taken from $10 \div 20$ data points across a 10 cm diameter wafer. Figure 8.7 illustrates an example of a radial profile of measured iron concentrations in an as-received FZ-grown wafer after iron contamination and diffusion [8.42]. The maximum variation of the results amount to almost 30%, which is well beyond the expected measurement error (about 10%) of DLTS, as determined by unpublished multilaboratory comparisons [8.44]. If a measurement error of about 10% is taken into account, the hatched area in Fig.8.7 suppresses the scattering of neighboring results. The hatched area then represents a W-shaped radial profile, as it is known, for instance, from radial profiles of resistivities or carbon concentrations in wafers. This experimental finding may offer a further indication of the presence of still unidentified gettering nuclei even in unprocessed crystals. The density of these gettering centers exhibits a quasirotational symmetry and therefore may depend on the local growing conditions during pulling of the crystal. The highest iron concentration in the radial profile (Fig.8.7) amounts to $2.45 \cdot 10^{14}$ cm^{-3} and is much closer to the iron solubility in silicon at $980°C$ tabulated in the literature, namely $(2.7 \div 2.9) \cdot 10^{14} cm^{-3}$, than the mean value of all 20 measurements, namely $2.12 \cdot 10^{14}$ cm^{-3}. This is close to the average reference value of $2.05 \cdot 10^{14}$ cm^{-3}, taken in our experiments as a reference concentration for electrically-active iron in samples without the gettering effect.

Beside the radial iron profile the axial profile provides additional indications for the presence of still unidentified gettering centers in the crystal before and after thermal processing. In Fig.8.8 the gettering efficiencies determined by the application of the iron test on p-type wafers before and

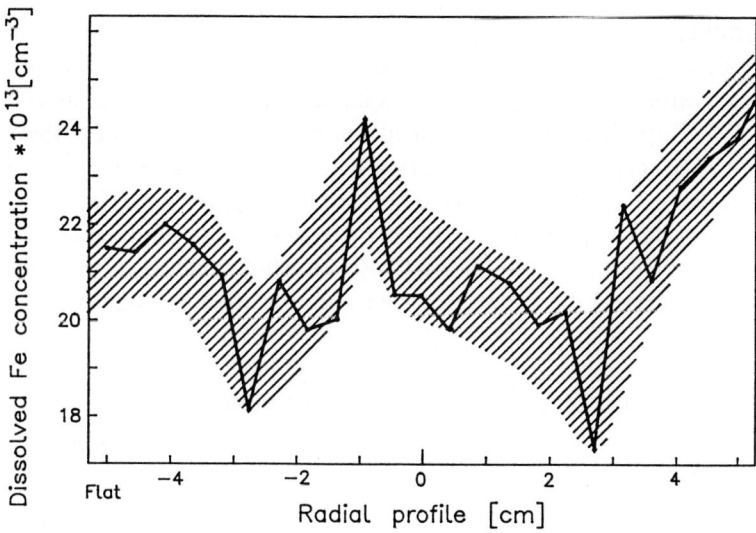

Fig. 8.7. Iron test. Radial profile of the iron concentration in a wafer without gettering effect. The hatched area is expected to compensate measurement errors of $\pm 5\%$ and represents a W-shaped, quasi-rotational symmetry. Pull velocity 900 mm/min

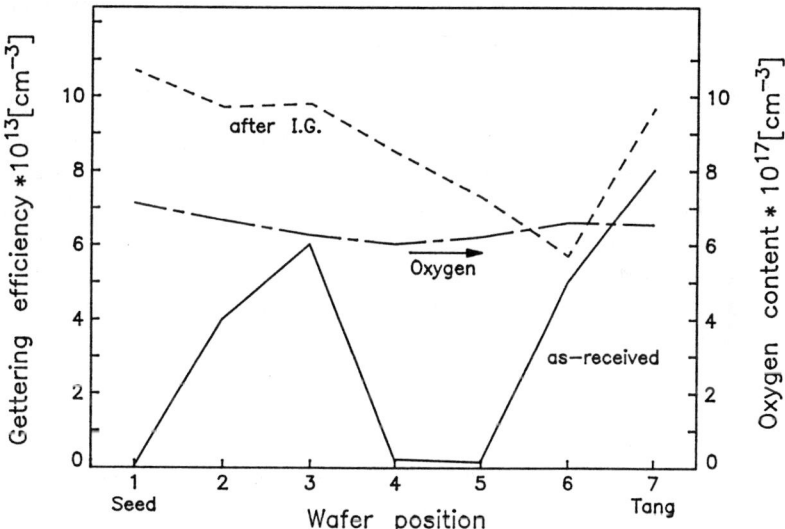

Fig. 8.8. Iron test. Gettering efficiencies (*left-hand scale*) as a function of the wafer position in the original crystal rod measured on as-received wafers (*full line*) and after internal gettering pre-annealing (*dashed line*). Pull velocity 900 mm/min. Oxygen concentrations measured in the same wafers (*right-hand scale*) for correlation with gettering effects

221

after performing internal gettering processes are plotted versus the wafer position in the original crystal rod [8.42, 43]. Instead of the measured iron concentrations after intentional iron contamination and diffusion of the sample at 980°C, which were depicted in the previous figures, now their differences to the mean iron concentration without gettering effect (2.05×10^{14} cm^{-3}) is plotted. It has been defined earlier as the gettering efficiency. In as-received wafers these gettering efficiencies increase with the distance from the seed end of the crystal starting with vanishing values. In the middle region of the crystal the values decrease and towards the tang end they increase once more, as shown by the solid lower curve in Fig. 8.8. The axial profile of the respective oxygen concentrations has been determined, in addition. The results are indicated by the broken line in the same figure. It is obvious that the decreasing, increasing, and finally slightly decreasing oxygen contents do not correlate with the respective gettering efficiencies measured in the as-received wafers.

The upper dashed curve in Fig. 8.8 represents the results for the gettering efficiencies determined after a high-low-high-temperature internal gettering pre-annealing cycle had been applied to the as-received wafers. The gettering efficiencies averaged over all wafer positions increased considerably compared to the respective values in the as-received wafers [8.42, 43]. In addition, deviations from a mean efficiency value are considerably reduced. Again, the gettering efficiencies do not correlate with the oxygen content in the as-received wafers. However, there is a general decreasing tendency in the gettering efficiency from the seed end towards the tang end of the crystal, which would correlate with the thermal history of the crystal. During crystal pulling the seed end of the crystal is annealed much longer than the tang end. Therefore, the density of thermal donors at the seed end, for instance, is much higher than at the tang end of the crystal. Since it was found that the gettering efficiency after performing internal gettering pre-annealing is essentially determined by the density of bulk stacking faults [8.30], it could be argued that these bulk stacking faults may correlate with the thermal history of the crystal [8.43].

Beside the tendency of the gettering efficiency to decrease with increasing distance from the seed end after having performed the internal gettering pre-annealing, there are strongly marked deviations, especially at the tang end of the crystal. These deviations increase in clarity if the mean value, indicated in Fig. 8.8 at the wafer position 3, which results from measurements on three different wafers, is replaced by the mean of the two wafers that exhibits both high gettering efficiencies, while omitting the results of the third wafer with the very low gettering efficiency. This is illustrated in Fig. 8.9 (dashed line). Now, it is more evident that the gettering efficiency, after having applied internal gettering, may consist of two different processes:

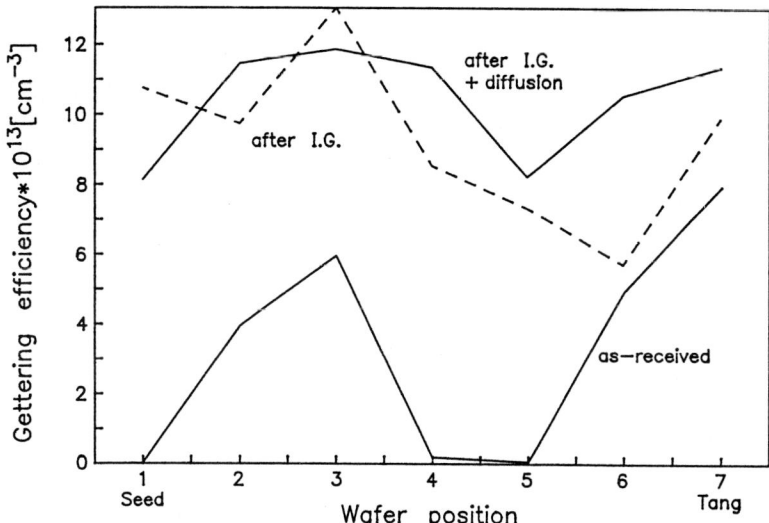

Fig. 8.9. Iron test. Comparison of gettering efficiencies as a function of the wafer positions in as-received wafers (*lower full line*), after internal gettering pre-annealing (*medium dashed line*), and after an additional buried layer diffusion (*upper full line*). Pull velocity 900 mm/min. Corrected measurement point in position 3 after internal gettering

(i) The tendency to decrease is correlated to the thermal history of the crystal and possibly caused by gettering of bulk stacking faults.

(ii) The gettering efficiencies observed for as-received wafers, with the maxima at the positions 3 and 7 due to gettering at unidentified intrinsic defect clusters which had been generated during the pulling of the crystal [8.43].

In order to control the assumption of two different, independent, but competitive gettering phenomena, an additional, simulated buried-layer diffusion process at 1110°C for four hours has been performed on one of the same wafers followed by new iron tests. For ease of comparison, the new results are indicated in Fig. 8.9 (upper solid curve) [8.43]. After this HT treatment, the decreasing tendency disappeared, and the mean gettering efficiency averaged over all positions still increased slightly. This appears reasonable since the memory effect due to the thermal history of the crystal is suppressed by extended subsequent heat treatments. At the same time the influence of gettering by unidentified centers in the as-received wafers is reduced, but it is still visible. This reduction may be caused by the increased gettering effect due to bulk stacking faults, or by the shrinkage of bulk intrinsic defect clusters.

It has been shown that the gettering efficiencies before and after performing internal gettering pre-annealing do not correlate with the respec-

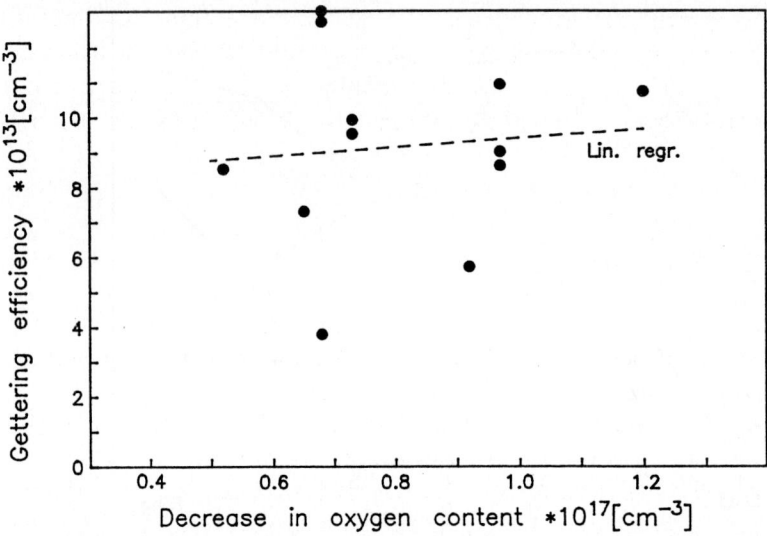

Fig. 8. 10. Gettering efficiencies from previous results on various positions in the crystal rod after internal gettering as a function of the decrease in oxygen content during the pre-annealing process. Linear regression

tive oxygen content in the wafers (Fig. 8.8). In addition, it can be shown that they do not correlate with the decrease in oxygen content either. This decrease in oxygen concentration during the pre-annealing is a measure of the amount of precipitated oxygen generated during these processes. Figure 8.10 presents the gettering efficiencies after pre-annealing for all wafer positions in the original crystal, including re-measurements on second and third samples as a function of the respective decrease in oxygen content. Although the amount of precipitated oxygen in the wafers investigated varies by more than a factor of two and the measured gettering efficiencies vary by more than a factor of three, a correlation of both parameters has not been observed. The calculated linear regression indicated in the figure is almost constant and independent of the amount of precipitated oxygen content. This experimental finding demonstrates that the amount of precipitated oxygen is not a suitable measure for the gettering efficiency in this case, in deviation of the wide-spread assumption repeatedly found in the literature. Corresponding results were also observed previously by other researchers [8.30, 46].

In conclusion, the unidentified gettering centers which are assumed to be present in as-received wafers are HT resistant and influence the gettering efficiency even after prolonged diffusion durations. However, their influence on the total efficiencies diminishes with increasing diffusion time [8.43, 46] and will be covered, at least, partly by the gettering effect of bulk stacking faults in our experiments. Their influence is still detectable after

HT heat treatments at 1100°C for 15 hours (pre-annealing) and an additional diffusion at 1110°C for 4 hours. All these results could be obtained by the application of a quantitative method to determine the gettering efficiency. It was also shown that the amount of precipitated oxygen is not a suitable measure to determine the internal gettering efficiency which has proven to be independent of the original and of the precipitated oxygen content.

In a recent paper on gettering in silicon in a sequence of simulated CMOS processes, *Falster* and *Bergholz* [8.46] found that gettering takes place even without the application of internal-gettering pre-annealing and also without the presence of bulk stacking faults. These researchers applied a combination of haze tests due to line-shaped Fe, Co, Ni, Cu, and Pd contaminations that were performed simultaneously on the same wafer. Although they found a rather low threshold oxygen concentration for the precipitation, they did not observe a correlation between the gettering effectiveness and the oxygen content or the amount of precipitated oxygen. This experimental finding agrees with our results presented above. We now assume that the gettering centers which were effective in their experiments were identical with the unidentified centers which we expect to be already present in the as-received wafers and which were found to be still effective after several extended HT processes.

The nature and the structure of the unidentified gettering centers are not known in detail. To date, only a small amount of information about their properties is available. Since the centers affect gettering of transition metals they can be decorated by these metals. Copper and nickel were found to be the metals for easiest gettering, and cobalt and palladium are progressively more difficult [8.46]. Iron is gettered with varying efficiencies, as shown by applying the iron test [8.43]. The gettering effectiveness of the unidentified centers is comparable to that of bulk stacking faults and is maintained even after prolonged HT heat treatments of the sample. The densities of these gettering centers vary in the crystal both radially and axially, and probably depend on the pulling parameters during crystal growing. The densities are neither correlated to the oxygen content nor to the amount of precipitated oxygen. It should be possible to determine the densities of these centers by applying TEM on iron-decorated centers. But these experiments have not yet been performed.

The crystal studied so far was grown several years ago. Therefore, additional investigations were performed on wafers with known positions in the crystal rod that originate from different suppliers and were recently delivered. Indeed, the results differed from those presented thus far in the amount of gettering efficiency. They also differed from one another. But all crystals exhibited gettering effects in the as-received state, which changed considerably after performing internal-gettering pre-annealing cycles and after subsequent simulated diffusion processes. At present it cannot be decided whether the different results are characteristic for the respective individual crystal or for the respective supplier.

In order to check the influence of the iron test itself upon the gettering efficiency of the sample to be investigated, several measurements have been performed on p-type wafers after performing a three-step internal gettering pre-annealing cycle followed by a simulated buried-layer diffusion. Employing a diffusion temperature of 980°C in the previous measurements proved that the gettering efficiency was very homogeneous within one wafer and from wafer to wafer. Thereafter, the influence of different diffusion temperatures upon the gettering effect has been investigated. The resulting concentrations of dissolved iron are plotted in Fig.8.11 as a function of the iron diffusion temperature [8.44]. The iron concentrations, as calculated from the solubility of iron (Fig.3.1), are shown in addition. From the experimental results it can be deduced that the concentrations of dissolved

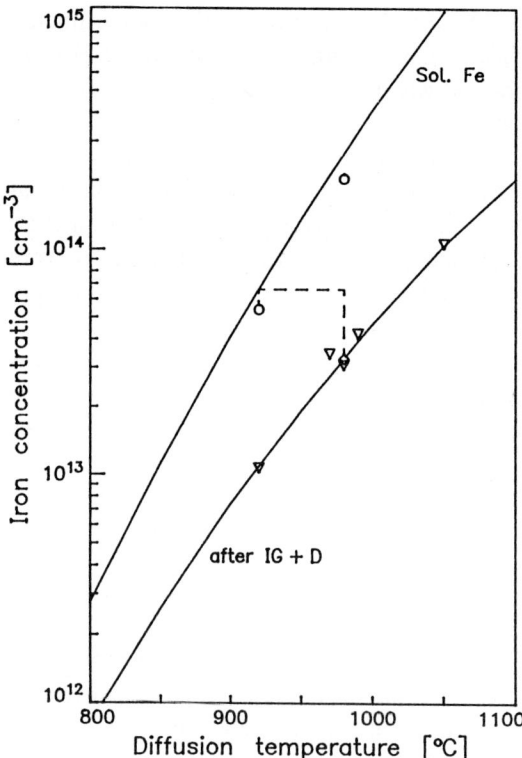

Fig. 8.11. Iron concentration after internal gettering followed by a simulated buried layer diffusion as a function of the diffusion temperature. Pull velocity 900 mm/min. Comparison to the solubility of iron (*upper line*). The gettering effect measured on a sample with reduced iron concentration (920° C) is also indicated at 980° C (*dashed line*, see text)

Table 8.1. Solubility and gettering of iron as a function of the diffusion temperature T

T [°C]	Solubitity [cm^{-3}]	Fe after gettering [cm^{-3}]	Gettering efficiency [cm^{-3}]	Rel. gettering efficiency [%]
900	$4.15 \cdot 10^{13}$	$7.32 \cdot 10^{12}$	$3.42 \cdot 10^{13}$	82.4
950	$1.36 \cdot 10^{14}$	$1.90 \cdot 10^{13}$	$1.17 \cdot 10^{14}$	86.0
1000	$4.08 \cdot 10^{14}$	$4.59 \cdot 10^{13}$	$3.62 \cdot 10^{14}$	88.7
1050	$1.12 \cdot 10^{15}$	$1.04 \cdot 10^{14}$	$1.02 \cdot 10^{15}$	90.7

iron after gettering exhibit an exponential slope at constant cooling rates (pull velocity after iron diffusion: 900 mm/min), as indicated in the figure.

The iron gettering efficiency defined as the difference in iron content before and after gettering now turns out to be strongly temperature dependent with increasing values for increasing diffusion temperatures. However, a **relative gettering efficiency** which can be defined as the gettering efficiency divided by the initial iron concentration is much less temperature dependent. Both the gettering efficiency and, to some extent, the relative gettering efficiency increase with increasing sample temperature, and consequently with increasing iron concentration in the sample after its diffusion. This is evident from Table 8.1 in which the solubilities of iron, the iron concentrations after gettering, the gettering efficiencies and the relative gettering efficiencies are listed for four different diffusion temperatures. All results were obtained via a pull velocity of 900 mm/min at the end of the iron diffusion process.

Whereas the gettering efficiencies increase by about a factor of 30 with increasing diffusion temperature from 900° to 1050°C, the increase of the relative gettering efficiencies remains small, exhibiting absolute values between about 82% and 90%. Taking into account that the solubility and the iron concentration after gettering are experimental results which were obtained by means of different measurement methods, and the gettering efficiencies are the differences of two large values, the relative gettering efficiency seems to be almost temperature-independent and amounts to about 85% in this example.

An increasing diffusion temperature increases the solubility of iron but it also increases its diffusivity. As a consequence, an increasing amount of iron atoms can be gettered at higher temperatures since the diffusion length of the iron atoms is likewise enhanced for a defined cooling rate. In this manner, the diffusion temperature determines the upper limit of the residual iron content in a sample at a fixed pull velocity with which the sample is withdrawn from the furnace tube. This upper limit of the iron concentration after gettering is a technologically important parameter since it characterizes the maximum deterioration of the electrical parameters of devices after gettering at the end of manufacturing. As demonstrated, it is temperature dependent although marked less strongly than the solubility of iron. The relative gettering efficiency, on the other hand, is almost temperature independent but it is less important for the technological application of gettering in device production since the impurity concentration in a device is rarely known during the various stages of technological processing.

A further question is whether the amount of initial impurity concentration influences the gettering efficiency. So far we investigated only the gettering effect on maximum soluble iron in the sample which, in general, exceeds by far common impurity contents in wafers during device fabrication. In order to measure the gettering effect for lower iron contents the sample used in the former investigations was iron-diffused at 920°C and quenched to RT. The respective iron concentration was determined, as indicated in Fig. 8.11 (open circle). Its deviation from the solubility of iron at this temperature corresponds to that obtained before at 980°C and is likewise indicated in the figure. Then, the residual iron silicide was removed from the sample surface by etching. Subsequently, the same sample was annealed once more at the higher temperature of 980°C and cooled to RT with a pull velocity of 900 mm/min, as in the experiments performed earlier. Now, the mean iron concentration in the bulk of the sample at HT amounted to 6.5×10^{13} cm^{-3} instead of about $2.0 \cdot 10^{14}$ cm^{-3} in the earlier experiments. The results obtained after this second annealing of the sample at 980°C is indicated in Fig. 8.11 and connected by a dashed line with the previous iron content at 920°C. The new result agrees well with those obtained on samples which exhibited much higher iron concentrations due to the solubility of iron at 980°C.

This experiment reveals that the upper limit of the iron concentration in a sample which exhibits a defined concentration of gettering centers is independent of the original iron concentration before gettering. On the other hand, this upper limit of the impurity concentration strongly depends upon the diffusion temperature and on the cooling rate applied (not shown in Fig. 8.11). As a consequence, we have to learn that low iron contents cannot be reduced further by internal gettering beyond this upper limit of iron content valid for defined conditions during the gettering process (diffusion temperature and pull velocity). This upper limit can be reduced by applying lower cooling rates but it can also be reduced drastically at lower diffusion temperatures, as deduced from Fig. 8.11.

This upper limit of the iron content is a measure for the number of gettering centers within the bulk of the sample, which can be reached by the diffusing iron atoms during the cooling period. Thus, it is independent of the iron concentration before gettering but it depends strongly on the sample temperature and the cooling rate since the diffusion length of the iron atoms depend on these parameters. Internal gettering in a sample can be characterized by this upper limit, by the gettering efficiency, or by the relative gettering efficiency as defined earlier. In any case, the iron concentration to be gettered must exceed the upper limit valid for the selected diffusion temperature and cooling rate. However, the results obtained cannot be transferred to other conditions such as a lower impurity content, various diffusion temperatures or different cooling rates. An iron gettering efficiency of, for example, 10^{14} cm^{-3} cannot be interpreted as the maximum iron-impurity concentration which is gettered. This result must be related to the solubility of iron at the respective diffusion temperature in order to calculate the upper limit for the residual iron concentration in the respective sample at the cooling rate employed at the end of the last HT process.

In conclusion, a final model for the physical mechanism of internal gettering cannot yet be proposed in detail. It seems likely that the upper limit of the residual iron content in a sample is determined by the density and the distribution of the gettering centers in the sample related to the temperature of the diffusion process and the applied cooling rate. This upper limit of the residual iron content is the amount of iron atoms per cm^3 which could not reach gettering centers during the cooling period because of the far distances (low densities of nucleation points, extended thickness of the denuded zone) or short diffusion times (high cooling rates).

In a final example for the application of the iron test it is shown that its results agree well with those obtained by utilizing the palladium test on the same sample. CZ-grown p-type wafers were treated by a three-step internal gettering pre-annealing cycle. Subsequently, they were arsenic-diffused at 1110°C while half of the wafers remained oxide-masked. The gettering efficiencies employing the palladium test were found to be considerably more effective in the arsenic-diffused region than in the oxide-masked region (8.1), as was shown in Fig. 8.4. The corresponding results applying the iron test are depicted in Fig. 8.12. The iron concentrations measured in the radial direction cross almost perpendicularly to the arsenic diffused-oxide-masked boundary near the center of the samples. The iron tests were performed on three different wafers with three different cooling rates applied at the end of the iron diffusion. From the graph it is evident that the iron concentrations in the arsenic-diffused regions exhibit gettering efficiencies higher than the oxide-masked regions, which is in agreement with the results obtained with the palladium test (8.1). Whereas the palladium test (Fig. 8.4) revealed preferably the difference between the two gettering efficiencies in the palladium-contaminated region near the boundary, the iron test (Fig. 8.12) presents the gettering efficiency over the whole diameter of the wafer, and additionally reveals more or less strongly marked decreasing gettering efficiencies at both edges of the wafers. Similar effects were also

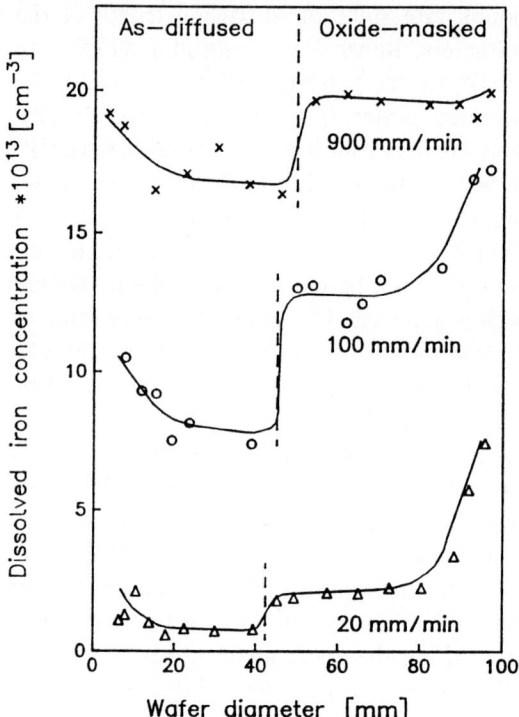

Fig. 8.12. Radial iron concentration profiles in wafers after internal gettering pre-annealing and subsequent arsenic diffusion with half wafers remaining oxide-masked. Different wafers with different cooling rates applied during iron-tests (compare with Fig. 8.4)

noted in Figs. 8.4, 5 by less gettered unintentional impurity contaminations near the wafer edges. The strong unintentional impurity contamination leading to a uniform haze on the whole wafer surface (Fig. 8.4, lower right quarter) took place during the arsenic diffusion. The reason for the reduced gettering efficiency at the edge of the wafers is not yet known but deviations near to wafer edges were observed repeatedly in both directions, and exhibit increasing or decreasing gettering efficiencies. A possible reason may be the different radial profiles of the unidentified intrinsic defect clusters which are assumed to play a major role in gettering.

A sudden change in the gettering efficiencies between the arsenic-diffused and the oxide-masked region is observed in all three samples depicted in Fig. 8.12, which differ in their cooling rates (the pull velocities are indicated). The levels of the mean iron concentrations differ considerably and decrease drastically with decreasing cooling rates. The relative change of the gettering efficiency at the boundary of an arsenic-doped/oxide-masked region depends strongly on the respective cooling rate. This is not revealed very well by the linear scale used in Fig. 8.12. This relative change exhibits the highest values (73 %) for the smallest cooling rate (20 mm/min), medium values (38 %) for the medium cooling rate (100 mm/min), and finally lowest values (15 %) for the highest cooling rate (900 mm/min). The bulk stacking-fault densities formed during the preceding temperature processes in all samples were determined by angle-lapping, preferential etching and counting with the aid of a microscope. They were found to be clearly lower in the oxide-masked region compared to the arsenic-diffused region, which is in full agreement with the experimental finding of different gettering efficiencies. This discrepancy can be explained by the enhanced emission of excess-silicon self-interstitials during the arsenic diffusion. These self-interstitials annihilate by precipitating and form bulk stacking faults. Therefore, it seems likely that gettering in this case is effected predominantly by bulk stacking faults.

During the cooling period when applying the high cooling rate of 900 mm/min the iron atoms in the oxide-masked region do not reach the low-density getter centers, therefore, maximum iron concentrations of $2 \cdot 10^{14}$ cm^{-3} are maintained. In the arsenic-diffused region the stacking-fault density is higher and a small quantity of iron atoms are precipitated. For the medium cooling rate of 100 mm/min, the gettering in both regions increased since the iron atoms have more time to reach even far away stacking faults. According to Fig. 8.6 gettering due to unidentified intrinsic defect clusters is still almost inoperative at this cooling rate. The absolute discrepancy in the gettering efficiency between both regions is at its highest ($\approx 5 \cdot 10^{13}$ cm^{-3}). Finally, for the low cooling rate of 20 mm/min, two gettering effects overlap, as was shown and discussed in connection with Fig. 8.6. Almost one half of the iron atoms should be gettered by the unidentified gettering centers which are assumed to be present in both regions. The second half of the gettering effect should be due to gettering by bulk stacking faults which together result in an almost quantitative gettering in the arsenic-diffused region and in a lower gettering efficiency in the oxide-masked region. The absolute difference in gettering efficiency between both regions is lowest.

In conclusion, the characterization of gettering by applying the palladium test, on the one hand, and the iron test, on the other hand, exhibit the same results. Thus, palladium and iron are gettered by the same centers and in the same manner, at least qualitatively. This may also be expected for the other main impurities nickel and copper, which can be introduced into samples during device production.

This, however, is not self-evident since during the precipitation of iron, silicon atoms are absorbed and during the precipitation of palladium, silicon atoms are emitted (Table 3.10). Furthermore, *Falster* and coworkers [8.52] pointed out recently that the precipitation of oxygen is drastically enhenced in silicon crystals that exhibit excess vacancies in contrast to most crystal regions which are characterized by excess self-interstitials. During oxygen precipitation self-interstitials are emitted (one silicon atom per two precipitated oxygen atoms similar to Pd precipitation). The emitted silicon atoms can easily recombine with excess vacancies and by this way approach thermal equilibrium, which may be the driving force for the enhanced precipitation of oxygen. By a sophisticated tailoring of the vacancy distribution in silicon wafers *Falster* succeeded even in annihilating the memory effect of the crystal-growing conditions at very high temperatures and forming a distribution of oxygen precipitates in the wafer which otherwise would require the well-known and time-consuming three-step internal gettering pre-annealing.

Equal results obtained by applying the palladium test, on the one hand, and the iron test, on the other hand, as reported above, were limited to the gettering effect of stacking faults. The gettering of still undetermined centers were so far only observed by applying the iron test because the palladium test does not allow a sufficient local resolution. It would be interesting to investigate iron precipitation in wafers with a controlled vacancy or self-interstitial concentration which is now possible by applying *Falster's* technique. It is assumed that the still unknown gettering centers could be detected by this way. It is further assumed that the striking inhomogeneities of the iron concentrations in the lateral (Fig. 8.7) and in longitudinal directions (Fig. 8.8) of the crystal will disappear.

The detailed presentation of several examples for the determination of the gettering efficiency should demonstrate that the application of these methods could enlighten the behavior of impurity metals in samples during heat treatments. On the other hand, the investigations enabled detection of the presence and effectiveness of non-identified gettering centers [8.43] which must be assumed to explain the experimental findings. Furthermore, these methods should help to optimise processes, to achieve sufficiently high gettering effects, and to maintain them during subsequent processes [8.30, 46]. Since gettering is directly correlated to specific impurities which should be reduced in concentration within the electrically active zone of devices, the properties of the respective impurities must be known and taken into account. In addition, the respective gettering parameters, especially the temperature and the cooling rate in the case of precipitation gettering, must be adapted to the requirements and to the defined impurity to be gettered. It is expected that the knowledge on gettering mechanisms will increase substantially in the near future, although several gettering techniques have already been applied in device manufacturing for a long period of time.

9. Conclusion and Future Trends

The present monograph reports on the properties and the behavior of transition-metal impurities in silicon as far as they are known and relevant for manufacturing devices in industrial applications or sample preparation in scientific laboratories. In general, only reliable values and facts are reported which facilitates the use of the data for any application. Data which have been reported only once in the literature are termed as single results. Many contradictory results were omitted if there is no reasonable explanation.

Most of the knowledge on the properties of the transition metals has been accumulated in the last two decades, but only a portion of it could be utilized in this monograph to give a comprehensive overview on the present state of the art. Too many details which may not help to solve practical problems would obscure the survey on important facts which should be known today for any kind of defect engineering and trouble shooting in device manufacturing. The increasing knowledge on deep energy level impurities present in wafers before and after processing leads to a considerably cleaner ingot material, cleaner handling and processing, and consequently enhanced yields which are required especially for a successful fabrication of integrated circuits of high complexity, as well as LSI, VLSI and finally ULSI memories.

Although much work has been done in recent years to collect important data and to understand the behavior of the impurities during various processes, there are, of course, a variety of blanks in the tables and many deficiencies in our knowledge. It will be a task of future investigations to close these gaps in order to present complete data collections to aid engineers and scientists in solving their practical problems. A remarkable lack of reliable data is still present in the properties of the 4d and 5d transition metals, although in this second edition most of the blanks concerning their deep energy levels could be replaced by at least single results. Several of these metals will gain increasing importance since the silicides of the refractive metals are now applied to solve modern interconnecting problems. Hence, these elements may also be found as unintentional impurities in wafers because of cross contamination during processing.

A special problem is the lack of reliable data for charge-carrier capture cross-sections of electrons and holes where data, if available at all, may sometimes scatter by several orders of magnitude. This is mainly due to the experimental problem of transferring short and rectangular electrical fill pulses to the sample which is mounted in a cryostat for the DLTS measurements. With the improved cleanness of materials and processes, the minority-carrier lifetimes in the devices increase and, as a consequence, the switching times increase likewise. Therefore, carrier-lifetime doping becomes necessary for more devices which formerly exhibited short switching times because of the presence of unintentional impurities. With an intentional lifetime doping the reverse current will increase depending on the special properties of the diffused recombination centers. So far, only a few techniques are applied to reduce the minority-carrier lifetime, such as gold or platinum diffusion and high-voltage electron or proton irradiation. It may be expected that more suitable materials exist but their selection is difficult because of missing reliable data for the respective capture cross-sections.

Wide areas of investigation can still be found in the theoretical treatment of the properties of the transition metals, such as solubilities, diffusivities, activation energies of the deep levels and carrier capture cross-sections. A first success was achieved in calculating the activation energies for the 3d transition metals which now agree fairly well with the experimental data. It was mainly the merit of the theoretical results that scientists no longer searched further for substitutional iron since it is unstable, as predicted by theory. A comparison of theoretical and experimental activation energies of the 4d and 5d transition metals is not yet satisfactory. This comparison could be useful to decide whether deep levels observed in experiments originate from interstitial or from substitutional defects. So far, this can only be assumed from other properties, especially if the impurities do not exhibit unpaired spins and are therefore Electron Paramagnetic Resonance (EPR) inactive.

In this 2nd edition a first tentative correlation between the position of the transition metal within the periodic table and its position within the silicon host lattice, on interstitial or on substitutial site, is postulated according to a proposal of *Lemke* (Table 3.3). Although this correlation has been verified so far only for the 3d transition metals the conspicuous similarities within one group of equal sum of outer s- and s-electrons concerning the number of deep energy levels formed in the band gap of silicon, their charge states, and even their activation energies enhance the probability that this correlation could hold also for the 4d and 5d transition metals.

First attempts have been published to calculate the diffusivities of the 3d transition metals as a function of their atomic radii, yielding promising results. For further improvements the model must be refined. To date, a theory to calculate the solubilities of the transition metals has not been developed, although theoretical predictions would be very helpful for various applications. Theoretical predictions for carrier capture cross-sections are still very poor and do not enable a critical selection of published experimental results. First attempts have also been undertaken to study impurity pairs theoretically. Pairing seems to be a wide-spread mechanism to overcome the problem of supersaturation, and it is expected that numerous impurity pairs are still not identified. To enable their detection, a prediction of their properties could give useful indications.

In this 2nd edition we include recently investigated hydrogen-transition metal complexes because they can considerably deteriorate the detection of several transition metals by DLTS. As mentioned before (Sect.2.2.2) hydrogen is introduced into the surface layer of silicon during etching the sample at RT and this is the same region where DLTS measurements are performed. By avoiding the introduction of hydrogen especially the three deep levels of palladium and platinum could be reliably detected what formerly was problematic. So far only a few metals were investigated in detail concerning their hydrogen complexes (Ti, Co, Ag, Pt, Au) and it is assumed that there is still a large number of unknown complexes to be studied. Furthermore, their structures have not yet been determined. Hence, a lot of investigations have to be done in the years to come.

Pairing of impurities within the host lattice belongs to a newly established impurity chemistry in silicon which is still in its infancy. From the multitude of deep energy levels which have been detected in silicon and published in the literature, only a small fraction could be identified. It is not certain whether the identifications are always reliable. There are still many defects which are termed as "impurity-correlated". Since there is no possibility to identify a defect from its deep energy-level activation energy, a complete listing of these data is almost useless. More investigations must go the other way by forming impurity compounds from definite components and subsequently analyzing them. Furthermore, there is a lack of suitable methods for analyzing defect structures on a routine basis, they would yield many more results than are available today. A reasonable number of experts must be trained to do this job in order to advance the knowledge of impurity chemistry as far as it is required to solve technical problems in the near future.

The increasing knowledge of materials science obtained by the few groups systematically investigating the formation mechanisms and properties of silicides should closely collaborate with the experts for TEM investigations of the formation of silicides during the precipitation of impurities. There is still a lack of knowledge of the precipitation mechanism, and the composition and structure of impurity precipitates in the silicon host lattice as a function of the preceding processes. Gettering mechanisms which are correlated to impurity precipitation must be investigated further. A suitable and simple gettering technique should be developed which can replace the

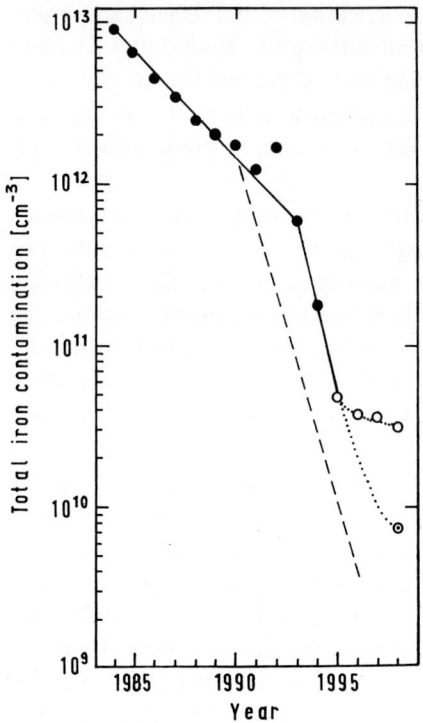

Fig. 9.1. Average of total iron concentration is as-received polished wafers after diffusion of surface cintaminants into the bulk of the wafer as a function of the year of delivery. Expected requirements for future memory devices (*dashed line*). (*Full dots*: results obtained by DLTS, *open circles*: results recalculated from carrier lifetimes in $5 \div 20 \, \Omega \cdot$ cm CZ-Si, *centered open circle*: recalculated from lifetimes in high-resistivity FZ-Si wafers)

phosphorus diffusion gettering in the case of phosphorus implantation followed by rapid thermal processing.

It is expected that the requirements for an improved cleanness of ingot material and processes will still continue in the future. The development of the iron content as a tracer impurity on the surface of as-received polished wafers originating from different suppliers in the past is illustrated in Fig. 9.1. The graph depicted in the first edition represents results that were obtained between 1984 and 1991. They showed a linear decrease of the mean total iron content of purchased polished silicon wafers after diffusion of the surface contamination into the volume of the wafer. The dashed line indicated the requirements for the production of future, highly integrated memory devices. In order to meet these requirements the decrease of the iron concentration with time had to become much steeper. After completing the graph for this second edition from 1992 to the present it revealed that the wafter suppliers, in fact, succeeded in diminishing the iron contamination, as has been required. There is only a retardation of about two years, necessary to adapt the production line.

Full data points signify mean results obtained by DLTS measurements. At iron contents below $5 \cdot 10^{10}$ cm^{-3} the sensivity of the DLTS technique reaches its lower limit for standard production material. However, at this time it was already known that the minority-carrier lifetime in p-type silicon is mainly determined by the iron content if this is electrically active on interstitial sites or in FeB pairs. Therefore, the time-consuming DLTS measurements were replaced by lifetime mappings of the whole wafer and the corresponding iron concentrations were recalculated. The open circles in Fig.9.1 are average values obtained on many wafers of Czochralski-grown (CZ) crystals that originated from three suppliers. It seems that the recalculated iron contents approach a level of about $2 \cdot 10^{10}$ cm^{-3}. However, recent control measurements on wafers of float-zone (FZ) crystals yielded in much lower lifetimes (centered open circle). So far the reason for this discrepancy is not yet known. There are three possibilities:

i) The iron content in FZ-silicon wafers of 15 cm diameter is lower than in CZ-silicon wafers of 10 cm diameter since the production of small diameters will be closed in the near future.

ii) The lifetime in CZ-grown crystals with iron contents in the region of about $2 \cdot 10^{10}$ cm^{-3} is no longer determined only by the iron content but also by other defects, for instance, its oxygen content in the form of precipitates or even intrinsic defect clusters.

iii) The iron concentration in FZ-grown crystals, in general, is much lower than in CZ-grown crystals. It will be a future task to investigate the real cause of this discrepancy.

To meet the future requirements for very low impurity concentrations in devices, an increased number of control measurements will be necessary, which, in turn, requires detection methods which are fast, inexpensive and enable a high number of results achievable by a single trained person. For this purpose automatic reading systems must be used. Furthermore, the suppliers of peripheral equipment have to check the cleanness of new instruments before their installation in a production line and should not wait for customer complaints, as has frequently been observed in the past. This once more requires experts on defect engineering even in peripheral industries. Despite all precautions and enhanced knowledge there will always be the possibility of new cross contaminations if processes are improved or replaced by other techniques, as was the case in ion implantation, dry etching techniques, rapid thermal processing, or sputtering equipment installed for the first time.

At present, process modeling with its increasing importance for the design of new devices mostly neglects the presence of unintentional metal impurities. If sufficient data were available, these impurities could be taken into account in future and could result in interesting information concerning the prediction of switching times with its increasing importance in future high-purity devices.

In conclusion, the importance of a better understanding of the behavior of metal impurities in silicon is expected to increase in future. The cleanness of wafers and processes has improved considerably in the past. This monograph is expected to become a basis for understanding the reasons for this improvement, and to offer a sufficient amount of presently known actual data which should be corrected and completed in reasonable periods of time. From the metals forming deep-energy-level impurities in silicon, the groups of alkaline and earth alkaline metals are still excluded from the discussion. The main reason is the poor knowledge to date on their properties and their behavior in silicon, this may be improved in the near future.

Appendix

A.1 Activation energy ΔE_A and majority-carrier capture cross-section σ_M of transition metals in alphabetic order

Metal	i/s	d/a	ΔE_A [eV]	E_∞ [eV]	σ_M [cm²]	T_{Dis}	Comment
Ag	s	a	− 0.55		$7.2 \cdot 10^{-17}$		
Ag	s	d	+0.38		$7.0 \cdot 10^{-16}$		
AgH$_x$	s-i		− 0.09				
AgH$_x$	s-i		− 0.45				
AgH$_x$	s-i		+0.28				
AgH$_x$	s-i		+0.38				
AgAg	s-i		− 0.28				
Au	s	a	− 0.55		$1.4 \cdot 10^{-16}$		
Au	s	d	+0.34		$2.5 \cdot 10^{-15}$		
AuAu	s-i		− 0.44				
AuH	s-i	a	− 0.56		$1.0 \cdot 10^{-16}$	300 K	
AuCr	s-i	d	+0.35				
AuCu	s-i		+0.42/0.32				bistable
AuFe	s-i	a	− 0.35		2.10^{-15}		
AuFe	s-i	d	+0.43		$2.6 \cdot 10^{-15}$		
AuMn	s-i	a	− 0.24				
AuMn	s-i	d	+0.57				
AuNi	s-i		+0.35/0.42				bistable
AuV	s-i	a	− 0.20				
AuV	s-i	d	+0.42				
Cd	s	aa	− 0.45				
Cd	s	a	+0.5				
Co	s?	d	+0.23		$1.8 \cdot 10^{-14}$	870 K	
Co	s	d	+0.41		$5.0 \cdot 10^{-18}$		
Co	s	a	− 0.41		$2.2 \cdot 10^{-15}$		
CoH$_x$		a?	− 0.07		$5.0 \cdot 10^{-14}$	400 K	
CoH$_x$		a?	− 0.17		$\approx 10^{-14}$	400 K	
CoH$_x$		a?	− 0.22		$\approx 10^{-15}$	400 K	
CoH$_x$		a?	− 0.26		$\approx 10^{-15}$	400 K	
CoH$_x$			− 0.39		$\approx 10^{-13}$	400 K	

Metal	i/s	d/a	ΔE_A [eV]	E_∞ [eV]	σ_M [cm^2]	T_{Dis}	Comment
CoH$_x$			-0.40		$\approx 10^{-14}$	400 K	
CoH			$+0.22$		$\approx 10^{-15}$		bistable
CoH			$+0.17$		$\approx 10^{-14}$		
CoH$_2$			$+0.09$		$\approx 10^{-14}$	600 K	
Cr	i	d	-0.22	0.00	$7.3 \cdot 10^{-15}$		
CrAl	i-s	d	$+0.45$				
CrAu	i-s	d	$+0.35$				
CrB	i-s	d	$+0.28$		$2.0 \cdot 10^{-14}$		
CrGa	i-s	d	$+0.48$				
CrZn	i-s	a	-0.1				
Cu	s	a	$+0.46$		$1.5 \cdot 10^{-14}$		
Cu	s	aa	-0.16		$1.9 10^{-17}$ *		
Cu	s	d	$+0.22$		$3.0 \cdot 10^{-14}$		
CuAu	i-s		0.42/0.32				bistable
CuB	i-s	...	...		...		el. inact.
CuCu	i-s	d	$+0.09$		$3.5 \cdot 10^{-15}$		
Cu$_x$		a	$+0.21$				(M-CENTER) metast.
Fe	i	d	$+0.39$	0.043	$3.8 \cdot 10^{-17}$ *		
FeAl 1	i-s	d	$+0.20$		$1.1 \cdot 10^{-15}$		
FeAl 2	i-s	d	$+0.13$				metast.RT
FeAu	i-s	a	-0.35		$2 \cdot 10^{-15}$		
FeAu	i-s	d	$+0.43$		$2.6 \cdot 10^{-15}$		
FeB 1	i-s	a	-0.27		$1.6 \cdot 10^{-14}$	400 K	
FeB 1	i-s	d	$+0.10$		$6.4 \cdot 10^{-15}$	400 K	
FeB 2	i-s						metast. <250 K
FeGa 1	i-s	d	$+0.25$		$4.0 \cdot 10^{-15}$		
FeGa 2	i-s	d	$+0.14$				metast. RT
FeIn 1	i-s	d	$+0.27$				metast. RT
FeIn 2	i-s	d	$+0.15$				
FePd	i-s		-0.32				
FeZn	i-s	a	-0.47				
Hf	i	a	-0.10		$>2 \cdot 10^{-14}$		
Hf	i	d	-0.40		$>2 \cdot 10^{-14}$		
Hf	i	dd	$+0.32$		$>5 \cdot 10^{-18}$		
Ir	s	a	-0.24		$9.1 \cdot 10^{-15}$		
Ir	s	d	-0.62		$7.2 \cdot 10^{-14}$		
Mn	i	a	-0.12		$3.1 \cdot 10^{-15}$		
Mn	i	d	-0.42	0.00	$3.1 \cdot 10^{-15}$		
Mn	i	dd	$+0.27$	0.064	$2.0 \cdot 10^{-18}$ *		

Metal	i/s	d/a	ΔE_A [eV]	E_∞ [eV]	σ_M [cm^2]	T_{Dis}	Comment
Mn	s	a	− 0.43		$9.0 \cdot 10^{-17}$		
Mn	s	d	+0.34		$2.0 \cdot 10^{-16}$		
Mn$_4$	i	d	− 0.28				
MnAl	i-s	d	− 0.45		$5.0 \cdot 10^{-15}$		
MnAu	i-s	a	− 0.24				
MnAu	i-s	d	+0.57				
MnB	i-s	d	− 0.55		$9.0 \cdot 10^{-14}$		
MnGa	i-s	d	− 0.42				
MnZn	i-s	d	+0.18				
Mo	i	d	+0.28		$6.0 \cdot 10^{-16}$		
Nb	i	a	− 0.28		$7.5 \cdot 10^{-18}$		
Nb	i	d	− 0.62		$>6 \cdot 10^{-16}$		
Nb	i	dd	+0.18		$3.8 \cdot 10^{-16}$		
Ni	s	aa	− 0.07		$5.4 \cdot 10^{-18}$ *		
Ni	s	a	− 0.41		$1.2 \cdot 10^{-16}$		
Ni	s	d	+0.17		$5.4 \cdot 10^{-15}$		
NiAu	i-s		+0.35/0.48				bistable
Os	i?	a	− 0.22		$4.6 \cdot 10^{-17}$		
Os	i?	d	+0.30		$8.0 \cdot 10^{-16}$		
Pd	s	a	− 0.21		$1.6 \cdot 10^{-15}$		
Pd	s	d	+0.30		$5.6 \cdot 10^{-16}$		
Pd	s	dd	+0.12		$6.5 \cdot 10^{-17}$		
PdFe	s-i		− 0.32				
Pt	s	a	− 0.23		$2.9 \cdot 10^{-14}$		
Pt	s	d	+0.32		$8.4 \cdot 10^{-15}$ *		
Pt	s	dd	+0.08		$3.5 \cdot 10^{-17}$		
PtH-rel.		d	− 0.18		$9.0 \cdot 10^{-16}$		
PtH-rel.		d	− 0.50		$3.0 \cdot 10^{-15}$		
PtH-rel.			+0.40		$1.0 \cdot 10^{-16}$		
PtH$_x$		a	+0.31		$8.0 \cdot 10^{-15}$		
Pt$_s$Pt$_i$	s-i		+0.43				
Re	i?	a	− 0.07		$8.7 \cdot 10^{-16}$		
Re	i?	d	− 0.35		$5.1 \cdot 10^{-16}$		
Rh	s	a	− 0.32		$5.6 \cdot 10^{-15}$		
Rh	s	d	− 0.58		$2.0 \cdot 10^{-14}$		
Ru	i?	a	− 0.14		$1.1 \cdot 10^{-16}$		
Ru	i?	d	+0.26		$9.2 \cdot 10^{-16}$		
Sc	i	d	− 0.21		$3.0 \cdot 10^{-14}$		
Sc	i	dd	− 0.50		$2.0 \cdot 10^{-14}$		

Metal	i/s	d/a	ΔE_A [eV]	E_∞ [eV]	σ_M [cm^2]	T_{Dis}	Comment
Sc	i	ddd	+0.20		$1.1 \cdot 10^{-19}$		
Ta	i?	a	− 0.22		$2.2 \cdot 10^{-17}$		
Ta	i?	d	− 0.58		$>4 \cdot 10^{15}$		
Ta	i?	dd	+0.15		$6.0 \cdot 10^{-17}$		
Ti	i	a	− 0.08		$3.5 \cdot 10^{-14}$		
Ti	i	d	− 0.27	0.004	$1.3 \cdot 10^{-14}$		
Ti	i	dd	+0.28	0.038	$1.9 \cdot 10^{-16}$ *		
TiH			− 0.31		$7.0 \cdot 10^{-16}$	570 K	
TiH			− 0.55		$1.4 \cdot 10^{-13}$	570 K	
V	i	a	− 0.18		$1.6 \cdot 10^{-16}$		
V	i	d	− 0.45		$2.0 \cdot 10^{-15}$		
V	i	dd	+0.32	0.14	$2.2 \cdot 10^{-18}$ *		
VH			− 0.50		$1.6 \cdot 10^{-13}$		
VAu	i-s	a	− 0.20				
VAu	i-s	d	+0.42				
VZn	i-s	d	+0.29		$1.0 \cdot 10^{-16}$		
W	i?	d	+0.40		$5.0 \cdot 10^{-16}$		
Zn	s	aa	− 0.53		$2.5 \cdot 10^{-15}$		
Zn	s	a	+0.32				
ZnB?	i-s		+0.09				
ZnCr	s-i	a	− 0.1		$3.0 \cdot 10^{-15}$		
ZnFe	s-i	a	− 0.47		$1.0 \cdot 10^{-16}$		
ZnMn	s-i	d	+0.18		$2.4 \cdot 10^{-15}$		
ZnV	s-i	d	+0.29		$1.0 \cdot 10^{-16}$		
Zr	i	a	− 0.13		$>10^{-14}$		
Zr	i	d	− 0.42		$>10^{-14}$		
Zr	i	dd	+0.32		$1.3 \cdot 10^{-17}$		

A.2 Activation energies (ΔE_A) of transition metals in numeric order

ΔE_A [eV]	E_∞ [eV]	σ_M [cm²]	Comment/T_{Dis}	Metal	i/s	d/a
− 0.62		$7.2\cdot10^{-14}$		Ir	s	d
− 0.62		$>5\cdot10^{-16}$		Nb	i	d
− 0.58		$2.0\cdot10^{-14}$		Rh	s	d
− 0.58		$>4\cdot10^{-15}$		Ta	i?	d
− 0.56		$1.0\cdot10^{-16}$	300 K	AuH	s-i	a
− 0.55		$7.2\cdot10^{-17}$		Ag	s	a
− 0.55		$1.4\cdot10^{-16}$		Au	s	a
− 0.55		$9.0\cdot10^{-14}$		MnB	i-s	d
− 0.55		$1.4\cdot10^{-13}$	570 K	TiH		
− 0.53		$2.5\cdot10^{-15}$		Zn	s	aa
− 0.50		$2.0\cdot10^{-14}$		Sc	i	dd
− 0.50		$1.6\cdot10^{-13}$		VH		
− 0.50		$3.0\cdot10^{-15}$		PtH-rel.		d
− 0.47		$1.0\cdot10^{-16}$		FeZn	i-s	a
− 0.45				Cd	s	aa
− 0.45		$5.0\cdot10^{-15}$		MnAl	i-s	d
− 0.45		$2.0\cdot10^{-15}$		V	i	d
− 0.44				AuAu	s-i	
− 0.43		$9.0\ 10^{-17}$		Mn	s	a
− 0.42	0.00	$3.1\cdot10^{-15}$		Mn	i	d
− 0.42				MnGa	i-s	d
− 0.42		$>1\cdot10^{-14}$		Zr	i	d
− 0.41		$2.2\cdot10^{-15}$		Co	s	a
− 0.41		$1.2\cdot10^{-16}$		Ni	s	a
− 0.40		$\approx10^{-14}$	400 K	CoH$_x$		
− 0.40		$>2\cdot10^{-14}$		Hf	i	d
− 0.39		$\approx10^{-13}$	400 K	CoH$_x$		
− 0.35		$2.0\cdot10^{-15}$		FeAu	i-s	a
− 0.35		$5.1\cdot10^{-16}$		Re	i?	d
− 0.32				FePd	i-s	
− 0.32		$5.6\cdot10^{-15}$		Rh	s	a
− 0.31		$7.0\cdot10^{-16}$	570 K	TiH		
− 0.28				AgAg	s-i	
− 0.28				Mn$_4$	i	d
− 0.28		$7.5\cdot10^{-18}$		Nb	i	a
− 0.27		$1.6\cdot10^{-15}$	400 K	FeB1	i-s	a
− 0.27	0.004	$1.3\cdot10^{-14}$		Ti	i	d

ΔE_A [eV]	E_∞ [eV]	σ_M [cm^2]	Comment/T_{Dis}	Metal	i/s	d/a
− 0.26		$\approx 10^{-15}$	400 K	CoH$_x$		a?
− 0.24		$9.1 \cdot 10^{-15}$		Ir	s	a
− 0.24				MnAu	i-s	a
− 0.23		$2.9 \cdot 10^{-14}$		Pt	s	a
− 0.22		$\approx 10^{-15}$	400 K	CoH$_x$		a?
− 0.22		$7.3 \cdot 10^{-15}$		Cr	i	d
− 0.22		$4.6 \cdot 10^{-17}$		Os	i?	a
− 0.22		$2.2 \cdot 10^{-17}$		Ta	i?	a
− 0.21		$1.6 \cdot 10^{-15}$		Pd	s	a
− 0.21		$3.0 \cdot 10^{-14}$		Sc	i	d
− 0.20				VAu	i-s	a
− 0.18		$1.6 \cdot 10^{-16}$		V	i	a
− 0.18		$9.0 \cdot 10^{-16}$		PtH-rel.	d	
− 0.17		$\approx 10^{-14}$	400 K	CoH$_x$		a?
− 0.16		$1.9 \cdot 10^{-17}$ *		Cu	s	aa
− 0.14		$1.1 \cdot 10^{-16}$		Ru	i?	a
− 0.13		$> 10^{-14}$		Zr	i	a
− 0.12		$3.1 \cdot 10^{-15}$		Mn	i	a
− 0.10		$> 2 \cdot 10^{-14}$		Hf	i	a
− 0.10				CrZn	i-s	a
− 0.08		$3.5 \cdot 10^{-14}$		Ti	i	a
− 0.07		$5.0 \cdot 10^{-14}$	400 K	CoH$_x$		a?
− 0.07		$5.4 \cdot 10^{-18}$ *		Ni	s	aa
− 0.07		$8.7 \cdot 10^{-16}$		Re	i?	a
+0.08		$3.5 \cdot 10^{-17}$		Pt	s	dd
+0.09		$\approx 10^{-14}$	600 K	CoH$_2$		
+0.09		$3.5 \cdot 10^{-15}$		CuCu	i-s	d
+0.09				ZnB?	i-s	
+0.10		$6.4 \cdot 10^{-14}$	400 K	FeB 1	i-s	d
+0.12		$6.5 \cdot 10^{-17}$		Pd	s	dd
+0.13			metast. RT	FeAl 2	i-s	d
+0.14			metast. RT	FeGa 2	i-s	d
+0.15				FeIn 2	i-s	d
+0.17		$\approx 10^{-14}$		CoH		
+0.17		$5.4 \cdot 10^{-15}$		Ni	s	d
+0.18				MnZn	i-s	d
+0.18		$3.8 \cdot 10^{-16}$		Nb	i	dd
+0.19		$6.0 \cdot 10^{-17}$		Ta	i?	dd
+0.20		$1.1 \cdot 10^{-15}$		FeAl 1	i-s	d
+0.20		$1.1 \cdot 10^{-19}$		Sc	i	ddd

ΔE_A [eV]	E_∞ [eV]	σ_M [cm^2]	Comment/T_{Dis}	Metal	i/s	d/a
+0.21			(M-cent) metast	Cu$_x$		a
+0.22		$\approx 10^{-15}$	bistable	CoH		
+0.22		$3.0 \cdot 10^{-14}$		Cu	s	d
+0.23		$1.8 \cdot 10^{-14}$	870 K	Co	s?	d
+0.25		$4.0 \cdot 10^{-15}$		FeGa 1	i-s	d
+0.26		$9.2 \cdot 10^{-16}$		Ru	i?	d
+0.27			metast. RT	FeIn 1	i-s	d
+0.27	0.064	$2.0 \cdot 10^{-18}$ *		Mn	i	dd
+0.28		$2 \cdot 10^{-14}$		CrB	i-s	d
+0.28		$6.0 \cdot 10^{-16}$		Mo	i	d
+0.28	0.038	$1.9 \cdot 10^{-16}$ *		Ti	i	dd
+0.29		$1.0 \cdot 10^{-16}$		VZn	i-s	d
+0.30		$8.0 \cdot 10^{-16}$		Os	i?	d
+0.30		$5.6 \cdot 10^{-16}$		Pd	s	d
+0.31		$8.0 \cdot 10^{-15}$		PtH$_x$		a
+0.32		$>5 \cdot 10^{-18}$		Hf	i	dd
+0.32		$8.4 \cdot 10^{-15}$ *		Pt	s	d
+0.32	0.14	$2.2 \cdot 10^{-18}$ *		V	i	dd
+0.32				Zn	s	a
+0.32		$1.3 \cdot 10^{-17}$		Zr	i	dd
+0.34		$2.5 \cdot 10^{-15}$		Au	s	d
+0.34		$2.0 \cdot 10^{-16}$		Mn	s	d
+0.35				CrAu	i-s	d
+0.35/0.42			bistable	AuNi	s-i	
+0.38		$7.0 \cdot 10^{-16}$		Ag	s	d
+0.39	0.043	$3.8 \cdot 10^{-17}$ *		Fe	i	d
+0.40		$1.0 \cdot 10^{-16}$		PtH/H$_2$-rel.		
+0.40		$5.0 \cdot 10^{-16}$		W	i?	d
+0.41		$5.0 \cdot 10^{-18}$		Co	s	d
+0.42				VAu	i-s	d
+0.43				PtPt	i-s	
+0.43/32			bistable	CuAu	i-s	
+0.43		$2.6 \cdot 10^{-15}$		FeAu	i-s	d
+0.45				CrAl	i-s	d
+0.46		$1.5 \cdot 10^{-14}$		Cu	s	a
+0.48				CrGa	i-s	d
+0.50				Cd	s	a
+0.57				MnAu	i-s	d

References

Chapter 1

1.1   D.V. Lang: Space-charge spectroscopy in semiconductors, in *Thermally Stimulated Relaxation in Solids*, ed. by. P. Bräunlich, Topics Appl. Phys., Vol.34 (Springer, Berlin, Heidelberg 1979) Chap.3

1.2   A.G. Milnes: *Deep Impurities in Semiconductors* (Wiley, New York 1973)

1.3   W. Schröter, M. Seibt, D. Gilles: High-temperature properties of 3d transition elements in silicon, in *Electronic Structure and Properties of Semiconductors*, ed. by W. Schröter (VHC, Weinheim 1991) p.539

1.4   M. Lannoo, J. Bourgoin: *Point Defects in Semiconductors I*, Springer Ser. Solid-State Sci., Vol.22 (Springer, Berlin, Heidelberg 1981)
       J. Bourgoin, M. Lannoo: *Point Defects in Semiconductors II*, Springer Ser. Solid-State Sci., Vol.35 (Springer, Berlin, Heidelberg 1989)

1.5   L. Reimer: *Transmission Electron Microscopy*, 4th edn., Springer Ser. Opt. Sci., Vol. 36 (Springer, Berlin, Heidelberg 1997)

1.6   M. Seibt, K. Graff: MRS Proc. **104**, 215 (1988)

Chapter 2

2.1   W. Schröter, M. Seibt, D. Gilles: High-temperature properties of 3d transition elements in silicon, in *Electronic Structure and Properties of Semiconductors*, ed. by W. Schröter (VHC, Weinheim 1991) p.539

2.2   K. Graff, H. Pieper: The behavior of transition and noble metals in silicon crystals, in *Semiconductor Silicon 1981*, ed. by H.R. Huff, J. Kriegler, Y. Takeishi (Electrochem. Soc., Pennington, NJ 1981) p.331

2.3   E. Ohta, K. Kakishita, H.Y. Lee: J. Appl. Phys. **65**, 3928 (1989)

2.4   F.D. Auret, R. Kleinhenz, C. P. Schneider: Appl. Phys. Lett. **44**, 209 (1984)

2.5   O. Kumagai, K. Kamko: J. Appl. Phys. **51**, 5430 (1980); ibid. **52**, 5143 (1981)

2.6   O.V. Aleksandrov, V.V. Koslovskii, V.V. Popov, B.E. Samorukov: Phys. Status Solidi (a) **110**, K61 (1988)

2.7   A. Van Wieringen, N. Warmholtz: Physica **22**, 849 (1956)

2.8   W. Jost: Kapazitätsspektroskopie an Übergangsmetall-Wasserstoffkomplexen in Silizium. Dissertation, University of Stuttgart (1996)

2.9   J.I. Pankove, D.E. Carlson, J.E. Berkeyheiser, R.O Wance: Phys. Rev. Lett. **51**, 2224 (1983)

2.10   S.J. Pearton, J.W. Corbett, M. Stavola: *Hydrogen in Crystalline Semiconductors*, Springer Ser. Mater. Sci., Vol.16 (Springer, Berlin, Heidelberg 1992)

2.11   H. Feichtinger, E. Sturm: Mater. Sci. Forum **143-147**, 111 (1994)

2.12   J.U. Sachse, E.Ö. Sveinbjörnsson, W. Jost, J. Weber, H. Lemke: Rhys. Rev. B
        55, 16176 (1997)
2.13   T. Sadoh, M. Watanabe, H. Nakashima, T. Tsurushima: Mater. Sci. Forum
        143-147, 939 (1994)
2.14   E.Ö. Sceinbjörnsson, O. Engström: Phys. Rev. B **52**, 4884 (1995)
2.15   P.F. Schmidt: J. Electrochem. Soc. **130**, 196 (1983)
2.16   B.O. Kolbesen, W. Bergholz, H. Wendt: Mater. Sci. Forum **38-41**, 1 (1989)
2.17   K. Graff, H. Fischer: Carrier lifetime in silicon and its impact on solar cell charac-
        teristics, in *Solar Energy Conversion*, ed. by B.O. Seraphin, Topics Appl. Phys.,
        Vol.31 (Springer, Berlin, Heidelberg 1979) p.173
2.18   K.P. Lisiak, A.G. Milnes: J. Appl. Phys. **46**, 5229 (1975)
2.19   M. Kittler, J. Lärz, W. Seifert, M. Seibt, W. Schröter: Appl. Phys. Lett. **58**, 911
        (1991)
2.20   T.I. Chappel, P.W. Chye, M.A. Tavel: Solid State Electr. **26**, 33 (1983)
2.21   M. Seibt, K. Graff: MRS Proc. **104**, 215 (1988)

Chapter 3

3.1   F. Beeler, O.K. Andersen, M. Scheffler: Phys. Rev. Lett. **55**, 1498 (1985)
3.2   F. Beeler, M. Scheffler: Mater. Sci. Forum **38-41**, 257 (1989)
3.3   F. Beeler, O.K. Andersen, M. Scheffler: Phys. Rev. B **41**, 1603 (1990)
3.4   J. Utzig: J. Appl. Phys. **65**, 3868 (1989)
3.5   E.R. Weber: Appl. Phys. A **30**, 1 (1983)
3.6   K. Graff: Unpublished (1980-90)
3.7   S.P. Murarka: *Silicides for VLSI Application* (Academic, Orlando 1983)
3.8   S. Hocine, D. Mathiot: Mater. Sci. Forum **38-41**, 725 (1989)
3.9   D. Gilles, W. Bergholz, W. Schröter: J. Appl. Phys. **59**, 3590 (1986)
3.10   D. Grünebaum, T. Czekalla, N.A. Stolwijk, H. Mehrer, I. Yonenaga, K. Sumino:
        Appl. Phys. A **53**, 65 (1991)
3.11   H. Bracht, H, Overhof: Phys. Status Solidi (a) **158**, 47 (1996)
3.12   R.N. Hall, H. Racette: J. Appl. Phys. **35**, 379 (1964)
3.13   D. Gilles: Einfluß der elektronischen Struktur auf Diffusion, Löslichkeit und Paar-
        bildung von 3d-Übergangselementen in Silizium, Dissertation, University of Göt-
        tingen (1987)
3.14   D. Gilles, W. Schröter, W. Bergholz: Phys. Rev. B **41**, 5770 (1990)
3.15   D.A. van Wezep, C.A.J. Ammerlaan: In *13th Int'l Conf. Defects in Semicond.*, ed.
        by L.C. Kimerling, J.M. Parsey (Metallugical Soc. of AIME, Warrendale, PA
        1984) p.863
3.16   H. Lemke: Phys. Status Solidi (a) **83**, 637 (1984)
3.17   G.W. Ludwig, H.H. Woodbury: *Solid State Phys.* **13**, 223 (Academic, New York
        1962)
3.18   F.C. Frank, D. Turnbull: Phys. Rev. **104**, 617 (1956)
3.19   U. Gösele, F. Morehead, W. Frank, A. Seeger: Appl. Phys. Lett. **38**, 157 (1981)
3.20   U. Gösele, T.Y. Tan: The role of vacancies and self-interstitials in diffusion and
        agglomeration phenomena in silicon, in *Aggregation Phenomena of Point Defects in
        Silicon*, ed. by E. Sirtl, J. Goorissen, P. Wagner (Electrochem. Soc., Penning-
        ton, NJ 1983) p.17

3.21   J.G. Hauber: Diffusion and Löslichkeit von Platin in Silizium. Dissertation, University of Stuttgart (1986)

3.22   H. Nakashima, T. Sadoh: Mater. Sci. Forum **143-147**, 761 (1994)

3.23   N.T. Bendik, V.S. Garnyk, L.S. Milkovski: Sov. Phys. - Solid State **12**, 150 (1970)

3.24   T. Heiser, A. Mesli: Phys. Rev. Lett. **68**, 978 (1992)

3.25   J. Utzig, D. Gilles: Mater. Sci. Forum **38-41**, 729 (1989)

3.26   M.K. Bakhadyrkhanov, S. Zainalidinov, A. Khanidov: Sov. Phys. - Semicond. **14**, 243 (1980)

3.27   A. Mesli, T. Heiser, E. Mulheim: Mater. Sci. Eng. B **25**, 141 (1994)

3.28   W. Schröter, M. Seibt, D. Gilles: High-temperature properties of 3d transition elements in silicon, in *Electronic Structure and Properties of Semiconductors*, ed. by W. Schröter (VHC, Weinheim 1991) p. 539

3.29   K. Graff, H. Pieper: The behavior of transition and noble metals in silicon crystals, in *Semiconductor Silicon* 1981, ed. by H.R. Huff, J. Kriegler, Y. Takeishi (Electrochem. Soc., Pennington, NJ 1981), p. 331

3.30   J. Weber, H. Bauch, R. Sauer: Phys. Rev. B **25**, 7688 (1982)

3.31   H. Lemke: Characterization of transition metal-doped silicon crystals prepared by float zone technique, in *Semiconductor Silicon 1994*, ed. by H.R. Huff, W. Bergholz, K. Sumino (Electrochem. Soc., Pennington, NJ 1994) p. 695

3.32   D.V. Lang: J. Appl. Phys. **45**, 3023 (1974)

3.33   E. Schibli, A.G. Milnes: Mater. Sci. Eng. **2**, 173 (1967)

3.34   B.L. Sharma: *Diffusion in Semiconductors* (Trans. Tech., Claustal-Zellerfeld 1970)

3.35   L.C. Kimerling, J.L. Benton, J.J. Rubin: Transition metal impurities in silicon, in *Defects and Radiation Effects in Semiconductors 1980*, Inst. Phys. Conf. Ser. **59**, 217 (1981)

3.36   J.W. Chen, A.G. Milnes, A. Rohatgi: Solid State Electron. **22**, 801 (1979)

3.37   J. Utzig: J. Appl. Phys. **64**, 3629 (1988)

3.38   G.G. DeLeo, G.D. Watkins, W.B. Fowler: Phys. Rev. B **23**, 1851 (1981); ibid. **25**, 4962 (1982)

3.39   P. Vogl: Private commun. (1981)

3.40   A. Zunger, U. Lindefelt: Phys. Rev. B **26**, 5989 (1982); ibid B **27**, 1191 (1983); ibid. B **28**, 3628 (1983)

3.41   S.D. Brotherton, P. Bradley, A. Gill: J. Appl. Phys. **57**, 1941 (1985)

3.42   H. Lemke: Phys. Status Solidi (a) **91**, 649 (1985)

3.43   H. Lemke: Mater. Sci. Forum **196-201**, 683 (1995)

3.44   H. Kitagawa, H. Nakashima, K. Hashimoto: In *Memoirs of the Faculty of Engineering*, Kyushu Univ. **46**, 119 (1986)

3.45   W. Jost, J. Weber, H. Lemke: Semicond. Sci. Techn. **11**, 22 (1996); ibid. **11**, 525 (1996)

3.46   N. Achtziger, T. Licht, U. Reislöhner, M. Rüb, W. Witthuhn: *The Physics of Semiconductors*, ed. by M. Scheffler, R. Zimmermann (World Scientific, Singapore 1996) Vol. 4, p. 2717

3.47   J. Weber: Priv. commun. (1997)

3.48   K. Graff, H. Pieper: J. Electrochem. Soc. **128**, 669 (1981)

3.49   H. Lemke: Phys. Status Solidi (a) **76**, 223 (1983); ibid. (a) **76**, K193 (1983)

3.50   H. Wendt, H. Cerva, V. Lehmann, W. Pamler: J. Appl. Phys. **65**, 2402 (1989)

3.51   L.C. Kimerling: Defect characterization by junction spectroscopy, in *Defects in Semiconductors*, ed. by J. Narajan, T.Y. Tan (Elsevier, New York 1981) p. 85

3.52 W. Bergholz, G. Zoth, H. Wendt, S. Sauter, G. Asam: Siemens Forsch. Entwickl. Ber. **16**, 241 (1987)

3.53 A. Chantre, D. Bois: Phys. Rev. B **31**, 7979 (1985)

3.54 H. Feichtinger, J. Oswald, R. Czaputa, P. Vogl, K. Wünstel: 13th Int'l Conf. on Defects in Semiconductors Coronado 1984, ed. by L.C. Kimerling, J.M. Parsey (Metallurgical Soc. of AIME, Warrendale, PA 1984) p.855

3.55 H. Lemke: Phys. Status Solidi (a) **72**, 177 (1982); ibid. **75**, 473 (1983)

3.56 S.D. Brotherton, P. Bradley, A. Gill: J. Appl. Phys. **55**, 952 (1984); ibid. **57**, 1783 (1985)

3.57 A. Van Wieringen, N. Warmholtz: Physica **22**, 849 (1956)

3.58 W. Jost: Kapazitätsspektroskopie an Übergangsmetall-Wasserstoffkomplexen in Silizium. Dissertation, University of Stuttgart (1996)

3.59 N.M. Johnson, C. Herring: Mater. Sci. Forum **143-147**, 867 (1994)

3.60 S.J. Pearton, J.W. Corbett, M. Stavola: Hydrogen in Crystalline Semiconductors, Springer Ser. Mater. Sci., Vol. 14 (Springer, Berlin, Heidelberg 1992)

3.61 J.I. Pankove, D.E. Carlson, J.E. Berkeyheiser, R.O. Vance: Phys. Rev. Lett. **51**, 2224 (1983)

3.62 T. Sadoh, M. Watanabe, H. Nakashima, T. Tsurushina: Mater. Sci. Forum **143-147**, 939 (1994)

3.63 J.W. Sachse, E.Ö. Sweinbjörnsson, W. Jost, J. Weber, H. Lemke: Phys. Rev. B **55**, 16176 (1997)

3.64 H. Feichtinger, E. Sturm: Mater. Sci. Forum **143-147**, 111 (1994)

3.65 W. Jost, J. Weber: Phys. Rev. B **54**, R11038 (1996)

3.66 E.Ö. Sweinbjörnsson, O. Engström: Appl. Phys. Lett. **61**, 2323 (1992)

3.67 H. Lemke: Substitional transition metal defects in silicon grown-in by the float zone technique, in High Purity Silicon IV, Vol.96-13, ed. by C.L. Claeys, P. Rai-Choudhury, P. Stallhofer, J.E. Maurits (Electrochem. Soc., Pennington, NJ 1996) p.272 H. Lemke: Mater. Sci. Forum **196-201**, 683 (1995)

3.68 M. Kittler, J. Lärz, W. Seifert, M. Seibt, W. Schröter: Appl. Phys Lett. **58**, 911 (1991)

3.69 M. Seibt, K. Graff: J. Appl. Phys. **63**, 4444 (1988)

3.70 M. Seibt: Elektronenmikroskopische Untersuchungen des Ausscheidungsverhaltens von Nickel in Silizium, Dissertation, University of Göttingen (1986)

3.71 M. Seibt: Homogeneous and heterogeneous precipitation of copper in silicon, in Semiconductor Silicon 1990, ed. by H.R. Huff, K.G. Barraclough, J.I. Chikawa (Electrochem. Soc., Pennington, NJ 1990) p.636

3.72 M. Seibt, W. Schröter: Philos. Mag. A **59**, 337 (1989)

3.73 M. Seibt, W. Schröter: Lokalisierung und Identifizierung von Mikrodefekten. Forschungsbericht, Bundesministerium für Forschung und Technologie FRG (1989)

3.74 U. Gösele, T.Y. Tan: In Mater. Sci. &Techn. IV, ed. by R.W. Cahn, P. Haasen, E. Kramer (VCH, Weinheim 1991) p.197

3.75 H. Lemke: Priv. commun. (1998)

3.76 R. Falster, D. Gambaro, M. Olmo, M. Cornara, H. Korb: MRS Proc. **510**, 27 (1998)

3.77 K. Graff: Unpubl. (1985)

3.78 K. Graff: The precipitation of transition metals in silicon crystal wafers, in Aggregation Phenomena of Point Defects in Silicon, ed. by E. Sirtl, J. Goorissen, P. Wagner (Electrochem. Soc. Pennington, NJ 1983) p.121

3.79 E. Sirtl, A. Adler: Z. Metallkunde **52**, 529 (1961)

3.80   K.H. Yang: J. Electrochem. Soc. **131**, 1140 (1984)
3.81   K. Graff, P. Heim, H. Pieper:: J. Electrochem. Soc. **141**, 2821 (1994)
3.82   M. Wright Jenkins: J. Electrochem. Soc. **124**, 757 (1974)
3.83   F. Secco d'Aragona: J. Electrochem. Soc. **119**, 948 (1972)
3.84   D.G. Schimmel: J. Electrochem. Soc. **123**, 734 (1976)
3.85   T.C. Chandler: J. Electrochem. Soc. **137**, 944 (1990)
3.86   *Annual Book of ASTM Standards* Vol. 10.05 (ASTM, Philadelphia, PA 1992) F1049

Chapter 4

4.1   F. Beeler, O.K. Andersen, M. Scheffler: Phys. Rev. Lett. **55**, 1498 (1985)
4.2   E.R. Weber: Appl. Phys. A **30**, 1 (1983)
4.3   K. Graff: Unpublished (1980-90)
4.4   H. Nakashima, T. Sadoh: Mater. Sci. Forum **143-147**, 761 (1994)
4.5   T. Heiser, A. Mesli: Phys. Rev. Lett. **68**, 978 (1992)
4.6   Average value calculated from results of
        K. Graff, H. Pieper: The behavior of transition and noble metals in silicon crystals, in *Semiconductor Silicon 1981*, ed. by H.R. Huff, J. Kriegler, Y. Takeishi (Electrochem. Soc., Pennington, NJ 1981) p.331
        H. Lemke: Phys. Status Solidi (a) **64**, 215 (1981)
        K. Wünstel, P. Wagner: Appl. Phys. A **27**, 207 (1982)
        S.D. Brotherton, P. Bradley, A. Gill: J. Appl. Phys. **57**, 1941 (1985)
        H. Nakashima, T. Sadoh: Mater. Sci. Forum **143-147**, 761 (1994)
4.7   K. Graff. H. Pieper: J. Electrochem. Soc. **128**, 669 (1981)
4.8   L.C. Kimerling: Defect characterization by junction spectroscopy, in *Defects in Semiconductors*, ed. by J. Narajan, T.Y. Tan (Elsevier North Holland, New York 1981) p.85
4.9   A. Chantre, L.C. Kimerling: Mater. Sci. Forum **10-12**, 387 (1986)
4.10   Average values were calculated with results presented in
        H. Lemke: Phys. Status Solidi (a) **76**, 223 (1983); ibid. (a) **76**, K193 (1983)
        H. Nakashima, T. Sadoh, T. Tsurushima: J. Appl. Phys. **73**, 2803 (1993)
        S.D. Brotherton, P. Bradley, A. Gill: J. Appl. Phys. **55**, 952 (1984)
        G. Zoth, W. Bergholz: Fall meeting, Electrochem. Soc. (Phoenix, AZ, 1991) Ext. abstracts, p.643
        S. Ghatnekar-Nilsson, M. Kleverman, P. Amanuelsson, H.G. Grimmeiss: Mater. Sci. Forum **143-147**, 171 (1994)
        W. Gehlhoff, U. Reese: In *Defect Engineering in Semiconductor Growth, Processing and Device Technology*, ed. by S. Ashok, J. Chevallier, K. Sumino, E. Weber. MRS Proc. **262**, 507 (Mater. Res. Soc., Pittsburgh, PA 1992)
4.11   H. Nakashima, T. Sadoh, T. Tsurushima: Phys. Rev. B **49**, 16983 (1994)
4.12   Average values were calculated with results presented in
        H. Feichtinger, J. Oswald, R. Czaputa, P. Vogl, K. Wünstel: In *Defects in Semiconductors*, ed. by L.C. Kimerling, J.M. Parsey (Metallurgical Soc. of AIME, Warrendale, PA 1984) p.855
        K. Wünstel, P. Wagner: Appl. Phys. A **27**, 207 (1982)
        H. Lemke: Phys. Status Solidi (a) **76**, 223 (1983); ibid. (a) **76**, K193 (1983)
        H. Conzelmann, K. Graff, E.R. Weber: Appl. Phys. A **30**, 168 (1983)
        A. Chantre, L.C. Kimerling: Mater. Sci. Forum **10-12**, 387 (1986)
        A. Chantre, D. Bois: Phys. Rev. B **31**, 7979 (1985)

4.13   Average values were calculated with results presented in
        K. Wünstel, P. Wagner: Appl. Phys. A **27**, 207 (1982)
        H. Lemke: Phys. Status Solidi (a) **76**, 223 (1983); ibid. (a) **76**, K193 (1983)
        H. Feichtinger, J. Oswald, R. Czaputa, P. Vogl, K. Wünstel: In *Defects in Semi-conductors*, ed. by L.C. Kimerling, J.M. Parsey (Metallurgical Soc. of AIME, Warrendale, PA 1984) p.855
        A. Chantre, L.C. Kimerling: Mater. Sci. Forum **10-12**, 387 (1986)
4.14   A. Chantre, D. Bois: Phys. Rev. B **31**, 7979 (1985)
4.15   S.D. Brotherton, P. Bradley, A. Gill: J. Appl. Phys. **55**, 952 (1984); ibid. **57**, 1941 (1985)
4.16   H. Lemke: Phys. Status Solidi (a) **72**, 177 (1982); ibid. **75**, 473 (1983)
4.17   R. Czaputa: Appl. Phys. A **49**, 431 (1989)
4.18   G. Zoth, W. Bergholz: Fall meeting (Electrochem. Soc., Phoenix, AZ 1991) Ext. abstracts, p.643
4.19   W. Bergholz, G. Zoth, H. Wendt, S. Sauter, G. Asam: Siemens Forsch.- u. Entwickl.-Ber. **16**, 241 (1987)
4.20   A.G. Cullis, L.E. Katz: Phil. Mag. **30**, 1419 (1974)
        P.D. Augustus, J. Knights, L.W. Kennedy: J. of Microsc. **118**, 315 (1980)
4.21   M. Seibt, W. Schröter: Lokalisierung und Identifizierung von Mikrodefekten. Forschungsbericht, Bundesministerium für Forschung und Technologie FRG (1989)
4.22   K. Honda, T. Nakanishi, A. Ohsawa, N. Toyokura: J. Appl. Phys. **62**, 1960 (1987)
4.23   K. Graff, H. Pieper: The behavior of transition and noble metals in silicon crystals, in *Semiconductor Silicon 1981*, ed. by H.R. Huff, J. Kriegler, Y. Takeishi (Electrochem. Soc., Pennington, NJ 1981) p.331
        K. Graff: The precipitation of transition metals in silicon crystal wafers, in *Aggregation Phenomena of Point Defects in Silicon*, ed. by E. Sirtl, J. Goorissen, P. Wagner (Electrochem. Soc., Pennington, NJ 1983) p.121
        K. Graff, P. Heim, H. Pieper: J. Electrochem. Soc. **141**, 2821 (1994)
4.24   M.K. Bakhadyrkhanov, S. Zainalidinov, A. Khanidov: Sov. Phys. - Semicond. **14**, 243 (1980)
4.25   P. Bonzel: Phys. Status Solidi **20**, 493 (1967)
4.26   K.H. Yoon, L.L. Levenson: J. Electr. Mat. **4**, 1249 (1975)
4.27   D. Gilles: Einfluß der elektronischen Struktur auf Diffusion, Löslichkeit und Paarbildung von 3d-Übergangselementen in Silizium, Dissertation, University of Göttingen (1987)
        D. Gilles, W. Schröter, W. Bergholz: Phys. Rev. B **41**, 5770 (1990)
        J. Utzig, D. Gilles: Mater. Sci. Forum **38-41**, 729 (1989)
4.28   L. Zhong, F. Shimura: Jpn. J. Appl. Phys. **32**, L1113 (1993)
4.29   M.-A. Nicolet, S.S. Lau: Formation and characterization of transition-metal silicides in materials and process characterization. *VLSI Electronics* **6**, 330 (Academic, New York 1983)
4.30   S.P. Murarka: *Silicides for VLSI Application* (Academic, Orlando, FL 1983)
4.31   H. Lemke: Phys. Status Solidi (a) **99**, 205 (1987)
        H. Lemke: Characterization of transition metal-doped silicon crystals prepared by float zone technique, in *Semiconductor Silicon 1994*, ed. by H.R. Huff, W. Bergholz, K. Sumino (Electrochem. Soc., Pennington, NJ 1994) p.695
4.32   Average values were calculated with results presented in
        H. Lemke: Characterization of transition metal-doped silicon crystals prepared by float zone technique, in *Semiconductor Silicon 1994*, ed. by H.R. Huff, W. Berg-

holz, K. Sumino (Electrochem. Soc., Pennington, NJ 1994) p.695

G. Chiavarotti, M. Conti, A. Messina: Solid-State Electron. **20**, 907 (1977)

H. Nakashima, T. Sadoh: Mater. Sci. Forum **143-147**, 761 (1994)

H. Lemke: Phys. Status Solidi (a) **99**, 205 (1987)

M. Yoshida, K. Saito: Jpn. J. Appl. Phys. **6**, 573 (1967)

H. Lemke: Characterization of transition metal-doped silicon crystals prepared by float zone technique, in *Semiconductor Silicon 1994*, ed. by H.R. Huff, W. Bergholz, K. Sumino (Electrochem. Soc., Pennington, NJ 1994) p.695

4.33 Average values were calculated with results presented in
H. Lemke: Phys. Status Solidi (a) **99**, 205 (1987)
H. Nakashima, T. Sadoh: Mater. Sci. Forum **143-147**, 761 (1994)
S.J. Pearton, A.J. Tavendale: J. Appl. Phys. **54**, 1375 (1983)

4.34 H. Lemke: Mater. Sci. Forum **196-201**, 683 (1995)

4.35 H. Lemke: Phys. Status Solidi (a) **99**, 205 (1987)

4.36 M. Seibt, K. Graff: MRS Proc. **104**, 215 (1988)

4.37 M. Seibt, K. Graff: J. Appl. Phys. **63**, 4444 (1988)

4.38 M. Seibt: Elektronenmikroskopische Untersuchungen des Ausscheidungsverhaltens von Nickel in Silizium, Dissertation, University of Göttingen (1986)

4.39 M. Seibt, W. Schröter: Philos, Mag. A **59**, 337 (1989)

4.40 T. Hosoya, Y. Ozaki, K. Hirata: J. Electrochem. Soc. **132**, 2436 (1985)

4.41 P.F. Schmidt: J. Electrochem. Soc. **130**, 196 (1983)

4.42 J.K. Solberg: Acta crystallog. A **34**, 684 (1978)

4.43 R.N. Hall, H. Racette: J. Appl. Phys. **35**, 379 (1964)

4.44 A. Mesli, T. Heiser: Interstitial defect reactions in silicon: The case of copper, in *Defect and Diffusion* Forum Vols. 131-132 (Scitec Publications, Switzerland 1996) p.89

4.45 T. Zundel, J. Weber, B. Benson: Appl. Phys. Lett. **53**, 1426 (1988)

4.46 A. Mesli, T. Heiser, E. Mulheim: Mater. Sci. Eng. B **25**, 141 (1994)

4.47 T. Prescha, T. Zundel, J. Weber, H. Prigge, P. Gerlach: Mater. Sci. Eng. B **4**, 79 (1989)

4.48 T. Heiser, A. Mesli: Appl. Phys. A **57**, 325 (1993)

4.49 M.O. Aboelfotoh, B.G. Svensson: Phys. Rev B **44**, 12742 (1991)

4.50 M. Seibt: Metal impurity precipitation in silicon, presented at 192 nd Meeting, Electrochem. Soc. (Paris 1997)

4.51 J. Weber, H. Bauch, R. Sauer: Phys. Rev. B **25**, 7688 (1982)

4.52 H.B. Erzgräber, K. Schmalz: J. Appl. Phys. **78**, 4066 (1995)

4.53 Average values were calculated with results presented in
H. Lemke: Phys. Status Solidi (a) **83**, 637 (1984)
H. Lemke: Characterization of transition metal-doped silicon crystals prepared by float zone technique, in *Semiconductor Silicon 1994*, ed. by H.R. Huff, W. Bergholz, K. Sumino (Electrochem. Soc., Pennington, NJ 1994) p.695
S.D. Brotherton, I.R. Ayres, A. Gill, H.W. van Kesteren, F.J.A.M. Greidanus: J. Appl. Phys. **62**, 1826 (1987)

4.54 Average values were calculated with results presented in
H. Lemke: Phys. Status Solidi (a) **83**, 637 (1984)
H. Lemke: Characterization of transition metal-doped silicon crystals prepared by float zone technique, in *Semiconductor Silicon 1994*, ed. by H.R. Huff, W. Bergholz, K. Sumino (Electrochem. Soc., Pennington, NJ 1994) p.695
H. Lemke: Substitutional transition metal defects in silicon grown-in by the float

zone technique. Mater. Sci. Forum **196-201**, 683 (1995)

S.D. Brotherton, I.R. Ayres, A. Gill, H.W. van Kesteren, F.J.A.M. Greidanus: J. Appl. Phys. **62**, 1826 (1987)

N. Toyama: Solid-State Electron. **26**, 37 (1983)

4.55 Average values were calculated with results presented in

H. Lemke: Phys. Status Solidi (a) **83**, 637 (1984)

H. Lemke: Characterization of transition metal-doped silicon crystals prepared by float zone technique, in *Semiconductor Silicon 1994*, ed. by H.R. Huff, W. Bergholz, K. Sumino (Electrochem. Soc., Pennington, NJ 1994) p.695

H. Lemke: Substitutional transition metal defects in silicon grown-in by the float zone technique. Mater. Sci. Forum **196-201**, 683 (1995)

S.D. Brotherton, I.R. Ayres, A. Gill, H.W. van Kesteren, F.J.A.M. Greidanus: J. Appl. Phys. **62**, 1826 (1987)

N. Toyama: Solid-State Electron. **26**, 37 (1983)

4.56 Average values were calculated with results presented in

J. Weber, H. Bauch, R. Sauer: Phys. Rev. B **25**, 7688 (1982)

H.B. Erzgräber, K. Schmalz: J. Appl. Phys. **78**, 4066 (1995)

S.D. Brotherton, I.R. Ayres, A. Gill, H.W. van Kesteren, F.J.A.M. Greidanus: J. Appl. Phys. **62**, 1826 (1987)

H. Lemke: Substitutional transition metal defects in silicon grown-in by the float zone technique, in *High Purity Silicon IV*, Vol.96-13, ed. by C.L. Claeys, P. Rai-Choudhury, P. Stallhofer, J.E. Maurits (Electrochem. Soc., Pennington, NJ 1996) p.272

A. Mesli, T. Heiser: Interstitial defect reactions in silicon: The case of copper, in *Defect and Diffusion* Forum, Vols. 131-132 (Scitec Publications, Switzerland 1996) p.89

K. Graff, H. Pieper: The behavior of transition and noble metals in silicon crystals, in *Semiconductor Silicon 1981*, ed. by H.R. Huff, J. Kriegler, Y. Takeishi (Electrochem. Soc., Pennington, NJ 1981) p.33

4.57 R. Czaputa: Appl. Phys. A **49**, 431 (1989)

4.58 H. Wendt, H. Cerva, V. Lehmann, W. Pamler: J. Appl. Phys. **65**, 2402 (1989)

4.59 A. Rohatgi, R.H. Hopkins, J.R. Davis, R.B. Campbell, H.C. Mollenkopf: Solid-State Electron. **23**, 1185 (1980)

4.60 H. Lemke: Phys. Status Solidi (a) **76**, K193 (1983)

4.61 L.J. Cheng, D.C. Leung: in *Electronic and Optical Properties of Polycrystalline or Impure Semiconductors and Novel Silicon Growth Methods*, ed. by K.V. Ravi, B. O'Mara (Electrochem. Soc., Pennington, NJ 1980) p.46

4.62 K. Graff, H.A. Hefner, H. Pieper: MRS Proc. 36, 19 (1985)

4.63 W. Frank: Private commun. 1989

4.64 F.C. Frank, D. Turnbull: Phys. Rev. **104**, 617 (1956)

4.65 U. Goesele, F. Morehead, W. Frank, A. Seeger: Appl. Phys. Lett. **38**, 157 (1981)

4.66 S.J. Pearton, E.E. Haller: J. Appl. Phys. **54**, 3613 (1983)

4.67 H. Lemke: Phys. Status Solidi (a) **86**, K39 (1984)

4.68 J.W. Sachse, W. Jost, J. Weber, H. Lemke: Appl. Phys. Lett. **71**, 1379 (1997)

4.69 H. Lemke: 18th Int'l. Conf. on Defects in Semiconductors (Sendai, Jpn. 1995)

M. Pugnet, J. Barbolla, J.C. Brabant, F. Saint-Yves, M. Brousseau: Phys. Status Solidi (a) **35**, 533 (1976)

4.70 M. Pugnet, J. Barbolla, J.C. Brabant, F. Saint-Yves, M. Brousseau: Phys. Status Solidi (a) **35**, 533 (1976)

4.71   J.A. Pals: Solid-State Electron. **17**, 1139 (1974)

4.72   L. So, S.K. Ghandi: Solid-State Electron. **20**, 113 (1977)

4.73   Z. Jie, R. Shengyang, H. Hong, G. Weikun, J. Xiujiang, L. Shuying: Mater. Sci. Forum **10-12**, 723 (1986)

4.74   A. Mirzaev, S. Makhamov: Sov. Phys. Semicond. **22**, 746 (1988)

4.75   J.G. Hauber: Diffusion und Löslichkeit von Platin in Silizium. Dissertation, University of Stuttgart (1986)

4.76   S. Braun, H.G. Grimmeiss, K. Spann: J. Appl. Phys. **48**, 3883 (1977)

4.77   W. Jost: Kapazitätsspektroskopie an Übergangsmetall-Wasserstoff-Komplexen in Silizium, Dissertation, University of Stuttgart (1996)

4.78   J.W. Sachse, E.Ö. Sweinbjörnsson, W. Jost, J. Weber, H. Lemke: Phys. Rev. B **55**, 16176 (1997)

4.79   H. Lemke: Substitutional transition metal defects in silicon grown-in by the float zone technique, in *High Purity Silicon IV*, Vol.96-13, ed. by C.L. Claeys, P. Rai-Choudhury, P. Stallhofer, J.E. Maurits (Electrochem. Soc., Pennington, NJ 1996) p.272

4.80   K.P. Lisiak, A.G. Milnes: J. Appl. Phys. **46**, 5229 (1975)

4.81   A.O. Evwaraye, E. Sun: J. Appl. Phys. **47**, 3172 (1976)

4.82   H. Bracht, H. Overhof: Phys. Status Solidi (a) **158**, 47 (1996)
       H. Bracht: Defects and Diffusion Forum **143-147**, 979 (1997)
       A.R. Schachtrup, H. Bracht, I. Yonenaga, H. Mehrer: Defects and Diffusion Forum **143-147**, 1021 (1997)

4.83   R. Falster: Appl. Phys. Lett. **46**, 737 (1985)

4.84   A.A. Lebedev, N.A. Sobolev, B.M. Urunbaev: Sov. Phys. - Semicond. **15**, 880 (1981)

4.85   S.D. Brotherton, P. Bradley: J. Appl. Phys. **53**, 1543 (1982)

4.86   J.C.M. Henning, E.C.J. Egelmeers: Phys. Rev. B **27**, 4002 (1983)

4.87   Average values were calculated with results presented in
       S.D. Brotherton, P. Bradley: J. Appl. Phys. **53**, 1543 (1982)
       S.D. Brotherton, P. Bradley, J. Bicknell: J. Appl. Phys. **50**, 3396 (1979)
       A.O. Evwaraye, E. Sun: J. Appl. Phys. **47**, 3172 (1976)
       Y.K. Kwon, I. Ishikawa, H. Kuwano: J. Appl. Phys. **61**, 1055 (1987)
       A.A. Lebedev, N.A. Sobolev, B.M. Urunbaev: Sov. Phys. - Semicond. **15**, 880 (1981)
       M. Pugnet, J. Barbolla, J.C. Brabant, F. Saint-Yves, P.M. Sandow, M.B. Das, J. Stach: J. Electrochem. Soc. **7**, 687 (1978)
       W. Jost: Kapazitätsspektroskopie an Übergangsmetall-Wasserstoff-Komplexen in Silizium, Dissertation, University of Stuttgart (1996)
       H. Lemke: Phys. Status Solidi (a) **76**, K193 (1983)
       L.J. Cheng, D.C. Leung: In *Electronic and Optical Properties of Polycrystalline or Impure Semiconductors and Novel Silicon Growth Methods*, ed. by K.V. Ravi, B. O'Mara (Electrochem. Soc., Pennington, NJ 1980) p.46
       J.W. Sachse, E.Ö. Sweinbjörnsson, W. Jost, J. Weber, H. Lemke: Phys. Rev. B **55**, 16176 (1997)

4.88   Average values were calculated with results presented in
       S.D. Brotherton, P. Bradley: J. Appl. Phys. **53**, 1543 (1982)
       S.D. Brotherton, P. Bradley, J. Bicknell: J. Appl. Phys. **50**, 3396 (1979)
       M.D. Miller, H. Schade, C.J. Nuese: J. Appl. Phys. **47**, 2569 (1976)
       S.J. Pearton, E.E. Haller: J. Appl. Phys. **54**, 3613 (1983)

M. Pugnet, J. Barbolla, J.C. Brabant, F. Saint-Yves, P.M. Sandow, M.B. Das, J. Stach: J. Electrochem. Soc. **7**, 687 (1978)

W. Jost: Kapazitätsspektroskopie an Übergangsmetall-Wasserstoffkomplexen in Silizium, Dissertation, University of Stuttgart (1996)

H. Lemke: Substitutional transition metal defects in silicon grown-in by the float zone technique, in *High Purity Silicon IV*, Vols.96-113, ed. by C.L. Claeys, P. Rai-Choudhury, P. Stallhofer, J.E. Maurits (Electrochem. Soc., Pennington, NJ 1996) p.272

J.W. Sachse, E.Ö. Sweinbjörnsson, W. Jost, J. Weber, H. Lemke: Phys. Rev. B **55**, 16176 (1997)

4.89 Average value calculated from results of:

W. Jost: Kapazitätsspektroskopie an Übergangsmetall-Wasserstoffkomplexen in Silizium, Dissertation, University of Stuttgart (1996)

H. Lemke: Substitutional transition metal defects in silicon grown-in by the float zone technique, in *High Purity Silicon IV*, ed. by C.L. Claeys, P. Rai-Choudhury, P. Stallhofer, J.E. Maurits (Electrochem. Soc., Pennington, NJ 1996) Proc. Vols. 96-113, p.272

J.W. Sachse, E.Ö. Sweinbjörnsson, W. Jost, J. Weber, H. Lemke: Phys. Rev. B **55**, 16176 (1997)

H. Zimmermann, H. Ryssel: Appl. Phys. Lett. **59**, 1209 (1991)

4.90 H. Lemke: Substitutional transition metal defects in silicon grown-in by the float zone technique, in *High Purity Silicon IV*, ed. by C.L. Claeys, P. Rai-Choudhury, P. Stallhofer, J.E. Maurits (Electrochem. Soc., Pennington, NJ 1996) Proc. Vols. 96-113, p.272

4.91 W. Jost: Kapazitätsspektroskopie an Übergangsmetall-Wasserstoffkomplexen in Silizium, Dissertation, University of Stuttgart (1996)

J.W. Sachse, E.Ö. Sweinbjörnsson, W. Jost, J. Weber, H. Lemke: Phys. Rev. B **55**, 16176 (1997)

4.92 H. Zimmermann, H. Ryssel: Appl. Phys. Lett. **59**, 1209 (1991)

4.93 W.R. Wilcox, T.J. LaChapelle: J. Appl. Phys. **35**, 240 (1964)

4.94 N.A. Stolwijk, B. Schuster, J. Hölzl, H. Mehrer, W. Frank: Physica B **116**, 335 (1983)

4.95 F.H. Baumann: Ausscheidungsverhalten von Gold in Silizium bei hoher Übersättigung, Dissertation, University of Göttingen (1988)

F.H. Baumann, W. Schröter: Phil. Mag. Lett. **57**, 75 (1988); also Phys. Rev. B **43**, 6510 (1991)

4.96 Average value calculated from results of:

D.V. Lang, H.G. Grimmeiss, E. Mejer, M. Jaros: Phys. Rev. B **22**, 3917 (1980)

S.D. Brotherton, J. Bicknell: J. Appl. Phys. **49**, 667 (1978)

S.D. Brotherton, J.E. Lowther: Phys. Rev. Lett. **44**, 606 (1980)

I. Dudeck, R. Kassing: Solid State Electron. **20**, 1033 (1977)

D. Engström, H.G. Grimmeiss: J. Appl. Phys. **46**, 831 (1975)

J.M. Fairfield, B.V. Gokhale: Solid State Electron. **8**, 685 (1965)

V. Kalyanarann, V. Kumar: Phys. Status Solidi (a) **70**, 317 (1982)

L.S. Lu, T. Nishida, C.T. Sah: J. Appl. Phys. **62**, 4773 (1987)

L.S. Lu, C.T. Sah: J. Appl. Phys. **59**, 173 (1986)

J.A. Pals: Solid State Electron. **17**, 1139 (1974)

F. Richou, G. Pelous, D. Lecrosnier: J. Appl. Phys. **51**, 6252 (1980)

C.T. Sah, L. Forbes, L.L. Rosier, A.F. Tasch: Solid State Electron. **13**, 759 (1970)

R.H. Wu, A.R. Peaker: Solid State Electron. **25**, 643 (1982)

H. Lemke: Characterization of transition metal-doped silicon crystals prepared by float zone technique, in *Semiconductor Silicon 1994*, ed. by H.R. Huff, W. Bergholz, K. Sumino (Electrochem. Soc., Pennington NJ 1994) p.695

E.Ö. Sveinbjörnsson, O. Engström: Appl. Phys. Lett. **61**, 2323 (1992)

J.A. Davidson, J.H. Evans: Semicond. Sci. Technol. **11**, 1704 (1996)

4.97 Average value calculated from results of:

G. Bemski: Phys. Rev. **111**, 1515 (1958)

W.D. Davis: Phys. Rev. **114**, 1006 (1959)

J.M. Fairfield, B.V. Gokhale: Solid State Electron. **8**, 685 (1965)

L.S. Lu, T. Nishida, C.T. Sah: J. Appl. Phys. **62**, 4773 (1987)

J.A. Pals: Solid State Electron. **17**, 1139 (1974)

C.T. Sah, L. Forbes, L.L. Rosier, A.F. Tasch: Solid State Electron. **13**, 759 (1970)

R.H. Wu, A.R. Peaker: Solid State Electron. **25**, 643 (1982)

H. Lemke: Characterization of transition metal-doped silicon crystals prepared by float zone technique, in *Semiconductor Silicon 1994*, ed. by H.R. Huff, W. Bergholz, K. Sumino (Electrochem. Soc., Pennington NJ 1994) p.695

E.Ö. Sveinbjörnsson, G.I. Andersson, O. Engström: Phys. Rev. B **49**, 7801 (1994)

J.A. Davidson, J.H. Evans: Semicond. Sci. Technol. **11**, 1704 (1996)

4.98 E.Ö. Sveinbjörnsson, O. Engström: Appl. Phys. Lett. **61**, 2323 (1992)

4.99 J.A. Davidson, J.H. Evans: Semicond. Sci. Technol. **11**, 1704 (1996)

4.100 E.Ö. Sveinbjörnsson, O. Engström: Phys. Rev. B **52**, 4884 (1995)

4.101 E.Ö. Sveinbjörnsson, G.I. Andersson, O. Engström: Phys. Rev. B **49**, 7801 (1994)

4.102 L.A. Ledebo, Z.G. Wang: Appl. Phys. Lett. **42**, 680 (1983)

J. Utzig, W. Schröter: Appl. Phys. Lett. **45**, 761 (1984)

4.103 D.V. Lang, H.G. Grimmeiss, E. Mejer, M. Jaros: Phys. Rev. B **22**, 3917 (1980)

Chapter 5

5.1 H. Lemke: Characterization of transition metal-doped silicon crystals prepared by float zone technique, in *Semiconductor Silicon 1994*, ed. by H.R. Huff, W. Bergholz, K. Sumino (Electrochem. Soc., Pennington, NJ 1994) p.695

5.2 N. Achtziger: J. Appl. Phys. **80** 1, (1996)

5.3 K. Graff: Mat. Sci. Eng. B **4**, 63 (1989)

5.4 A.A. Lebedev, N.A. Sultanov, P. Yuspov: Sov. Phys. - Semicond. **14**, 340 (1980)

5.5 M.-A. Nicolet, S.S. Lau: Formation and characterization of transition-metal silicides. *VLSI Electron.* **6**, 330 (Academic, New York 1983)

5.6 S. Hocine, D. Mathiot: Mater. Sci. Forum **38-41**, 725 (1989)

5.7 S. Kuge, H. Nakashima: Jpn. J. Appl. Phys. **30**, 2659 (1991)

H. Nakashima, T. Sadoh: Mater. Sci. Forum **143-147**, 761 (1994)

5.8 Average value calculated from results of:

K. Graff, H. Pieper: The behavior of transition and noble metals in silicon crystals, in *Semiconductor Silicon 1981*, ed. by H.R. Huff, J. Kriegler, Y. Takeishi (Electrochem. Soc., Pennington, NJ 1981) p.33

H. Nakashima, T. Sadoh: Mater. Sci. Forum **143-147**, 761 (1994)

H. Lemke: Substitutional transition metal defects in silicon grown-in by the float zone technique. Mater. Sci. Forum **196-201**, 683 (1995)

N. Achtziger: J. Appl. Phys. **80**, 1 (1996)

N. Achtziger, T. Licht, U. Reislöhner, M. Rüb, W. Witthuhn: In *The Physics of Semiconductors*, Vol.4, ed. by M. Scheffler, R. Zimmermann (World Scientific, Singapore 1996) p.2717

W. Jost, J. Weber: Phys. Rev. B **54**, R11038 (1996)

W. Jost: Kapazitätsspektroskopie an Übergangsmetall-Wasserstoffkomplexen in Silizium, Dissertation, University of Stuttgart (1996)

5.9   Average value calculated from results of:

Chen, A.G. Milnes, A. Rohatgi: Solid State Electron. **22**, 801 (1979)

J.R. Morante, J.E. Carceller, P. Cartujo, J. Barbella: Solid State Electron. **26**, 1 (1983)

A. Rohatgi, R.H. Hopkins, J.R. Davis: IEEE Trans. ED-**28**, 103 (1981)

A.C. Wang, C.T. Sah: J. Appl. Phys. **56**, 1021 (1984)

H. Nakashima, T. Sadoh: Mater. Sci. Forum **143-147**, 761 (1994)

N. Achtziger: J. Appl. Phys. **80**, 1 (1996)

N. Achtziger, T. Licht, U. Reislöhner, M. Rüb, W. Witthuhn: In *The Physics of Semiconductors*, Vol.4, ed. by M. Scheffler, R. Zimmermann (World Scientific, Singapore 1996) p.2717

W. Jost, J. Weber: Phys. Rev. B **54**, R11038 (1996)

W. Jost: Kapazitätsspektroskopie an Übergangsmetall-Wasserstoffkomplexen in Silizium, Dissertation, University of Stuttgart (1996)

5.10   Average value calculated from results of:

J.W. Chen, A.G. Milnes, A. Rohatgi: Solid State Electron. **22**, 801 (1979)

L.J. Cheng, D.C. Leung: In *Electronic and Optical Properties of Polycrystalline or Impure Semiconductors and Novel Silicon Growth Methods*, ed. by K.V. Ravi, B. O'Mara (Electrochem. Soc., Pennington, NJ 1980) p.46

K. Graff, H. Pieper: The behavior of transition and noble metals in silicon crystals, in *Semiconductor Silicon 1981*, ed. by H.R. Huff, J. Kriegler, Y. Takeishi (Electrochem. Soc., Pennington, NJ 1981) p.33

A.M. Salama, L.J. Cheng: J. Electrochem. Soc. **127**, 1164 (1980)

A.C. Wang, C.T. Sah: J. Appl. Phys. **56**, 1021 (1984)

H. Nakashima, T. Sadoh: Mater. Sci. Forum **143-147**, 761 (1994)

N. Achtziger: J. Appl. Phys. **80**, 1 (1996)

N. Achtziger, T. Licht, U. Reislöhner, M. Rüb, W. Witthuhn: In *The Physics of Semiconductors*, ed. by M. Scheffler, R. Zimmermann (World Scientific, Singapore 1996) Vol.4, p.2717

W. Jost, J. Weber: Phys. Rev. B **54**, R 11038 (1996)

W. Jost: Kapazitätsspektroskopie an Übergangsmetall-Wasserstoffkomplexen in Silizium, Dissertation, University of Stuttgart (1996)

5.11   Average value calculated from results of:

W. Jost, J. Weber: Phys. Rev. B **54**, R 11038 (1996)

W. Jost: Kapazitätsspektroskopie an Übergangsmetall-Wasserstoffkomplexen in Silizium, Dissertation, University of Stuttgart (1996)

N. Achtziger: J. Appl. Phys. **80**, 1 (1996)

N. Achtziger, T. Licht, U. Reislöhner, M. Rüb, W. Witthuhn: In *The Physics of Semiconductors*, ed. by M. Scheffler, R. Zimmermann (World Scientific, Singapore 1996) Vol.4, p.2717

5.12   W. Jost, J. Weber: Phys. Rev. B **54**, R 11038 (1996)

5.13   D.A. van Wezep, C.A.J. Ammerlaan: In *13th Int'l Conf. Defects in Semicond.*, ed.
       by L.C. Kimerling, J.M. Parsey (Metallurgical Soc. of AIME, Warrendale, PA
       1984) p. 863

5.14   A. Rohatgi, J.R. Davis, R.H. Hopkins, P. Rai-Choudhury, P.G. McMullin: Solid
       State Electron. **23**, 415 (1980)

5.15   A.M. Salama, L.J. Cheng: J. Electrochem. Soc. **127**, 1164 (1980)

5.16   K. Leo, R. Schindler, J. Knobloch, B. Voss: J. Appl. Phys. **62**, 3472 (1987)

5.17   T. Sadoh, H. Nakashima: Appl. Phys. Lett. **58**, 1653 (1991)

5.18   G.W. Ludwig, H.H. Woodbury: In *Solid State Physics* 13, 223 (Academic, New
       York 1962)

5.19   Average value calculated from results of:
       E. Ohta, M. Sakata: Solid State Electron. **23**, 759 (1980)
       E. Ohta, T. Kunio, T. Sato, M. Sakata: J. Appl. Phys. **56**, 2890 (1984)
       K. Graff, H. Pieper: The behavior of transition and noble metals in silicon crystals,
       in *Semiconductor Silicon 1981*, ed. by H.R. Huff, J. Kriegler, Y. Takeishi
       (Electrochem. Soc., Pennington, NJ 1981) p. 33
       H. Lemke: Phys. Status Solid (a) **64**, 549 (1981)
       H. Nakashima, T. Sadoh: Mater. Sci. Forum **143-147**, 761 (1994)
       H. Lemke: Characterization of transition metal-doped silicon crystals prepared by
       float zone technique, in *Semiconductor Silicon 1994*, ed. by H.R. Huff, W. Berg-
       holz, K. Sumino (Electrochem. Soc., Pennington, NJ 1994) p. 695
       N. Achtziger, H. Gottschalk, T. Licht, I. Meir, M. Rüb, U. Reislöhner, W. Wit-
       thuhm: Appl. Phys. Lett. **66**, 2370 (1995)
       N. Achtziger: J. Appl. Phys. **80**, 1 (1996)

5.20   Average value calculated from results of:
       E. Ohta, M. Sakata: Solid State Electron. **23**, 759 (1980)
       E. Ohta, T. Kunio, T. Sato, M. Sakata: J. Appl. Phys. **56**, 2890 (1984)
       K. Graff, H. Pieper: The behavior of transition and noble metals in silicon crystals,
       in *Semiconductor Silicon 1981*, ed. by H.R. Huff, J. Kriegler, Y. Takeishi
       (Electrochem. Soc., Pennington, NJ 1981) p. 33
       H. Lemke: Phys. Status Solidi (a) **64**, 549 (1981)
       H. Nakashima, T. Sadoh: Mater. Sci. Forum **143-147**, 761 (1994)
       H. Lemke: Characterization of transition metal-doped silicon crystals prepared by
       float zone technique, in *Semiconductor Silicon 1994*, ed. by H.R. Huff, W. Berg-
       holz, K. Sumino (Electrochem. Soc., Pennington NJ 1994) p. 695
       N. Achtziger, H. Gottschalk, T. Licht, I. Meir, M. Rüb, U. Reislöhner, W. Wit-
       thuhm: Appl. Phys. Lett. **66**, 2370 (1995)
       N. Achtziger, T. Licht, U. Reislöhner, M. Rüb, W. Witthuhn: In *The Physics of
       Semiconductors*, ed. by M. Scheffler, R. Zimmermann (World Scientific,
       Singapore 1996) Vol. 4, p. 2717

5.21   Average value calculated from results of:
       E. Ohta, M. Sakata: Solid State Electron. **23**, 759 (1980)
       E. Ohta, T. Kunio, T. Sato, M. Sakata: J. Appl. Phys. **56**, 2890 (1984)
       K. Graff, H. Pieper: The behavior of transition and noble metals in silicon crystals,
       in *Semiconductor Silicon 1981*, ed. by H.R. Huff, J. Kriegler, Y. Takeishi
       (Electrochem. Soc., Pennington, NJ 1981) p. 33
       H. Lemke: Phys. Status Solidi (a) **64**, 549 (1981)
       H. Nakashima, T. Sadoh: Mater. Sci. Forum **143-147**, 761 (1994)
       H. Lemke: Characterization of transition metal-doped silicon crystals prepared by

float zone technique, in *Semiconductor Silicon 1994*, ed. by H.R. Huff, W. Bergholz, K. Sumino (Electrochem. Soc., Pennington NJ 1994) p.695

N. Achtziger, T. Licht, U. Reislöhner, M. Rüb, W. Witthuhn: In *The Physics of Semiconductors*, ed. by M. Scheffler, R. Zimmermann (World Scientific, Singapore 1996) Vol.4, p.2717

5.22   N. Achtziger, T. Licht, U. Reislöhner, M. Rüb, W. Witthuhn: In *The Physics of Semiconductors*, ed. by M. Scheffler, R. Zimmermann (World Scientific, Singapore 1996) Vol.4, p.2717

T. Sadoh, M. Watanabe, H. Nakashima, T. Tsurushina: Mater. Sci. Forum **143-147**, 939 (1994)

5.23   H. Lemke: Phys. Status Solidi (a) **75**, 473 (1983)

5.24   E.R. Weber: Appl. Phys. A **30**, 1 (1983)

5.25   N.T. Bendik, V.S. Garnyk, L.S. Milkovski: Sov. Phys. - Solid State **12**, 150 (1970)

5.26   H. Nakashima, T. Sadoh: Mater. Sci. Forum **143-147**, 761 (1994)

5.27   A. Mesli, T. Heiser, E. Mulheim: Mater. Sci. Eng. B **25**, 141 (1994)

T. Heiser, A. Mesli: Appl. Phys. A **57**, 325 (1993)

5.28   Average value calculated from results of:

H. Conzelmann, K. Graff, E.R. Weber: Appl. Phys. A **30**, 169 (1983)

H. Feichtinger, R. Czaputa: Appl. Phys. Lett. **39**, 706 (1981)

K. Graff, H. Pieper: The behavior of transition and noble metals in silicon crystals, in *Semiconductor Silicon 1981*, ed. by H.R. Huff, J. Kriegler, Y. Takeishi (Electrochem. Soc., Pennington, NJ 1981) p.33

T. Kunio, T. Nishino, E. Ohta, M. Sakata: Solid State Electron. **24**, 1087 (1981)

T. Kunio, T. Yamazaki, E. Ohta, M. Sakata: Solid State Electron. **26**, 155 (1983)

R.A. Muminov, K.K. Dzhuliev, S. Makhkamov, A.T. Mamadalimov: Sov. Phys. - Semicond. **16**, 376 (1982)

H. Lemke: Characterization of transition metal-doped silicon crystals prepared by float zone technique, in *Semiconductor Silicon 1994*, ed. by H.R. Huff, W. Bergholz, K. Sumino (Electrochem. Soc., Pennington NJ 1994) p.695

H. Nakashima, T. Sadoh: Mater. Sci. Forum **143 147**, 761 (1994)

N. Achtziger, T. Licht, U. Reislöhner, M. Rüb, W. Witthuhn: In *The Physics of Semiconductors*, ed. by M. Schettler, R. Zimmermann (World Scientific, Singapore 1996) p.2717

5.29   Average value calculated from results of:

H. Conzelmann, K. Graff, E.R. Weber: Appl. Phys. A **30**, 169 (1983)

H. Feichtinger, J. Oswald, R. Czaputa, P. Vogl, K. Wünstel: *13th Int'l Conf. On Defects in Semiconductors* (Coronado 1984), ed. by L.C. Kimmerling, J.M. Parsey (Metallurgical Soc. AIME, Warrendale, PA 1984) p.855

H. Lemke: Phys. Status Solidi (a) **75**, K49 (1983)

H. Nakashima, T. Sadoh: Mater. Sci. Forum **143-147**, 761 (1994)

N. Achtziger, T. Licht, U. Reislöhner, M. Rüb, W. Witthuhn: In *The Physics of Semiconductors*, ed. by M. Scheffler, R. Zimmermann (World Scientific, Singapore 1996) p.2717

5.30   Average value calculated from results of:

H. Lemke: Phys. Status Solidi (a) **75**, K49 (1983)

H. Feichtinger, J. Oswald, R. Czaputa, P. Vogl, K. Wünstel: *13th Int'l Conf. On Defects in Semiconductors* (Coronado 1984), ed. by L.C. Kimerling, J.M. Parsey (Metallurgical Soc. AIME, Warrendale, PA 1984) p.855

5.31   H. Lemke: Phys. Status Solidi (a) **75**, K49 (1983)

5.32   H. Lemke: Phys. Status Solidi (a) **72**, 177 (1982); ibid. **75**, 473 (1983)

5.33   N.T. Bendik, V.S. Garnyk, L.S. Milkovski: Sov. Phys. - Solid State **12**, 150 (1970)

5.34   D. Gilles, W. Bergholz, W. Schröter: J. Appl. Phys. **59**, 3590 (1986)

5.35   H. Lemke: Phys. Status Solidi (a) **83**, 637 (1984)

5.36   Average value calculated from results of:
R. Czaputa, H. Feichtinger, J. Oswald: Solid State Commun. **47**, 223 (1983)
K. Graff, H. Pieper: The behavior of transition and noble metals in silicon crystals, in *Semiconductor Silicon 1981*, ed. by H.R. Huff, J. Kriegler, Y. Takeishi (Electrochem. Soc., Pennington, NJ 1981) p.33
H. Lemke: Phys. Status Solidi (a) **64**, 549 (1981)
H. Lemke: Characterization of transition metal-doped silicon crystals prepared by float zone technique, in *Semiconductor Silicon 1994*, ed. by H.R. Huff, W. Bergholz, K. Sumino (Electrochem. Soc., Pennington, NJ 1994) p.695
H. Nakashima, T. Sadoh: Mater. Sci. Forum **143-147**, 761 (1994)

5.37   Average value calculated from results of:
M.M. Akhmedova, L.S. Berman, L.S. Kostina, A.A. Lebedev: Sov. Phys. - Semicond. **9**, 1516 (1976)
R. Czaputa, H. Feichtinger, J. Oswald: Solid State Commun. **47**, 223 (1983)
K. Graff, H. Pieper: The behavior of transition and noble metals in silicon crystals, in *Semiconductor Silicon 1981*, ed. by H.R. Huff, J. Kriegler, Y. Takeishi (Electrochem. Soc., Pennington, NJ 1981) p.33
H. Lemke: Phys. Status Solidi (a) **64**, 549 (1981)
H. Lemke: Characterization of transition metal-doped silicon crystals prepared by float zone technique, in *Semiconductor Silicon 1994*, ed. by H.R. Huff, W. Bergholz, K. Sumino (Electrochem. Soc., Pennington NJ 1994) p.695
H. Nakashima, T. Sadoh: Mater. Sci. Forum **143-147**, 761 (1994)

5.38   H. Lemke: Phys. Status Solidi (a) **71**, K215 (1982)

5.39   H. Feichtinger, J. Oswald, R. Czaputa, P. Vogl, K. Wünstel: *13th Int'l Conf. On Defects in Semiconductors* (Coronado 1984), ed. by L.C. Kimerling, J.M. Parsey (Metallurgical Soc. AIME, Warrendale, PA 1984) p.855

5.40   S.D. Brotherton, P. Bradley, A. Gill: J. Appl. Phys. **55**, 952 (1984); ibid. **57**, 1783 (1985)

5.41   D. Gilles: Einfluβ der elektronischen Struktur auf Diffusion, Löslichkeit und Paarbildung von 3d-Übergangselementen in Silizium, Dissertation, University of Göttingen (1987)

5.42   J. Utzig, D. Gilles: Mater. Sci. Forum **38-41**, 729 (1989)

5.43   M. Seibt, W. Schröter: Lokalisierung und Identifizierung von Mikrodefekten. Forschungsbericht, Bundesministerium für Forschung und Technologie, FRG (1989)

5.44   F. Beeler, O.K. Andersen, M. Scheffler: Phys. Rev. Lett. **55**, 1498 (1985)

5.45   Average value calculated from results of:
H. Kitagawa, H. Nakashima, K. Hashimoto: Jpn. J. Appl. Phys. **24**, 373 (1985); Memoirs of the Fac. Of Engin. Kyushu Univ. 46, 119 (1986)
V.M. Arutyunyan, I.A. Sarkisyan: Sov. Phys. - Semicond. **9**, 826 (1975)
H. Lemke: Phys. Status Solidi (a) **91**, 649 (1985)
H. Nakashima, T. Sadoh: Mater. Sci. Forum **143-147**, 761 (1994)
H. Lemke: Characterization of transition metal-doped silicon crystals prepared by float zone technique, in *Semiconductor Silicon 1994*, ed. by H.R. Huff, W. Bergholz, K. Sumino (Electrochem. Soc., Pennington NJ 1994) p.695
W. Jost, J. Weber, H. Lemke: Semicond. Sci. Techn. **11**, 22 (1996)

5.46　Average value calculated from results of:

H. Kitagawa, H. Nakashima, K. Hashimoto: Memoirs Fac. Eng. Kyushu Univ. **46**, 119 (1986)

H. Lemke: Phys. Status Solidi (a) **91**, 649 (1985)

W. Jost, J. Weber, H. Lemke: Semicond. Sci. Techn. **11**, 525 (1996)

H. Lemke: Characterization of transition metal-doped silicon crystals prepared by float zone technique, in *Semiconductor Silicon 1994*, ed. by H.R. Huff, W. Bergholz, K. Sumino (Electrochem. Soc., Pennington, NJ 1994) p.695

5.47　H. Ewe: Relaxationsinduziertes Gettern von Kobalt und Nickel in Silizium, Dissertation, University of Göttingen (1996)

5.48　W. Bergholz: Physica B **116**, 312 (1983)

J. Utzig: J. Appl. Phys. **64**, 3629 (1988)

5.49　W. Jost, J. Weber, H. Lemke: Semicond. Sci. Techn. **11**, 22 (1996)

5.50　N. Achtziger, T. Licht, U. Reislöhner, M. Rüb, W. Witthuhn: In *The Physics of Semiconductors*, ed. by M. Scheffler, R. Zimmermann (World Scientific, Singapore 1996) p.2717

5.51　W. Jost, J. Weber, H. Lemke: Semicond. Sci. Techn. **11**, 525 (1996)

5.52　H. Lemke: Mater. Sci. Forum **196-201**, 683 (1995)

5.53　H. Lemke: 18th Int'l. Conf. Defect in Semicond., Sendai, Jpn. (1995)

H. Lemke: Substitutional transition metal defects in silicon grown by the FZ technique, in *High Purity Silicon IV*, ed. by C.L. Claes, P. Rai-Choudhury, P. Stallhofer, J.E. Maurits (Electrochem. Soc., Pennington, NJ 1996) Vols. 96-113, p.272

5.54　D. Grünebaum, T. Czekalla, N.A. Stolwijk, H. Mehrer, I. Yonenaga, K. Sumino: Appl. Phys. A **53**, 65 (1991)

5.55　A.G. Milnes: *Deep Impurities in Semiconductors* (Wiley, New York 1973)

5.56　H. Bracht, H. Overhof: Phys. Status Solidi (a) **158**, 47 (1996)

5.57　H. Bracht, N.A. Stolwijk, H. Mehrer: Phys. Rev.B **52**, 16542 (1995)

5.58　N.A. Stolwijk, Ch. Poisson, J. Bernardini: J. Phys., Condens Matter **8**, 5843 (1996)

Ch. Poisson, A. Rolland, J. Bernardini, N.A. Stolwijk: J. Appl. Phys. **80**, 6179 (1996)

5.59　S. Weiss, R. Beckmann, R. Kassing: Private commun. (1989)

5.60　Average value calculated from results of:

J.M. Hermann, C.T. Sah: J. Appl. Phys. **44**, 1259 (1973)

J.W. Chen, A.G. Milnes: Ann. Rev. Mater. Sci. **10**, 157 (1980)

H .Lemke: Phys. Status Solidi (a) **72**, 177 (1982)

S. Weiss, R. Beckmann, R. Kassing: Private commun. (1989)

H. Bracht, N.A. Stolwijk, H. Mehrer: Phys. Rev.B **52**, 16542 (1995)

5.61　Average value calculated from results of:

J.M. Hermann, C.T. Sah: J. Appl. Phys. **44**, 1259 (1973)

J.W. Chen, A.G. Milnes: Ann. Rev. Mater. Sci. **10**, 157 (1980)

H. Lemke: Phys. Status Solidi (a) **72**, 177 (1982)

S. Weiss, R. Beckmann, R. Kassing: Private commun. (1989)

H. Bracht, N.A. Stolwijk, H. Mehrer: Phys. Rev.B **52**, 16542 (1995)

A. Dörnen, R. Kienle, K. Thonke, P. Stolz, G. Pensl, D. Grünebaum, N.A. Stolwijk: Phys. Rev B **40**, 12005 (1989)

P. Stolz, G. Pensl, D. Grünebaum, N.A. Stolwijk: Mater. Sci. Eng. B **4**, 31 (1989)

5.62　A. Altink: Solid State Commun. **75**, 1 (1990)

5.63   H. Lemke: Phys. Status Solidi (a) **122**, 617 (1990)

5.64   A. Rohatgi, J.R. Davis, R.H. Hopkins, P.G. McMullin: Solid State Electron. **26**, 1039 (1983)

5.65   K. Schmalz, H.G. Grimmeiss, H. Pettersson, L. Tilly: Mater. Sci. Forum **143-147**, 809 (1994)

5.66   R. Czaputa: Appl. Phys. A **49**, 431 (1989)

5.67   Average value calculated from results of:
        H. Lemke: Phys. Status Solidi (a) **85**, K137 (1984); ibid. **91**, 143 (1985)
        H. Lemke: Characterization of transition metal-doped silicon crystals prepared by float zone technique, in *Semiconductor Silicon 1994*, ed. by H.R. Huff, W. Bergholz, K. Sumino (Electrochem. Soc., Pennington, NJ 1994) p.695
        R. Czaputa: Appl. Phys. A **49**, 431 (1989)

5.68   Average value calculated from results of:
        K.P. Lisiak, A.G. Milnes: Solid State Electron. **19**, 115 (1976)
        H. Lemke: Phys. Status Solidi (a) **85**, K137 (1984); ibid. **91**, 143 (1985)
        H. Lemke: Characterization of transition metal-doped silicon crystals prepared by float zone technique, in *Semiconductor Silicon 1994*, ed. by H.R. Huff, W. Bergholz, K. Sumino (Electrochem. Soc., Pennington, NJ 1994) p.695
        R. Czaputa: Appl. Phys. A **49**, 431 (1989)

5.69   F. Rollert, N.A. Stolwijk, H. Mehrer: J. Physik D **20**, 1148 (1987)

5.70   K. Graff, H. Pieper: The behavior of transition and noble metals in silicon crystals, in *Semiconductor Silicon 1981*, ed. by H.R. Huff, J. Kriegler, Y. Takeishi (Electrochem. Soc., Pennington, NJ 1981) p.33

5.71   W.R. Wilcox, T.J. LaChapelle: J. Appl. Phys. **35**, 240 (1964)

5.72   H. Lemke: Phys. Status Solidi (a) **94**, K55 (1986)

5.73   J. Bollmann, H. Klose, A. Mertens: Phys. Status Solidi (a) **97**, K135 (1986)

5.74   K. Graff: The precipitation of transition metals in silicon crystal wafers, in *Aggregation Phenomena of Point Defects in Silicon*, ed. by E. Sirtl, J. Goorissen, P. Wagner (Electrochem. Soc., Pennington, NJ 1983) p.121

5.75   L.D. Yau, C.F. Smiley, C.T. Sah: Phys. Status Solidi (a) **13**, 457 (1987)

5.76   H. Feichtinger, E. Sturm: Mater. Sci. Forum **143-147**, 111 (1994)

5.77   M. Lang, G. Pensl, M. Gebhard, N. Achtziger: Appl. Phys. A **53**, 95 (1991)

5.78   W. Frank: Private commun. (1989)

5.79   K. Miyata, C.T. Sah: Solid State Electron. **19**, 611 (1976)

5.80   H.U. Habermeier, W. Frank: Private commun. (1987)

5.81   Average value calculated from results of:
        Y. Fujisaki, T. Ando, H. Kozuka, Y. Takano: J. Appl. Phys. **63**, 2304 (1988)
        H. Lemke: Phys. Status Solidi (a) **76**, K193 (1983)
        K. Miyata, C.T. Sah: Solid State Electron. **19**, 611 (1976

5.82   H. Lemke: Private commun. (1997)

5.83   K.P. Lisiak, A.G. Milnes: Solid State Electron. **19**, 115 (1976)

5.84   N.A. Stolwijk: Priv. commun. (1999)

Chapter 6

6.1   D.K. Schroder: *Semiconductor Material and Device Characterization* (Wiley, New York 1990)
        Analysis of Microelectronic Materials and Devices, ed. by M. Grasserbauer, H.W. Werner (Wiley, Chichester 1991)

6.2   P.F. Schmidt, C.W. Pearce: J. Electrochem. Soc. **128**, 630 (1981)

6.3   D.J. O'Connor, B.A. Sexton, R.St.C. Smart (eds.): *Surface Analysis Methods in Materials Science*, Springer Ser. Surf. Sci., Vol.23 (Springer, Berlin, Heidelberg 1992)

6.4   A. Corradi, M. Domenici, A. Guaglio: J. Cryst. Growth **89**, 39 (1988)

6.5   A. Shimazaki, H. Hiratsuka, Y. Matsushita, S. Yoshii: 16th Conf. on Solid State Devices and Materials (Jpn. Society of Applied Physics, Tokyo 1984) extended abstracts p.281

6.6   P. Eichinger, H.J. Rath, H. Schwenke: *Semiconductor Fabrication: Technology and Metrology*, ed. by D.C. Gupta (Am. Soc. for Testing and Materials, ASTM STP 990, Ann Arbor, MI 1989) p.314

6.7   C. Neumann, P. Eichinger: Spectrochim. Acta B **46**, 1369 (1991)

6.8   T. Shiraiwa, N. Fujino, S. Sumita, Y. Tanizoe: *Semiconductor Fabrication: Technology and Metrology*, ed. by D.C. Gupta (Am. Soc. for Testing and Materials, ASTM STP 990, Ann Arbor, MI 1989) p.314

6.9   D.V. Lang: J. Appl. Phys. **45**, 3023 (1974)
        D.V. Lang: Space-charge spectroscopy in semiconductors, in *Thermally Stimulated Relaxation in Solids*, ed. by R. Bräunlich, Topics Appl. Phys., Vol.37 (Springer, Berlin, Heidelberg 1979)

6.10   G. Ferenczi, J. Kiss: Acta Phys. Acad. Scient. Hung. **50**, 285 (1981)

6.11   L.C. Kimerling, J.L. Benton, J.J. Rubin: Transition metal impurities in silicon, in *Defects and Radiation Effects in Semiconductors 1980*, Inst. Phys. Conf. Ser. **59**, 217 (1981)

6.12   A.C. Wang, C.T. Sah: J. Appl. Phys. **56**, 1021 (1984)

6.13   E. Ohta, M. Sakata: Solid-State Electron. **23**, 759 (1980)
        H. Lemke: Phys. Status Solidi (a) **64**, 549 (1981)

6.14   T. Kunio, T. Nishino, E. Ohta, M. Sakata: Solid-State Electron. **24**, 1087 (1981)
        R.A. Muminov, K.K. Dzhuliev, S. Makhkamov, A.T. Mamadalimov: Sov. Phys. - Semicond. **16**, 376 (1982)
        H. Conzelmann, K. Graff, E.R. Weber: Appl. Phys. A **30**, 169 (1983)

6.15   R. Czaputa, H. Feichtinger, J. Oswald: Solid State Commun. **47**, 223 (1983)

6.16   H. Feichtinger, A. Gschwandtner, J. Waltl: Phys. Status Solidi (a) **53**, K71 (1979)
        V.B. Voronkov, A.A. Lebedev, A.T. Mamadalimov, B.M. Urunbaev, T.A. Usmanov: Sov. Phys. - Semicond. **14**, 1217 (1980)
        K. Wünstel, P. Wagner: Appl. Phys. A **27**, 207 (1982)
        S.D. Brotherton, P. Bradley, A. Gill: J. Appl. Phys. **57**, 1941 (1985)

6.17   G. Ferenczi, J. Boda, T. Pavelka: Phys. Status Solidi (a) **94**, K119 (1986)

6.18   K. Ikeda, H. Takaoka: Jpn. J. Appl. Phys. **21**, 462 (1982)

6.19   M. Okuyama, H. Takakura, Y. Hamakawa: Solid-State Electron. **26**, 689 (1983)

6.20   S. Weiss, R. Kassing: Solid-State Electron. **31**, 1723 (1988)

6.21   N.M. Johnson, D.J. Bartelink, R.B. Gold, J.F. Gibbon: J. Appl. Phys. **50**, 4828 (1979)

6.22   H.G. Grimmeiss, N. Kullendorff: J. Appl. Phys. **51**, 5852 (1980)

6.23   H. Lefevre, M. Schulz: IEEE Trans. ED-**24**, 973 (1977)

6.24   L. Mahdjoubi, M. Benmalek, M. Derdouri: Solid-State Electron. **25**, 925 (1982)

6.25   S. Duenas, M. Jaraiz, J. Vicente, E. Rubio, L. Bailon, J. Barbolla: J. Appl. Phys. **61**, 2541 (1987)

6.26   M. Takikawa, T. Ikoma: Jpn. J. Appl. Phys. **19**, L436 (1980)

6.27   S. Brehme, R. Pickenhain: Phys. Status Solidi (a) **90**, K119 (1985)

6.28   D. Pons: Appl. Phys. Lett. **37**, 413 (1980); J. Appl. Phys. **55**, 3644 (1984)

6.29   J.H. Zhao, J.C. Lee, Z.Q. Fang, T.E. Schlesinger, A.G. Milnes: J. Appl. Phys. **61**, 1063 (1987)

6.30   K. Maass, K. Irmscher, H. Klose: Phys. Status Solidi (a) **91**, 667 (1985)

6.31   P. Eichinger: Microelectronic **2**, 264 (1988)

6.32   R. Czaputa: Appl. Phys. A **49**, 431 (1989)

6.33   K. Graff, H. Fischer: Carrier lifetime in silicon and its impact on solar cell characteristics, in *Solar Energy Conversion*, ed. by B.O. Seraphin, Topics Appl. Phys., Vol. 31 (Springer, Berlin, Heidelberg 1979) p. 173

6.34   DIN 50440 part 1 (Beuth Verlag, Berlin, revised 1993)

6.35   Annual Book of ASTM Standards, Vol. 10.05, Electronics (II) F 391 (ASTM Philadelphia, PA 1989)

6.36   H. Föll, V. Lehmann, G. Zoth, F. Gelsdorf, B. Göttinger: *Analytical Techniques for Semiconductor Materials and Process Characterization*, ed. by B.O. Kolbesen, D.V. McCaughan, W. Vandervorst (Electrochem. Soc., Berlin 1989) Abstr. 03

6.37   W. Bergholz, V. Penka, G. Zoth: *Analytical Techniques for Semiconductor Materials and Process Characterization*, ed. by B.O. Kolbesen, D.V. McCaughan, W. Vandervorst (The Electrochem. Soc., Pennington, NJ 1989) p. 29

6.38   W. Bergholz, G. Zoth, H. Wendt, S. Sauter, G. Asam: Siemens Forsch. Entwickl. Ber. **16**, 241 (1987)

6.39   T.S. Horanyi, T. Pavelka, P. Tütö: Appl. Surface Sci. **63**, 306 (1993)

6.40   M. Schöfthaler, U. Rau, G. Langguth, M. Hirsch, R. Brendel, H. Werner: *Proc. 12th Europ. Solar Energy Conf.* (Amsterdam) (H.S. Stephens & Ass., Bedford 1994) p. 533

6.41   K. Graff: Techniques for metal contamination analysis and control, in *Analytical Techniques for Semiconductor Materials and Process Characterization II*, ed. by B.O. Kolbesen, C. Claeys, P. Stallhofer (Electrochem. Soc., Pennington, NJ 1995) p. 3

6.42   W. Arndt, K. Graff, P. Heim: Novel method to reduce the surface recombination velocity for carrier lifetime measurement in silicon wafers, in *Analytical Techniques for Semiconductor Materials and Process Characterization II*, ed. by B.O. Kolbesen, C. Claeys, P. Stallhofer (Electrochem. Soc., Pennington, NJ 1995) p. 44

6.43   K. Graff: Unpublished (1980-90)

6.44   M. Hourai, K. Murakami, T. Shigematsu, N. Fujino, T. Shiraiwa: Jpn. J. Appl. Phys. **28**, 2413 (1989)

Chapter 7

7.1   W. Bergholz, V. Penka, G. Zoth: *Analytical Techniques for Semiconductor Materials and Process Characterization*, ed. by B.O. Kolbesen, D.V. McCaughan, W. Vandervorst (The Electrochem. Soc., Pennington, NJ 1989) p. 29

7.2   K. Graff: *Semiconductor Processing and Equipment Symposium* Technical Proceedings (Semiconductor Equipment and Materials Institute, Zürich 1983) p. 9

7.3   A. Rohatgi, J.R. Davis, R.H. Hopkins, P. Rai-Choudhury, P.G. McMullin: Solid-State Electron. **23**, 415 (1980)

7.4   R. Iscoff: Semicond. Int'l Jan. 1991 p. 60

7.5   FORD Q 101 (1985). This internal report can be requested from Ford Company USA or from respective national representatives in various languages.

7.6   W. Bergholz, G. Zoth, H. Wendt, S. Sauter, G. Asam: Siemens Forsch. Entwickl. Ber. **16**, 241 (1987)

7.7   H. Föll, V. Lehmann, G. Zoth, F. Gelsdorf, B. Göttinger: *Analytical Techniques for Semiconductor Materials and Process Characterization*, ed. by B.O. Kolbesen, D.V. McCaughan, W. Vandervorst (Electrochem. Soc. Berlin 1989) Abstr.03

Chapter 8

8.1   A. Goetzberger, W. Shockley: J. Appl. Phys. **31**, 1821 (1960)

8.2   N. Momma, H. Tamiguchi, M. Ura, T. Ogawa: J. Electrochem. Soc. **125**, 963 (1978)

8.3   R. Sawada, T. Karaki, J. Watanabe: Jpn. J. Appl. Phys. **20**, 2097 (1981)

8.4   C.L. Reed, K.M. Mar: J. Electrochem. Soc. **127**, 2058 (1980)

8.5   R. Sawada: Jpn. J. Appl. Phys. **23**, 959 (1984)

8.6   G.H. Schwuttke: Phys. Status Solidi (a) **42**, 553 (1977)

8.7   R.J. Pressley: Laser process for gettering defects in semiconductor devices. US Patent H01L21/268

8.8   G.E.J. Eggermont, R.J. Falster, S.K. Hahn: Solid State Techn. **26**, 171 (1983)

8.9   T.M. Buck, J.M. Poate, K.A. Pickar, C.M. Hsieh: Surf. Sci. **35**, 362 (1973)

8.10   E. Bugiel, M. Kittler, A. Borchardt, H. Richter: Phys. Status Solidi (a) **84**, 143 (1984)

8.11   V.A. Atsarkin, E.Z. Mazel: Sov. Phys. - Solid State **2**, 1874 (1961)

8.12   R.J. Kriegler, Y.C. Cheng, D.R. Colton: J. Electrochem. Soc. **119**, 388 (1972)

8.13   T.A. Baginski, I.R. Monkowski: J. Electrochem. Soc. **132**, 2031 (1985)

8.14   P.H. Robinson, F.P. Heimann: J. Electrochem. Soc. **118**, 141 (1971)

8.15   D.R. Young, C.M. Osburn: J. Electrochem. Soc. **120**, 1578 (1973)

8.16   C.K. Chu, J.E. Johnson: J. Electrochem. Soc. **74**, 86 (1974)

8.17   P.M. Petrott, G.A. Rozgonyi, T.T. Sheng: J. Electrochem. Soc. **123**, 565 (1976)

8.18   Y. Mada: Jpn. J. Appl. Phys. **21**, L683 (1982)

8.19   V. Schlosser: Verfahren zum Gettern von Halbleiter-Bauelementen. Europatent EP 0092540, European Patent Office, Stockholm (1983)

8.20   T.Y. Tan, E.E. Gardner, W.K. Tice: Appl. Phys. Lett. **30**, 175 (1977)

8.21   G.A. Rozgonyi, C.W. Pearce: Appl. Phys. Lett. **32**, 747 (1978)

8.22   K. Nagasawa, Y. Matsushita, S. Kishino: Appl. Phys. Lett. **37**, 622 (1980)

8.23   L.E. Katz, P.F. Schmidt, C.W. Pearce: J. Electrochem. Soc. **128**, 620 (1981)

8.24   D. Lecrosnier, J. Paugam, G. Pelous, F. Richou, M. Salvi: J. Appl. Phys. **52**, 5090 (1981)

8.25   G.F. Cerofolini, M.L. Polignano: J. Appl. Phys. **55**, 579 (1984)

8.26   G.F. Cerofolini, M.L. Polignano, H. Bender, C. Claeys: Phys. Status Solidi (a) **103**, 643 (1987)

8.27   G. Keefe-Fraundorf, D.E. Hill, R.A. Craven: Backside gettering of defects in silicon. EMTAS Conf., Phoenix, SME Technical Paper EE83-**131**, 1 (1983)

8.28   H. Suga, Y. Shimanuki, K. Murai, K. Endo: *Semiconductor Processing* ASTM STP 850, ed. by D.C. Gupta (1984) p.241

8.29   L. Jastrzebski, R. Soydan, B. Goldsmith, J.T. McGinn: J. Electrochem. Soc. **131**, 2944 (1984)

8.30 K. Graff, H.A. Hefner, W. Hennerici: J. Electrochem. Soc. **135**, 952 (1988)

8.31 W. Schröter, M. Seibt, D. Gilles: High-temperature properties of 3d transition elements in silicon in *Electronic Structure and Properties of Semiconductors*, ed. by W. Schröter (VHC, Weinheim 1991) p.539

8.32 A. Ourmazd, W. Schröter: Appl. Phys. Lett. **45**, 781 (1984)

8.33 A. Bourret, W. Schröter: Ultramicroscopy **14**, 97 (1984)

8.34 M. Seibt, K. Graff: J. Appl. Phys. **63**, 4444 (1988)

8.35 T.Y. Tan, C.Y. Kung: J. Appl. Phys. **59**, 917 (1986)

8.36 K. Graff: Semicon Europa 1990 (Semiconductor Equipment and Materials Int'l, Zürich 1990) Techn. proc. p.2

8.37 T.I. Chappel, P.W. Chye, M.A. Tavel: Solid-State Electron. **26**, 33 (1983)

8.38 K. Graff, H. Fischer: Carrier lifetime in silicon and its impact on solar cell characteristics, in *Solar Energy Conversion*, ed. by B.O. Seraphin, Topics Appl. Phys., Vol.31 (Springer, Berlin, Heidelberg 1979) p.173

8.39 D. Gilles, W. Schröter, W. Bergholz: Phys. Rev. B **41**, 5770 (1990)

8.40 M. Seibt, W. Schröter: Lokalisierung und Identifizierung von Mikrodefekten. Forschungsbericht, Bundesministerium für Forschung und Technologie FRG (1989)

8.41 K. Graff, H.A. Hefner, H. Pieper: MRS Proc. **36**, 19 (1985)

8.42 K. Graff: Mat. Sci. Eng. B **4**, 63 (1989)

8.43 K. Graff: Internal gettering as function of the wafer position in the original crystal rod. ESPRIT'90, Brussels (Kluwer, Dordreche 1990) Proc. p.77

8.44 K. Graff: Unpublished (1980-90)

8.45 D. De Busk, J. Van Wagener, J. Chambers: *Semicond. Int'l* (Cahners Publ., Hoofdorp, Nederlands 1992) p.124

8.46 R. Falster, W. Bergholz: J. Electrochem. Soc. **137**, 1548 (1990)

8.47 E.R. Weber: Appl. Phys. A **30**, 1 (1983)

8.48 W. Bergholz, G. Zoth, H. Wendt, S. Sauter, G. Asam: Siemens Forsch. Entwickl. Ber. **16**, 241 (1987)

8.49 H. Führer: Private commun. (1990)

8.50 K. Graff, H. Pieper: J. Electrochem. Soc. **128**, 669 (1981)

8.51 K. Graff: Current understanding of the behavior of transition metal impurities in silicon, in *Semiconductor Silicon 1986*, ed. by H.R. Huff, T. Abe, B. Kolbesen (Electrochem. Soc., Pennington, NJ 1986) p.751

8.52 R. Falster, D. Gambaro, M. Olmo, M. Cornara, H. Korb: MRS Symp. **510**, 27 (1998)

Subject Index

Springer Series in
MATERIALS SCIENCE

Series Editors: U. Gonser · R. M. Osgood, Jr. · M. B. Panish · H. Sakaki
Founding Editor: H. K. V. Lotsch

* The 2nd edition is available as a textbook with the title: *Laser Processing and Chemistry*

Druck: Strauss Offsetdruck, Mörlenbach
Verarbeitung: Schäffer, Grünstadt